Table of Contents

PART TWO: ENVIRONMENTAL AND RESOURCES CONSTRAINTS

PART THREE: CONTRIBUTED PAPERS

Introduction

Shigeto Tsuru

It was deemed appropriate in the planning stage for the Fifth Congress of the International Economic Association to organize a specialized session for Japan's economic growth and her resource problems inasmuch as the site of the Congress was to be Japan and no less for the reason that the sustained high growth rate of Japan over the two decades, 1954-73, was in fact unmatched by any other country's performance. This was all the more remarkable, it was commented on by some, since Japan, probably more than any other industrialized country, was poorly endowed in terms of indigenous resources.

To set up a specialized session on Japan was decided upon in August 1975 at the executive committee meeting of the Association at Saltsjöbaden, Sweden and the planning of the session was entrusted in my hand. Administrative details apart, preparation for the session began in Japan immediately afterwards under the chairmanship of Professor Jokichi Uchida, head of the Division III of the Science Council of Japan.

In October 1975 Professor Uchida commenced his preparation by addressing letters of request to five constituent disciplinary associations[1] within the Union of National Economic Associations in Japan asking them to cooperate with him in selecting specific topics and active participants for the specialized session. A tentative plan was drawn up as a result with four specific topics as follows:

1. Past and Present of the Economic Development of Japan
2. Development of Industrial Structure of Japan
3. Economic Growth of Japan and Environmental Problems
4. Economic Development and International Relations

[1] The Japanese Association of Theoretical Economic (Riron-Keiryo Keizai Gakkai), Japan Society of Political Economy (Keizai Riron Gakkai), The Japan Society of International Economics (Kokusai Keizai Gakkai), Japan Economic Policy Association (Nihon Keizai Seisaku Gakkai) and Japan Society for the Study of Business Administration (Nihon Keiei Gakkai).

A list of names, both from Japan and abroad, was also drawn up as paper-writers and discussants. And this became the basis for the final plan which subsequently was decided upon by myself in my responsibility as chairman of the local organizing committee for the Congress.

The final program as carried out in the Congress was as follows:

31 August 1977

Morning Session: Chairman—Mark Perlman (U.S.A.)
Rapporteur—A. Hernadi (Hungary)

Kazushi Ohkawa (Japan): Past Economic Growth of Japan in Comparison with the Western Case: Trend Acceleration and Differential Structure
Discussant: *Kazuo Shibagaki (Japan)*

Martin Bronfenbrenner (U.S.A.): A "Marginal-Efficiency" Theory of Japanese Growth
Discussant: *Tuvia Blumenthal (Israel)*

Dale W. Jorgenson and Mieko Nishimizu (U.S.A.): U.S. and Japanese Economic Growth 1952–1973: An International Comparison
Discussant: *Hugh Patrick (U.S.A.)*

Kenzo Yukizawa (Japan): Relative Productivity of Labor in American and Japanese Industry and Its Change, 1958–1972
Discussant: *Masaru Saito (Japan)*

Afternoon Session: Chairman—Mark Perlman (U.S.A.)
Rapporteur—A. Hernadi (Hungary)

Ken'ichi Imai and Masu Uekusa (Japan): Industrial Organization and Economic Growth in Japan
Discussant: *Soichiro Giga (Japan)*

Roberto Zaneletti (Italy): Capital Accumulation and Economic Growth: A Comparison of Italy and Japan
Discussant: *Hisao Onoe (Japan)*

Hisao Kanamori (Japan): Japanese Economic Growth and Economic Welfare
Discussant: *Kimio Uno (Japan)*

Kazuo Sato (Japan): Did Technical Progress Accelerate in Japan?
Discussant: *Harry T. Oshima (U.S.A.)*

1 September 1977

Morning Session: Chairman—Ronald Dore (U.K.)
Rapporteur—Anne Hori Androuais (France)

Rémy Prud'homme (France): Appraisal of Environmental Policies in Japan
Discussant: *Nobuo Okishio (Japan)*

Ken'ichi Miyamoto (Japan): The Environmental Protection Policy in Japan
Discussant: *Jacques Godchot (France)*

Yasukichi Yasuba (Japan): Resources in Japan's Development
Discussant: *Akrasanee Narongchai (Thailand)*

E. Stuart Kirby (U.K.): Resource Potentials of Continental East Asia and Japan's Material Needs
Discussant: *Tsuneo Nakauchi (Japan)*

Afternoon Session: Chairman—Lynn Turgeon (U.S.A.)
Rapporteur—Anne Hori Androuais (France)

Masao Sakisaka (Japan): Economic Growth in Japan and Energy
Discussant: *Hirobumi Uzawa (Japan)*

Jun Nishikawa (Japan): "Resource Constraints": A Problem of the Japanese Economy
Discussant: *A. Hernadi (Hungary)*

D.J. Daly (Canada): Japanese Economic Developments, 1970–1976
Discussant: *Ippei Yamazawa (Japan)*

Yuzuru Hanayama (Japan): Urbanization and Land Prices: The Case of Tokyo
Discussant: *Giorgio Stefani (Italy)*

It is to be regretted that not all the official discussants were able to submit written comments for publication. But summaries of their comments as well as of discussions are given in the text by Mark Perlman and A. Hernadi for the first day and by Ronald Dore and Lynn Turgeon for the second day. For the sessions on Japan as a whole, I presented a general summary on the closing day of the Congress and it is printed in a separate volume for the Congress, edited by Edmond Malinvaud, incorporating the papers and discussions at plenary sessions. Here, therefore, it is probably more pertinent to raise, and comment on, a number of problems which still remain controversial in connection with the subject matter of this volume.

Firstly, a question might be raised as to how serious a constraint the lack of indigenous resources has been in the course of postwar economic growth of Japan. Reviewing the price trend of industrial raw materials and fuel in the postwar period, one is struck by the fact that most of them, notably petroleum, were in oversupply which put strong downward pressure on their prices. As a matter of fact, the world market price of petroleum in the latter half of 1960s was as low as one-eighth of the prewar level. Having had to depend almost entirely on imports for industrial fuel, Japan was, paradoxically as it were, in the position to take full advantage of the cheap source of Middle East petroleum supply, the price being around $1.60 per barrel. At the time, on the other hand, the United States, for example, was depending on the domestic petroleum supply to the extent of some 92 percent for which the tariff-protected price of $5.00 per barrel prevailed. The United Kingdom and West Germany, both of which then were, like Japan, very poorly endowed with domestic petroleum resources, could have taken advantage of the cheap Middle East petroleum supply; but they had their domestic coal mines to look after, and the cost of thermal electricity generation burning coal was said to be equivalent to using $7.00-per-barrel petroleum at the time. France, too, was constrained then to purchase the Algerian petroleum at the price of $3.95 per barrel on account of the center-periphery relation of political character.

Japan was untrammelled by any of these constraints in the purchase, not only of petroleum, but also of most of the industrial raw materials and was able to bask in the classical benefit of free trade, orienting her industrial sites and structures in such a way as to facilitate and economize the use of industrial input coming from abroad. Development of efficient ocean transportation systems and the siting of industrial plants on the reclaimed land along the seashore were pursued most energetically during the period of rapid growth. These factors, combined together, were instrumental in making Japan's manufacturing industries highly competitive with those in other industrialized countries. But, of course, the question remains if and how long such a favorable condition will last in face of the increasing trend of various impediments to free trade, including the cartel-like development of primary goods producers.

The second problem worth pursuing is: Granted the remarkable record of rapid rate of growth in terms of real GNP, what significance should we attach to it from the welfare point of view? Kanamori, who tackled this problem in the session, confined himself mainly to contrasting the Net National Welfare indicator (NNW) with the real GNP and concluded that even in terms of the former the average rate of growth during the two decades of 1955–75 was 7.4 percent per annum, which is remarkable enough. He did admit, however, that the NNW was meant to measure only a part of economic welfare and certainly should not be taken as corresponding to a broader concept of human welfare.

Economists probably should not be too ambitious in their attempt to perfect an aggregate measure of national welfare since there are so many factors in this world which are relevant to human welfare and yet defy any quantification. As a matter of fact, the course of events in the period of rapid growth in postwar Japan impresses us with multiplying instances of (A) the deterioration in amenities which cannot be taken into account in the estimation of the NNW indicator and (B) the dilation of those GNP components which, though devoid of welfare significance, are not deducted when the NNW is calculated. A typical example of the former (A) is provided by the ubiquitous instances of "Extending the area for *kitchen* by destroying the *garden*," or the reclaiming of shallow coast of the archipelago for factory sites all over the country effacing the access to beaches for the citizens in the nearby area. As for the latter (B), it may be sufficient to recall what Marx wrote more than one hundred years ago, pointing out that the criminal, by "producing" crimes, contributes towards expanding national income through necessitating various means of defense, both human and material, against criminal acts of burglary, forgery, etc. "The cost of life" type of consumption expenditures, as exemplified by the commuting expenses, is now quite proliferative; and the institutionalization of waste and "the interference of income," both of which are instance of "the suppliers' sovereignity," contribute to the inflating of real GNP without being deducted when the NNW is measured.

The experience of Japan of recording an inordinately high rate of growth in GNP in the postwar period leads us to wonder if a much more fruitful approach as an overall quantitative measure of economic welfare might not be resuscitation of Irving Fisher's concepts of capital and income. For Fisher, "income" consisted solely of services as received by ultimate consumers, whether from their material or from their human environment which together might be called "social wealth" or "capital." Social wealth consisted not only of producers' real capital such as plant and equipment but also of what are nowadays called "common property resources" as well as geological capital and consumers' real capital. In this scheme, "production" is defined as an addition to this social wealth and "consumption" as a subtraction from it. Since "income" is essentially proportional to the stock of social wealth, "consumption" would have a negative effect on "income," while "production" a positive one. It is worth recording that Pigou, who took issue with Fisher over this problem, conceded that Fisher's conceptual scheme would be appropriate if one were interested in "comparative amounts of economic welfare which a community obtains over a long series of years."[2]

It would be quite instructive if we were to make a list of pertinent items to be included in the "social wealth" of Japan and to attempt their statistical estimates for some benchmark years such as 1935, 1955 and 1975.

[2] A.C. Pigou, *The Economics of Welfare*, London, Macmillan, 1932, p. 36.

Although the index number of per capita real GNP grew from 100 in 1935 to 114 in 1955 and to 490 by 1975, it is highly doubtful if the stock of residential building per capita, for example, kept pace with such a rate of growth. No satisfactory statistical estimates of the stock of residential housing for those benchmark years exist yet; but according to Kazushi Ohkawa, the total stock of residential housing in Japan increased from 9,934 billion yen in 1938 to 10,832 billion yen in 1960, both in 1960 prices, that is by 9 percent, while the population increased by 32.3 percent over the same period. This means a decline by some 13 percent on the per capita basis. There do exist also estimates of per capita floor space in urban areas, and the comparison between the prewar and the present indicates at most an increase of 20 percent. What is more indicative of the discrepancy between the flow approach and the stock approach in the housing sector is the fact that the ratio between the acquisition cost of average urban houses and the average annual income of urban dwellers has been steadily rising, from 2.57 in 1960 to 4.75 in 1973, while the unit space of a house cum garden has been shrinking at the same time. The Fisherian approach is certainly called for in such circumstances.

The third problem to which we might address ourselves relates itself to the future. Most of the special factors which have been cited as accounting for the extraordinary rate of growth of the Japanese economy in the postwar period are now almost exhausted. Especially notable is the fact that Japan's catching-up process in technologies with the relatively more advanced countries of the West in the postwar period is now practically over; and at the same time some of the newly industrializing countries, such as the Republic of Korea, are already able to compete out Japanese products in the world market, not only in light industries but also in heavy industries such as shipbuilding and steel.

Opinions seem to diverge as to what rate of growth, and how long, Japan may be able to expect to maintain in the coming future. The most sanguinely optimistic forecast is offered by Herman Kahn, who cites the annual growth rate of 10 to 12 percent, followed by many of the Japanese academic and government economists, who consider the 6 to 7 percent rate to be not only feasible but also essential from the standpoint of maintaining a reasonable level of near-full employment. At the opposite pole are grouped a minority of economists in Japan who, for one reason or another, take the view that the rate of real GNP growth of the Japanese economy in the coming decade is likely to be 5 percent or less. In this latter group, however, are included two distinctly different viewpoints: one which regards the low rate of growth with pessimism and the other which considers the shift in the growth rate downwards to be in accord with the maturation of the Japanese economy accompanied by a gradual shift in the value systems prevalent in society.

This very last group often quotes a dictum of John Stuart Mill, who wrote: "It is scarcely necessary to remark that a stationary condition of capital and population implies no stationary state of human improvement. There would be as much scope as ever for all kinds of mental culture, and moral and social progress; as much room for improving the Art of Living, and much more likelihood of its being improved, when minds ceased to be engrossed by the art of getting on."[3] No one probably would quarrel with the idealism of Mill, just as we may agree with Keynes when he writes that: "I look forward, in days not so very remote, to the greatest change which has ever occurred in the material environment of life for human beings in the aggregate.... The course of affairs will simply be that there will be ever larger and larger classes and groups of people from whom problems of economic necessity have been practically removed." Once we are ushered in onto such a stage, Keynes wrote, "we shall once more value ends above means and prefer the good to the useful." "But beware!" he continued, "The time for all this is not yet. For at least another hundred years we must pretend to ourselves and to every one that fair is foul and foul is fair; for foul is useful and fair is not."[4]

Keynes wrote these words almost 50 years ago in the midst of the Great Depression; but the time dimension he had in mind for the coming of the stage of "ends being valued above means" may well have been too optimistic. Realistically speaking, a capitalist society like Japan, though partaking the character of a mixed economy in an increasing degree, still retains that basic motive force of *accumulation* which Marx spoke of. Schumpeter, too, did write that: "Unlike other economic systems, the capitalist system is geared to incessant economic change.... Whereas a stationary feudal economy would still be a feudal economy, and a stationary socialist economy would still be a socialist economy, stationary capitalism is a contradiction in terms."[5] How the "soft landing" of the high-growth-geared economy of Japan to a more or less stationary condition can be accomplished is a moot question, especially when we reflect upon the fact that the *intra*-industry coordination for the restraint of growth cannot be so smoothly done under competitive conditions and also the fact that the "natural rate of growth" (in Harrodian sense) for Japan is likely to be still fairly high in the coming decades. The adjustment of labor participation rate downwards and the reduction of working hours and workdays are both likely to face strong resistance yet for some time to come.

[3] J.S. Mill, *Principles of Political Economy*, Longmans, Green and Co. Ltd., London, 1926, p. 751.

[4] J.M. Keynes, *Essays in Persuasion*, 1931, p. 372.

[5] J.A. Schumpeter, "Capitalism in the Postwar World," in *Postwar Economic Problems*, edited by S.E. Harris, 1943, pp. 116–17.

There are a number of other factors which we will have to take into account if we are to venture predictions on the probable rate of GNP growth of the Japanese economy in the future; in particular, the development in international spheres will be highly relevant for the country like Japan, such as what is likely to be the trend of the terms of trade for Japan, how far will the protectionist tendency of the present persist, how quickly will the newly industrializing countries catch up with Japan in the latter's key export industries, and so on. In any case, it will not be a simple question of "confidence," as Herman Kahn suggests,[6] for Japan to be able to continue on her course of rapid economic growth.

* * *

In Part Three of this volume we have included two "contributed papers": one by Kimio Uno on "A Social Indicators Approach to Economic Development" and the other by Tomotaka Ishimine on "Ocean Resources: An Analysis of Conflicting Interests." Both of them were valuable contributions to the specialized session on Japan of the Congress, and we regret that we did not have enough time to discuss them on the floor of the session.

For the editing of this volume we are greatly indebted to Mrs. Drucilla Ekwurzel, the Assistant Editor for *Journal of Economic Literature*. Were it not for her most painstaking scrutiny over the manuscripts in minimizing inadequate English expressions and obscure sentences, this volume would not have attained the degree of intelligibility which it now has. To her and also to the staff of the *Asahi Evening News*, whose patient cooperation was essential in the course of printing of the volume, I should like to take this opportunity to express my profound gratitude.

[6] *Cf*. Herman Kahn and Thomas Pepper, *The Japanese Challenge*, Hudson Institute, 1978.

PART ONE

COMPARATIVE ANALYSIS OF ECONOMIC GROWTH, INDUSTRIAL ORGANIZATION AND GROWTH ACCOUNTING

Past Economic Growth of Japan in Comparison with the Western Case: Trend Acceleration and Differential Structure*

Kazushi Ohkawa *Hitotsubashi University*

Trend acceleration in the macro-output growth rate and a differential structure in sectoral productivities over time are conceptualized as the most distinct features drawn from an overall quantitative appraisal of Japan's long-term growth patterns for 1885–1965. Both are also found for several Western countries of economic backwardness. A somewhat generalized observation on the catch-up process in terms of social capability and technological-institution relevance is developed, and some suggestions are made for contemporary LDCs.

* In the quantitative appraisal of Japan's past growth, I should like to acknowledge the help I have had from all the members of our study group of the Long-Term Economic Statistics (LTES) of Japan. An English summary version of the 14 LTES volumes in Japanese (Toyo Keizai Shinpo-sha, 1965–) is forthcoming: Kazushi Ohkawa and Miyohei Shinohara eds. *Japanese Economic Growth: A Quantitative Appraisal* (forth.).

I must also acknowledge a special intellectual debt to my long collaboration with Professor Henry Rosovsky, the co-author of *Japanese Economic Growth: Trend Acceleration in the Twentieth Century* (1973) on which the basic idea of this paper depends.

INTRODUCTION

The purpose of this paper is to present an overview of Japan's long-term growth pattern in comparison with the Western experience. The comparison is of particular interest as Japan presents the only non-Western case of long-term modern economic growth. Starting from its low Asian levels a century ago, Japan's per capita income now ranks with the average European level. In this context one is naturally inclined to be interested in the analysis of the Japanese case, which may well be regarded as exceptional among advanced, industrialized countries.

Japan's case has been characterized by a number of unique features such as a high growth rate, a high proportion of savings and investment, speedy industrialization and trade expansion, as well as a dualistic economic structure which has marked not only the post–World War II period but also the longer prewar period. In what follows, I am not tempted to amplify these unique or exceptional aspects of Japan's growth pattern. Instead, I will try to elucidate Japan's experience in a way comparable to the Western record so as to enrich, as far as possible, our knowledge of growth and development in general. The difference or dissimilarity found between Japan and the West is assumed to be explainable if the relevant conditions and factors are systematically understood. I have increasingly felt the usefulness of this approach, though being aware of the demands it places on us. Furthermore, I feel this work to be of great comparative interest to the development pattern of contemporary LDCs, particularly in light of the facile talks we have had on the relevance or irrelevance of advanced countries' experience to the contemporary problems of LDCs. To be practical and effective, I believe, the case of Japan should be treated as a sort of bridge between the DCs and LDCs.

Out of a number of possible subjects, I take up "trend acceleration" and "differential structure" and their relationship. Trend acceleration is a specific pattern of output growth: it is characterized by a long-term trend of acceleration in the growth rate of per capita product (a counter pattern is a trend deceleration). It is a widely prevailing view that Japan belongs to the highest category of growth rate performance. Actually, Japan's rate of growth was rather moderate during its early phases of development; its average high rate is the result of trend acceleration. In any event, I am not concerned here with such an average observation, but rather interested in the pattern of over-time change in the growth rate.

Differential structure is a specific production structure featured by a growing gap of sectoral productivities (output per worker) over time. Dualism has often been talked about in Japan in terms of productivity, wages, and related terms. I am concerned here, not with the static, but rather with the dynamic aspect of production structure. The problem of the growing productivity gap pattern is equivalent to what Simon

Kuznets has called "diversion" as against the "conversion" of differences in output per worker between major sectors.

Both trend acceleration and differential structure are not abstract analytical concepts, but rather empirically derived terms, which attempt to describe the Japanese pattern of modern economic growth. The underlying mechanism and interrelations developed in our volume (1973) will be applied to interpreting these phenomena in the last section of this paper.

I. TREND ACCELERATION

Let us begin by identifying the historical pattern of Japan. The relevant statistics are summarized in Table 1. Because of the existence of long swings of the Kuznets type, special consideration is taken for comparing over-time changes to identify secular trends in the growth rate.

In Table 1, Panel A shows an alternation of upswings (U, odd-numbered lines) and downswings (D, even-numbered lines); Panel B covers full swings for successive troughs (T to T) and peaks (P to P); while Panel C presents secular trends, T to P with the exception of 1887–1930, which is T to T. Beginning with GNP performance in Panel A, swing regularity is perfect for the entire period. Period average growth rates tend to increase in each successive upswing phase (U) from I through VII; in downswing phases (D), they increase from II to IV, but from IV to VI, decrease. An extraordinarily small rate of growth for VI is caused, of course, by war dislocation. In saying this, we do not intend to ignore the actual negative effects brought on by historical events, but they are considered to be *ad hoc* decelerating factors. Admittedly, differing views may be possible on this issue. Nevertheless, we feel that the trend performance of the increasing growth rate can be positively identified with respect to upswings; the trend performance in downswings presents no compelling reason for denying operation of the trend accelerating factors. In Panel B the average growth rate increases in each P to P period (II′ to IV′ and IV′), whereas for T to T periods it is mixed: an increase for I′ to III′, but a decrease for III′ to V′, again reflecting war dislocation.

The secular growth performance illustrated in Panel C may serve to deepen understanding of Japan's experience, particularly if looked at in comparison with the figures in Panels A and B. For example, the postwar growth rate, 9.6 percent for 1953–69, is much higher than the 3.1 percent average for the entire prewar period (1887–1938). However, when the postwar period is treated as an upswing exhibiting a secular trend pattern, continuity between the two periods is more evident. Comparing the prewar period (1887–1938) and the entire span of modern growth (1887–1969) also places postwar growth in better perspective: the average

TABLE 1

LONG-TERM PATTERN OF AGGREGATE GROWTH RATES: GNP, POPULATION, AND PER CAPITA

(Average Annual Rates of Growth, Percentages)

Period (Length in Years)			*GNP* (*1*)	*Total Population* (*2*)	*Per Capita GNP*[a] (*3*)
A. Long-swing phases					
I	(U)	1887–1897 (10)	3.21	0.96	2.25
II	(D)	1897–1904 (7)	1.83	1.16	0.67
III	(U)	1904–1919 (15)	3.30	1.19	2.11
IV	(D)	1919–1930 (11)	2.40	1.51	0.89
V	(U)	1930–1938 (8)	4.88	1.28	3.60
VI	(D)	1939–1953 (15)	0.58	1.36	–0.78
VII	(U)	1953–1969 (16)	9.56	1.03	8.53
B. Trough to trough and peak to peak					
I′	(T)	1887–1904 (17)	2.64	1.04	1.60
II′	(P)	1897–1919 (22)	2.72	1.18	1.54
III′	(T)	1904–1930 (26)	2.92	1.32	1.60
IV′	(P)	1919–1938 (19)	3.44	1.35	2.09
V′	(T)	1930–1953 (23)	2.08	1.29	0.79
VI′	(P)	1938–1969 (31)	5.21	1.06	4.15
C. Secular trends					
		1887–1930 (43)	2.81	1.21	1.60
		1904–1938 (34)	3.26	1.25	2.01
		1887–1938 (51)	3.13	1.22	1.91
		1887–1969 (82)	3.92	1.21	2.71
		1904–1969 (65)	4.19	1.17	3.02

Source:

Real GNP: LTES, Vol. 1, Part III, Tables 18 and 19a, revised by the new data for export and import prices (LTES, Vol. 14, forthcoming). Population: LTES, Vol. 2, forthcoming.

Notes:

The annual rate of growth is calculated as a percentage increase from the preceding year in smoothed series of 7-year (5-year for postwar) moving averages. The period average growth rate is a simple average of individual years' growth rate. An exception is for 1938–53, calculated by a simple bridge between the two years due to the lack of consistent data.

[a] Column 2 minus column 1.

growth rate is 3.9 percent measured for the entire period 1887–1969, 3.1 percent up to 1938, and 5.2 percent for 1938–69. In other words, while getting back to the over-time trend is responsible for a good deal of postwar growth, there has been a clear acceleration of the secular growth rate including that of the postwar period. This long-term, historical pattern of progressive growth rate increases is what we may define as trend acceleration.

Turning again to Table 1, Panel A, we see that the population growth rate has no positive relation with the output growth rate swings. Rather, it shows a trend of moderate acceleration until some time in the 1930s. Since the 1930s, population growth has decelerated (Panels A and B), resulting in very stable average rates of increase in secular comparisons (Panel C). Thus, growth rate swings of per capita GNP, shown in column 3, Panel A, are essentially the same as those of GNP, being largely unaffected by the population growth pattern. As for the trend in GNP per capita, acceleration appears somewhat blurred for earlier years because of the rising rate of population increase. Furthermore, an abrupt population increase in VI due to postwar repatriation from overseas territories causes much of the negative value of this period. If we take the same view regarding the abnormality of the period as we did in the GNP case, the essence of trend acceleration can also be seen in per capita GNP. The performance for longer intervals (shown in Panel B) endorses this, though with somewhat stronger qualifications for T to T periods.

Secular growth rates shown in Panel C may appear far more moderate than usually expected for Japan's experience. For example, per capita GNP increased at an average rate of 2.7 percent for the entire period 1887–1969, which is less than one-third the postwar rate. Furthermore, the prewar average rate had been less than 2 percent. Notwithstanding the operation of growth accelerating factors, the depressed 1920s and war-afflicted 1940s contributed much to drawing down the average growth rate of per capita GNP.

This is, I believe, a new finding, which implies that the high postwar growth rate may be explained as a continuum in the secular trend of growth rate acceleration in Japan. In other words, what is asserted by trend acceleration is essentially a recognition of the progressive increase in growth rates from one phase to another. Nevertheless, some may not accept this explanation, emphasizing the discontinuity between pre- and postwar periods in the light of the internationally high rates of postwar growth, of which Japan presents a most remarkable case. Others may have serious reservations due to the relatively short (only a century) period being analyzed. Still others may hesitate to accept the secular trend explanation because such a trend pattern has not been found for Western countries.

Of the possible criticisms mentioned above, let us discuss the one relevant to the growth pattern of Western nations. Following Kuznets's findings (1971),[1] we have first to accept the fact that, internationally, there is no common pattern of trends. For the per capita product growth rate, this appears to imply that no significant acceleration or deceleration is found for the entire period of modern economic growth of various nations. However, there is a distinct systematic trend pattern for some countries. In Table 2, Kuznets's data for selected, relatively backward European countries are reproduced.

I have never examined in detail the historical performance of these nations. Yet, these records strongly suggest an acceleration pattern similar to Japan. I think the reason why one hesitates to recognize such a secular pattern may be due to the lack of comparative work in this area. I am not asserting, however, that all relatively backward nations (that is those whose initial dates of entry into modern economic growth is relatively late) tend to exhibit such a trend performance. Instead, my point is that the trend acceleration found for Japan (initial dates, 1870–80, GNP per capita: $140 for 1885–90, to be compared with the figures in Table 2) is not unique and that some of the conditions and factors relevant to this performance may be common to relatively backward nations.[2]

II. DIFFERENTIAL STRUCTURE

In Japan the macro phenomenon of trend acceleration took place at the same time that a differential structure was developing at the sectoral level. In Table 3, an overview of this pattern is shown in terms of the ratio of output per worker in agriculture (A), services (S), and their sum ($A + S$) to that of industry (I). The conventional procedure of taking the aggregate output per worker as the standard is not adopted because of the dualistic nature of growth in Japan. It should be noted here that the modern sectors are mostly represented by industry, although some modern sectors such as banking are contained in S.

A secular trend can be seen in the growing gap of output per worker in A/I, S/I, as well as $(A + S)/I$ extending into the postwar years until the mid-sixties. The trend is diluted somewhat because changes in relative prices of output were in favor of the A and S sectors. This is, of course, the result of the differential rates of output growth per worker between the major sectors, implying a different pattern of changes in the sectoral shares between output and labor force (and capital) as well as a divergent

[1] Specifically pp. 40–41.

[2] This finding may appear relevant to the Gerschenkron hypothesis (1962) that the later the entry of a country into the modern economic process, the higher the growth rate is likely to be. What he suggests seems to be more rigorous in connecting growth rates and the degree of economic backwardness. As far as the empirical data suggest, we cannot claim such an association.

TABLE 2

GROWTH PATTERN OF SELECTED EUROPEAN COUNTRIES OF ECONOMIC BACKWARDNESS

	Initial Dates (1)	Per Capita GNP at Initial Dates (1965$) (2)		Duration of Period (years) (3)	Ratio of Growth for Decade (%)		
					Total Product (4)	Population (5)	Product per Capita (6)
Norway	1865–69	287	1865–69 to 1885–94	22.5	18.8	7.2	10.8
			1885–94 to 1905–14	20	24.9	9.2	14.3
			1905–14 to 1925–29	17.5	32.5	9.2	21.4
			1925–29 to 1963–67	38	42.6	8.0	32.1
			1925–29 to 1950–54	25	38.8	7.5	29.1
			1950–54 to 1963–67	13	50.4	9.0	38.0
Sweden	1861–69	215	1861–69 to 1885–94	24.5	28.5	6.5	20.7
			1885–94 to 1905–14	20	38.8	7.1	29.6
			1905–14 to 1925–29	17.5	28.6	6.2	21.1
			1925–29 to 1963–67	38	47.0	6.5	38.0
			1925–29 to 1950–54	25	45.5	6.5	36.6
			1950–54 to 1963–67	13	49.9	6.5	40.8
Italy	1895–99	271	1895–99 to 1925–29	30	24.6	6.5	16.9
	1861–69	261	1925–29 to 1963–67	38	37.0	7.2	27.8
			1925–29 to 1951–54	25.5	22.7	7.4	14.3
			1951–54 to 1963–67	12.5	71.5	6.9	60.4

Source:
Simon Kuznets (1971), Table 2, p. 24 for (1)–(2) and Table 4, p. 39 for (3)–(6).

pace of technological-organizational progress between sectors, in particular between the modern and traditional sectors.

The statistical evidence for this is shown in Panel B of Table 3, in terms of both simple and weighted growth rates of output per worker for selected periods, in trough to trough and peak to peak ranges. The phenomena of trend acceleration can also be seen in the secular performance of the aggregate growth rates of output per worker. Sectorwise, a sharp contrast is seen between a more-than-4 percent growth rate of the *I* sector and a near zero (except for the postwar period) growth rate of the *S* sector. Between the high growth rate of the *I* sector and the low rate of the *S* sector, we see the relatively small growth rate of the *A* sector. The weighted growth rates reveal that the sectoral contribution to trend acceleration is made solely by the *I* sector, while the *A* and *S* sectors operated as decelerating factors except during the postwar period.

The data in Table 4 tend to invalidate the notion mentioned above that the differential structure phenomena must be unique to Japan. We see therein that a similar pattern is found in the development pattern of at least a few Western nations, though limited in number. What characterizes Japan's case is the pace of widening differentials, as is illustrated by the figures of average annual rate of change.[3]

III. THE CATCH-UP PROCESS FOR LATECOMERS

The term "catch-up" is used here in the purely economic sense that nations of later entry into modern economic growth eventually arrive, irrespective of political intentions, at an income level equivalent to that of advanced nations. The Japanese record presents a typical case of the process of long-term catch-up. Trend acceleration and differential structure are the representative characteristics of this process. I would like to interpret these two phenomena not as independent of each other, but by setting an integrated hypothesis composed of two basic concepts: social capability and relevance in technological-organizational transfer to countries of economic backwardness.

Japan's entry into modern economic growth in the nineteenth century as a latecomer has been explained in the form of a sizeable gap between a backward economy of Asian levels and the far less backward socio-political structure. As a result of this discrepancy, in its early phase of

[3] These ratios cannot directly be used for the so-called "divergence or convergence" discussion, particularly because of the higher ratio than unity often found for earlier phases. In this respect, different interpretations should be attached to *A/I* and *S/I* differences. For example, in Table 3, Panel A for Japan's case, the ratio *S/I* is only really relevant for differential structure formation since 1915–22 (current price series) and since 1927–33 (constant price series). Such demarcation is not possible in Table 4.

TABLE 3

OUTPUT PER WORKER BY MAJOR SECTOR: RATIOS AND RATES OF GROWTH

(A) Sectoral Ratios

Selected Periods	*A/I*		*S/I*		*(A + S)/I*	
	Current prices (1)	*Constant prices (2)*	*Current prices (3)*	*Constant prices (4)*	*Current prices (5)*	*Constant prices (6)*
1885–89	0.42	0.68	1.75	3.71	0.65	1.21
1894–99	0.43	0.55	1.42	3.12	0.63	1.07
1901–07	0.36	0.44	1.21	2.23	0.55	0.84
1908–14	0.34	0.38	1.03	1.66	0.50	0.68
1915–22	0.32	0.33	0.88	1.61	0.48	0.68
1927–33	0.25	0.27	0.92	0.99	0.47	0.51
1936–40	0.24	0.21	0.63	0.77	0.38	0.41
1955–60	0.29	0.23	0.78	0.71	0.52	0.45
1960–65	0.29	0.19	0.87	0.46	0.60	0.34

(B) Annual Growth Rates of Constant Price Series

	Simple			*Weighted*			*Aggregate*
	A	*I*	*S*	*A*	*I*	*S*	*(%)*
1887–1904	1.49	3.68	0.92	0.53	0.57	0.45	1.55
1897–1919	2.43	4.49	0.60	0.78	0.92	0.28	1.98
1904–1930	1.91	4.16	−0.19	0.49	1.28	−0.08	1.85
1919–1938	1.05	4.32	−0.03	0.21	1.80	−0.01	2.02
1934–1964	2.90	4.39	3.07	0.39	2.09	1.32	3.70

Source:

Panel A is reproduced from Table 3-5, p. 56 and Panel B from Table 3-4, p. 54 (with a new addition for 1938–64) in LTES, Vol. 1.

Notes:

(i) Output is domestic product net of capital depreciation.

(ii) Agriculture (*A*) includes forestry and fisheries; industry (*I*) is composed of manufacturing, mining, construction, transportation, communication and public utilities; and services (*S*) includes all others.

(iii) Constant prices are calculated in 1934–36 prices. Because of the lack of official price deflators by sector for the postwar years, these are our estimates. For the years since 1965, deflators for the *S* sector are not available.

TABLE 4

RATIOS OF OUTPUT PER WORKER BY SECTOR: SECULAR CHANGES IN WESTERN COUNTRIES

		Sectoral Ratios		*Average Annual Rate of Changes*	
		A/I	*S/I*	*A/I*	*S/I*
(A) In constant prices:					
Germany	1850– 59	1.22	3.10	−0.95	−1.58
(1913 prices)	1935– 38	0.52	0.73		
Sweden	1861– 70	0.56	1.60	−0.36	−0.57
(1959 prices)	1963– 67	0.39	0.89		
Italy	1861– 70	1.13	3.79	−0.76	−1.34
(1963 prices)	1963– 67	0.53	0.99		
Japan	1885– 89	0.68	3.71	−1.67	−2.74
(1934–36 prices)	1960– 65	0.19	0.46		
(B) In current prices:					
Sweden	1861– 70	0.51	1.14	−0.13	−0.24
	1963– 67	0.45	0.86		
Italy	1861– 70	1.09	2.33	−0.63	0.75
	1881–1900	0.96	2.72	−0.79	−1.20
	1963– 67	0.51	1.04		
Japan	1885– 89	0.42	1.75	−0.49	−0.91
	1960– 65	0.29	0.87		

Source:

Simon Kuznets (1971), Table 45, pp. 290–92 (except Japan). He gives the ratio for *A/I* and *S/I* as "sectoral product per worker relative to adjusted country-wide product per worker." The ratio *A/I* is converted from the above. *S/I* is cited directly from his table.

Notes:

Output is mostly GDP; specifically excluded is the output of financial sectors such as banking, insurance, and real estate and the income from dwellings. For Japan, NDP, imputed income for dwelling excluded.

development, Japan had considerable capability to exploit modern technology and organization. This initial gap may go a long way to explain Japan's initial phase of growth. But the real analytical challenge is to explain the continuum of trend acceleration throughout the twentieth century when economic backwardness was *decreasing*. Surely, a hypothesis of economic backwardness cannot cope with this challenge, except for the short-term catch-up process of the postwar years, which lasted until the mid-sixties.

For a latecomer there is always room for introducing advanced technologies of foreign origin. However, there is a set of limiting factors that we once called the amount and level of social capability. We use that term to designate those factors constituting a country's ability to import or engage in technological and organizational progress.[4] When Professor Moses Abramovitz says "countries with the proper apparatus of governmental and commercial institutions, with educated and skilled population and advanced technological capabilities," he seems to refer to, more or less, the same phenomena.[5] What we are particularly concerned with here is the process of increasing its amount and of raising its level because we hypothesize that these set the basic potentiality for trend acceleration in the growth rate.

Although the mechanism is still not sufficiently clear, it can reasonably be supposed that social capability changes as a function of time. This is relevant to what Professor Abramovitz defines as a gap between the potentiality for and the realization of the growth in productivity. In my opinion, the potentiality itself will not be "given" over the long term for a country of economic backwardness. As such a country's social capability rises, the amount and level of borrowable backlog of technological knowledge from advanced countries will be enlarged and/or raised up. As relative backwardness measured in terms of the productivity gap becomes narrowed, the productivity growth potential of latecomers will accordingly be increased. Here we can see the possibility of trend acceleration for latecomers. Viewed from the standpoint of comparison with advanced countries, this becomes a secular process of catch-up because empirically (from the beginning of this century at least) the growth rate of output per worker or per capita product of advanced countries, such as the United States, has continuously been modest by comparison.[6] With respect to the facilitating factors, the crucial element is capital

[4] Ohkawa and Rosovsky (1972), especially, Chapters 1 and 8.

[5] Moses Abramovitz, "Rapid Growth Potential and Its Realization: The Experience of Capitalist Economies in the Postwar Period" (1977)—a paper presented to this Congress. I have owed much to his clarification of the productivity growth process in terms of distinguishing the potentiality and its facilitating factors. However, the catch-up process in his context seems to be posed in shorter-terms in order to deal with the postwar growth rather than the longer-term catch-up process that I discuss.

[6] The interwar period is often said to be a "lower-than-average rate of growth" period, and in his paper, Abramovitz states (referring to Table 1) that "the large productivity gap of 1913 and its enlargement during the disturbed years between then and 1950 did, in fact, constitute a strong potentiality for rapid postwar growth in the less advanced countries" (1977). This does not mean, however, all the less advanced countries had failed to keep their growth rate of per capita product greater than that of the U.S. According to Kuznets's data (1972), during 1925–29 to 1950–54, Japan, Italy, Sweden, Norway, and Canada maintained greater growth rates than that of the U.S. It is true that the process of catch-up became weaker, but even the interwar period is not an exception for the basic pattern of growth of these countries.

formation. Japan's fixed investment (and savings) proportion of GNP has shown a pattern of secular trend increase: starting from a moderate proportion of a little more than 10 percent (in gross terms), it recently reached a height of 35 percent. This implies a secular increase in capital per worker, especially in the modern sectors. The association between this and the rate of productivity increase has been explained in a model of "interaction" between relevant factors in our volume (1973), chapter 8. I shall not repeat here our explanation of this aspect of the realization problem, except to restate one point. That is, we expect to see Japan enter into a phase of decelerating growth rate towards the end of the sixties due to the less vigorous operation of the facilitating factors.

So much for aggregate observation. For a country like Japan, which experienced dualistic growth heavily conditioned by the existence of surplus labor in the traditional sectors of the economy, the relevance of what previously has been mentioned must have been limited. Such sectors as *A* and *S*, as well as small scale enterprises in manufacturing (inseparable from the *I* sector in our statistics), need labor-intensive technology. In many lines of production activity, borrowable technological-organizational knowledge cannot yield sufficient gains in productivity. The classic example would be the low relevance of Western agricultural machinery to paddy field cultivation in Japan (and other Asian countries). What has been called the differential structure is one aspect of the failure of borrowing technology to achieve a broader degree of relevance because in the modern sector the relevance has been continuously increased in the process of its successful growth. This does not mean that no technological progress has been made in the traditional sectors. To the contrary, output per worker in the *A* sector, for example, increased at a pace at least comparable to the average growth rate in industrialized countries. The problem at issue is relative backwardness in expanding the relevance of borrowable technology within the national economy.

It is to be noted that similar performance can be seen, though to a lesser extent, in Western countries and that the sectoral difference in the relevance of modern technology and of the rate of increase of this difference over time can be more generally interpreted beyond Japan's case. With respect to the sectoral productivity differentials, a number of authors have discussed relevant conditions and causal factors, among which the most important may be the sectoral differences in quality of "human capital": in the lower productivity sectors a lower level of quality of human capital prevails.

If this is valid, our data suggest the quality differentials must have been widened over time in that sectoral differences widened. However, as far as the Japanese record is concerned, this hypothesis cannot be accepted. Furthermore, in an international comparison it is hard to say that Japan's relatively greater degree of differential structure can be explained by its greater degree of sectoral differences in human capital quality. A much more plausible explanation may be found in surplus labor or the over-occupied labor situation in agriculture and services, which constitutes the major factor of limiting the relevance to advanced technological progress. I believe this limitation is highly suggestive *vis-à-vis* the contemporary LDCs' pattern of development. For the modern sectors a trend acceleration pattern of growth rates can be possible in modernizing the production structure if the necessary conditions are realized after passing through a certain period of preparation. However, if surplus labor does exist more seriously in the traditional sectors than in Japan, we can expect stronger accompanying differentials in sectoral productivity growth. This is the area of greatest challenge for further analysis.

REFERENCES

Abramovitz, Moses. 1977. "Rapid Growth Potential and Its Realization: The Experience of Capitalist Economies in the Postwar Period." Paper presented to the Fifth World Congress of the International Economic Association, Tokyo, 29 August–3 September.

Gerschenkron, Alexander. 1962. *Economic Backwardness in Historical Perspective: A Book of Essays*. Cambridge, Mass.: Harvard University Press.

Kuznets, Simon. 1971. *Economic Growth of Nations: Total Output and Production Structure.* Cambridge, Mass.: Harvard University Press.

Ohkawa, Kazushi and Rosovsky, Henry. 1973. *Japanese Economic Growth: Trend Acceleration in the Twentieth Century.* Stanford, Calif.: Stanford University Press.

——— and Shinohara, Miyohei, eds. *Japanese Economic Growth: A Quantitative Appraisal.* New Haven: Yale University Press (forthcoming).

A "Marginal-Efficiency" Theory of Japanese Growth

Martin Bronfenbrenner *Duke University*

Three often neglected but (in my opinion) highly important additions or bold-facings to any laundry-list of causes for the two Japanese "economic miracles" (mid-Meiji and postwar-Showa periods) are presented in this paper. These are, in order of significance as I see it: (1) Governmental diminution of business risk for the larger companies, as reflected in higher marginal efficiency of investment; (2) the role of inflation at less than catastrophic rates in holding actual consumption below planned consumption—"forced frugality" if not precisely "forced saving"—and (3) the weakness of craft unionism, "bloody-mindedness," and similar labor pressures against reductions in capital coefficients.

I. STABLE OR UNSTABLE PATH?

The first question I should like to consider about Japanese growth as we approach 1978—the centennial year for much of the Japanese economic-statistical record—is whether the country's long-term growth path has been technically stable or unstable.

This is more than a pedantic or semantic quibble. The implication of a stable growth path is dominance by "market forces," with an additional implication that growth would have proceeded at approximately its actual rates without such government interference (positive or negative) as actually occurred. The implication of an unstable growth path is that government interference was vitally necessary, inasmuch as the market growth process, left to itself, is apt to shoot upwards from "knife-edge" equilibrium to hyperinflation, or downwards to some form of stagnation, or to explosive fluctuations between these two states.

I have been among those students, both Japanese and foreign, who have seen Japanese growth as essentially an unstable process, and who have seen Japanese government economic policy as holding the actual growth path significantly above its equilibrium level without permitting hyperinflation.[1] It seems to me that the majority of such students have taken some such position. (This does not mean that we have been right, but only that I disavow any claim to originality.)

But more recently, cliometric studies—applying the higher econometrics to the Japanese historical record—have come out strongly on the other side of this debate. The authors have been primarily American.[2] I accordingly find it relevant to discuss early on the principal factors—beyond hardening of my intellectual arteries—which keep me unconvinced and unconverted.

Let us consider a mini-model tracing the movement of two abstract economic variables x and y over time t. Let us more explicitly suppose that we have fitted two functions f_1 and f_2, such that the equation system or "model":

$$x = f_1(y, t, \alpha) \qquad y = f_2(x, t, \beta) \tag{1-2}$$

yields stable paths of both x and y over time t. But these observed (x, y) values include the effects of our "shift variables" α and β, which we suppose to represent public policies, but on which we do not have useful separate quantitative observations.

[1] M. Bronfenbrenner, "Economic Miracles and Japan's Income-Doubling Plan" (1965), restated and expanded in Bronfenbrenner, "The Japanese Growth Path: Equilibrium or Disequilibrium?" (1970).

[2] Allen C. Kelley and Jeffrey G. Williamson, *Lessons from Japanese Development* (1974). I am also grateful to Professor Williamson for showing me a book-length manuscript in which he (and Leo J. de Bever) has subsequently re-examined the Kelley-Williamson methodology and conclusions.

From the stable time paths of our (x, y) system, as I understand it, a cliometrician might surmise that the growth path was essentially stable aside from random shocks and that public policy made little observable difference beyond "sound and fury." But now suppose that in some "nightwatchman state" without public intervention—(x, y) unaffected by any (α, β)—we should have, instead of equations (1-2), the alternative system:

$$x = g_1(y, t) \qquad y = g_2(x, t) \tag{3-4}$$

an unstable system which "explodes" or "shakes itself to pieces." This possibility, as far as I am aware, has not been ruled out thus far by any of the cliometric experiments undertaken on Japanese growth in either the Meiji or the Showa eras.[3]

II. THE MARGINAL-EFFICIENCY "SWORD"

If we can agree that public policy, including growth policy, really matters for good or ill, and that economic growth is more than the extrapolation of a cliometric multi-equation model, the next question arises from further disagreement between what Dr. Kanamori calls the "one-sword" and "literary" schools.[4] The "one-sword" school—Kanamori credits the term to Professor Saburo Shinohara—"explains the situation with a theory that cuts through a vital point of [the] economy—a point so vital

[3] An example from a related branch of economic dynamics may be helpful here. Samuelson's multiplier-accelerator model of economic fluctuations starts with a three-equation system relating income Y, consumption C, and investment I. (Time is indicated by subscripts.)

$$Y_t = C_t + I_t \qquad C_t = aY_{t-1} \qquad I_t = b(C_t - C_{t-1})$$

which combine to a growth path in Y.

$$Y_t - a(1 + b)Y_{t-1} + abY_{t-2} = 0.$$

This path may be stable or explosive, monotone or cyclical, depending on the numerical values of the Keynesian multiplier $(1/a)$ and the Aftalion-Clark-Frisch accelerator b. Suppose that the econometrician fits a dynamic (recursive) equation to the observed values of Y, and obtains (a, b) estimates implying stability. But the Y values might have been quite different in the absence of public-policy interventions, with coefficients (a', b') derived from the "free-market" path of Y implying explosive or anti-damped cyclical behavior.

See Paul A. Samuelson, "Interactions Between the Multiplier and the Principle of Acceleration" (1939), reprinted in Gottfried Haberler (ed.) *Readings in Business Cycle Theory* (1944).

[4] Hisao Kanamori, "Nihon no Seicho-Ritsu wa Naze Takai ka?" (1970), translated as "What Makes Japan's Economic Growth Rate So High?" in *Japanese Economic Studies* (Fall, 1972). Kanamori also mentions an "econometric" school, which we call "cliometric."

that a push there would set the whole economy pulsating."[5] The "literary" or possibly "laundry-list" school contents itself arraying large numbers of "factors responsible" for the Japanese record. Some journalists have listed as many as 25 such factors; Kanamori himself, while professing adherence to this school, requires only six.[6]

I classify my own "line" as closer to the Shinohara (or perhaps Shimomura) than to the Kanamori position. My candidate for the role of the Shinohara sword is, in Keynesian terms, a high marginal efficiency of domestic capital investment in Japan. This has resulted in its turn from two fairly constant and consistent Japanese government policies: (1) essentially guaranteeing against both failure and financial stringency an important subset of large domestic *zaibatsu, zaikai, or keiretsu* firms—I think the distinctions are more exclusively terminological than many Japanese industrial economists do—and (2) holding foreign multinational competition at bay by various capital-import controls until these "chosen instrument" firms have established themselves in dominant positions on the Japanese domestic market.[7] And under heading (1) I should include a subheading or codicil, that the "chosen instrument" firms are expected in turn to support and guarantee their principal subcontractors, sales agencies, and similar affiliates, both formal affiliates *de jure* and less formal affiliates *de facto*.

Macro-economists will recall the textbook definition of the "marginal efficiency" of investment in a capital instrument as the static marginal productivity of that instrument averaged and discounted over its expected life with additional discount for uncertainty. I quote one sentence in which Lord Keynes stresses most particularly the importance of uncer-

[5] Kanamori (1970), p. 32, citing the example of Dr. Osamu Shimomura's stress on fixed investment in heavy industry.

[6] Kanamori (1970), p. 32 f., "I am often asked by foreigners why Japan's economic growth rate has been high, [but] by mentioning the highest rate of savings, the fast increase in fixed investment, the sufficient manpower resources to draw on, the high educational level, and the small military expenditures, I find most of my questioners become satisfied."

[7] Herman Kahn has gone so far as to put the term "risk capital" in quotation marks in describing Japan. This is because "under present-day Japanese conditions...the real risks are often actually low. The high growth rate usually cuts down the odds for loss, and...often it also makes it easy to...cover over any losses that do occur. Finally, a large Japanese firm in really serious trouble would...be bailed out by the government—either by arranging for a merger or for the banks to extend loans. In any case, the employees, stockholders, debtors...[would be] taken care of. Thus one of the reasons that many large modern Japanese firms can afford to operate in a seemingly risky way is that they know their government and society stands behind them....By American standards the Japanese seem to take too many risks. They are very expansionist....While this...can produce some mistakes, they...are the right kind of mistakes.... They are the kinds of mistakes that allow for high expansion rates and for overcoming...lethargy, inattention,...and rigidity...." *The Emerging Japanese Superstate* (1970), running quotation from pp. 107–08.

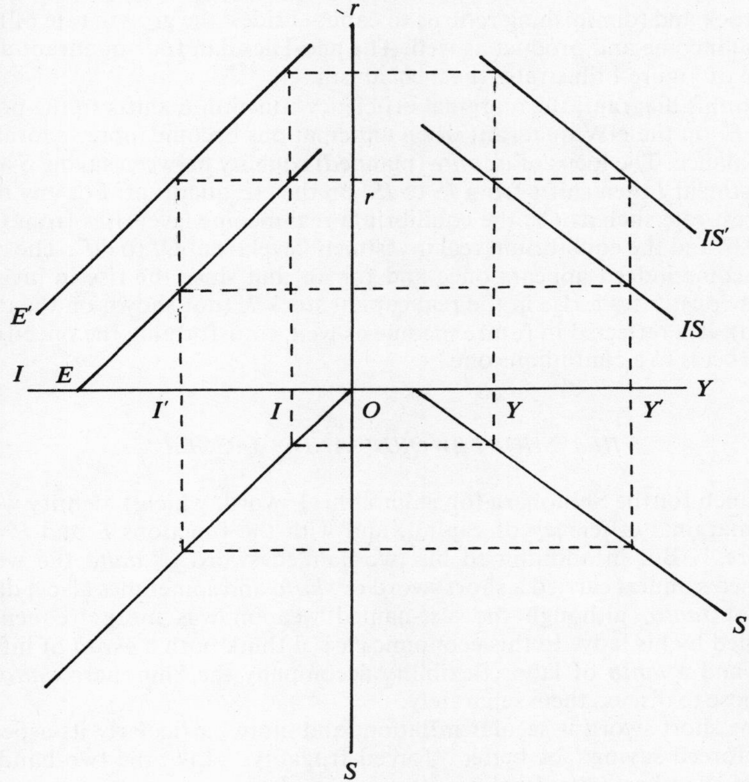

Figure 1

tainty:[8]

> It is important to understand the dependence of the marginal efficiency of a given stock of capital on changes in expectation, because it is chiefly this dependence which renders the marginal efficiency of capital subject to the somewhat violent fluctuations which are the explanation of the Trade Cycle.

It appears to follow that if uncertainty is reduced, and especially if the less sanguine "changes in expectation" are attenuated by public policy, the overall marginal efficiency of capital is increased. And therefore the level of investment is increased also, likewise the growth rate of the capi-

[8] John Maynard Keynes, *General Theory of Employment, Interest and Money* (1936), p. 143 f.

tal stock and (diminishing returns to capital aside)[9] the growth rate of national income and product as well. The neo-Hicksian four-quadrant diagram of Figure 1 illustrates the mechanism.

On this diagram, the marginal efficiency function E shifts to the position E' on the NW quadrant when anticipations become more favorable on balance. The locus of *ex ante* (planned) equality between saving S and investment I then shifts from IS to IS' on the NE quadrant. For any real interest rate such as Or, the equilibrium real income level rises from OY to OY' and the equilibrium real investment level from OI to OI'. The rise in income indeed appears once and for all, but since the rise in investment constitutes a rise in the real capital stock K (not shown on the diagram), it is reflected in future income as well, transforming the once-for-all process to a continuous one.

III. "SHORT SWORD" AND "DAGGER"

So much for the Shinohara (or Shimomura) sword, which I identify with the marginal efficiency of capital, and with the functions E and E' of Figure 1. But in addition to his two-handed sword or *daito* the well-dressed samurai carried a short sword or *shoto* and sometimes also a dagger or *tanto,* although the last-named weapon was more frequently wielded by his lady. In this economic case, I think both a *shoto* of inflation and a *tanto* of labor flexibility accompany the Shinohara *daito.* I propose to discuss them separately.

The short sword is secular inflation, and more particularly its aspects of "forced saving" or better "forced frugality." Like the two-handed broad-sword or *daito,* it relates to aggregate *demand.*

But the dagger relates to aggregate *supply,* by way of labor productivity in Lord Keynes's "efficiency units." To some extent, of course, the impressive annual rise of measured Japanese labor productivity per man-day or man-hour reflects only the increased and improved capital stock with which that labor works. But insofar as rising productivity, both average and marginal, also reflects improved human capital—in this case, both the ability and the willingness to update, broaden, or even transform old skills while learning new ones—it merits special and independent attention.

[9] I accept the argument of Frank H. Knight, "Diminishing Returns from Investment" (1944), for reasons set forth in M. Bronfenbrenner, *Income Distribution Theory* (1971), p. 316.

IV. THE DOUBLE ROLE OF INFLATION

A long-term inflationary bias—kept within "reasonable" bounds, which we shall not define—embodies two distinct mechanisms for encouraging and accelerating economic growth as economists conventionally measure it. One such mechanism, forced saving or forced frugality, we have mentioned already and will soon discuss in greater detail. The second mechanism is to focus saving upon physical and human capital, whose accumulation furthers continued growth, and away from cash balances, Keynes's "liquidity preferences," which tend rather to reduce growth. Let us explain these mechanisms in turn.

My favorite exposition of the forced-saving effect of inflation derives from the Swedish school of the 1930s (Erik Lindahl, Erik Lundberg, Bertil Ohlin).[10] In a closed economy, the *sources* of real social income (Y) are society's real expenditures for private consumption (C), for private investment (I), and for public goods and services (G). In real terms, we then have:

$$Y = C + I + G.$$

The planned or anticipated *uses* of this real income we may denote, following the Swedish exemplars, by Y^a, with the superscript denoting anticipation. At some pre-existing rate of price-level change, anticipated income Y^a is divided among three uses, anticipated consumption C^a, anticipated saving S^a, and anticipated tax payments T^a. Accordingly we have, still in real terms:

$$Y^a = C^a + S^a + T^a$$

and simple subtraction yields:

$$Y - Y^a = (C - C^a) + (I - S^a) + (G - T^a).$$

In the presence of unanticipated inflation or unanticipated acceleration of an existing inflation rate, we may suppose "money illusion" to induce increased supplies of both labor and raw materials, so that Y exceeds Y^a. We may also suppose inflation to reduce consumption C below its planned rate C^a, as consumers cannot now afford certain planned purchases or find their provisions for the future to be inadequate. (It is this inequality, $C < C^a$, which constitutes forced saving; there is no requirement that S exceed S^a, and indeed the opposite is more commonly the case.) We are left with the residual inequality $(I + G) > (S^a + T^a)$, which normally builds up capital and encourages growth by comparison with the non-inflationary case of $Y = Y^a$, $C = C^a$, and $(I + G) = (S^a + T^a)$.[11]

To present our second inflationary mechanism—the effect of inflation

[10] Ohlin, "Some Notes on the Stockholm Theory of Savings and Investment" (1937).

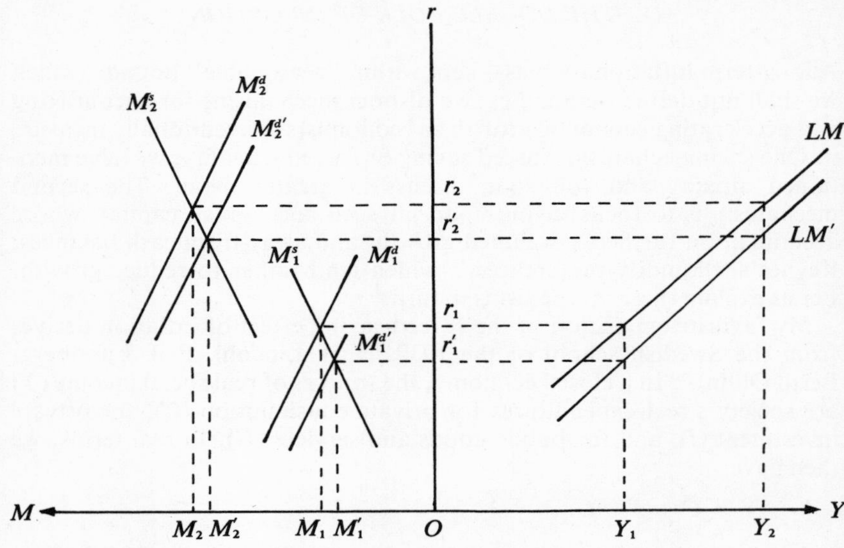

Figure 2

in shifting savings from cash balances to physical and human capital formation—we turn to Figures 2 and 3.[12] In Figure 2 we assume both the demand M^d and the supply M^s of real money M to rise with real income Y. We also postulate two effects of the real interest rate r, to reduce the demand for money M^d (by enhancing the attractiveness of alternative interest-bearing assets) and to increase the supply of money M^s (by increasing the profits of bankers from increasing loans and creating deposit currency). The combined outcome of shifts in (Y, r) upon (M^d, M^s) is another Hicksian locus, labeled *LM* and normally upward-sloping, connecting points for which the demand L for liquidity (our M^d) equals the money supply M (our M^s). The effect of unanticipated inflation or unanticipated increase in a pre-existing inflation rate is primarily to lower the M^d function for each income and interest rate on the left-hand panel of Figure 2. Thus, M_1^d, at income level OY_1, becomes $M_1^{d'}$; similarly

[11] If forced saving or forced frugality goes to finance private misinvestment or the "public component" component of *G*, this result will not occur. Sukarno's Indonesia is an apt Asian illustration of a case where it did not.

[12] Our argument depends on the analysis of James Tobin, "A Dynamic Aggregative Model" (1955) and "Money and Economic Growth" (1965). *See* also Edwin Burmeister and A. Rodney Dobell, *Mathematical Theories of Economic Growth* (1970), Chap. 6.

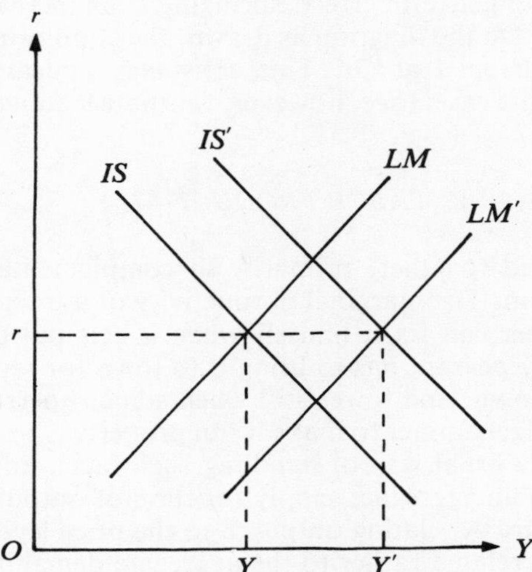

Figure 3

for M_2^d at income level OY_2. The eventual outcome is a rightward shift of *LM* to *LM'* on the right-hand panel of the figure.

Omitted from Figure 2, but also important, is inflation's supplementary contribution to the marginal efficiency of capital itself, and to the *IS* function. Inflation plays a (third) role in reducing one sort of risk in particular, namely, the risk of nominal capital loss from any cause whatever. This is a more significant element than it seems, particularly when (as in Japan) the representative firm's volume of fixed money debt is an "unsafely large" proportion of its total assets.

Figure 3 combines long and short swords. The joint effects of reduced uncertainty in general and inflation in particular upon the marginal efficiency of capital investment are represented on this figure by a rightward shift of *IS* to *IS'*, as on Figure 1. They are reinforced by the major effect of inflation upon the *LM* function of Figure 2, which becomes *LM'*. The rightward movements of these two Hicksian functions produce an equilibrium income level of *OY'* rather than *OY*.[13] The rise from *OY*

[13] Both this analysis and the diagrammatics of Figure 3 assume the initial situation at *OY* to be marked by sufficient unemployed or under-employed labor and other resources to permit movement to or beyond *OY'*.

to OY' combines once-for-all effects (which do not influence longer-term growth) with the longer-term effects springing from increased investment and capital stock. On the diagram as drawn, the equilibrium real interest rate remains unchanged at Or. This constancy appears as, and may actually be, a special case. (*See*, however, footnote 9 above.)

V. LABOR PRODUCTIVITY

Resources are used together, primarily in complementary rather than competitive fashion. The marginal productivity of a tractor is high when the superior farmer can fix it himself when it gets out of order, lower when the ordinary peasant has to bring it to town for repairs or call the repairman from town, and lower still when some superstitious illiterate beats it with a sledgehammer to make it run properly.

The economist's usual way of handling such facts, however, is to incorporate them in an aggregate supply function of output, like the function S of our Figure 4, relating output Y to the price level p, while marginal efficiency is related rather to the aggregate demand function D of the same variables on the same diagram. This practice we follow here, with the important *caveat* that D and S are less independent than they are drawn.

The supply function S of Figure 4 features an upward slope and upward concavity representing diminishing returns to fixed inputs (primarily "land"). It eventually becomes vertical when all resources are fully employed and no more income can be produced. This analysis ignores the possible availability of supplementary resources from abroad by gift, or by running down reserves of international resources.

An increase in the marginal efficiency of capital, as induced by the various considerations developed in the earlier sections of this paper, is reflected in upward and rightward shifts of a conventional aggregate demand function D. The increasing capital stock to which it leads is reflected in an outward and rightward shift of the aggregate supply function S—thereby depicting marginal efficiency at one remove. The inflationary trend manifest in the Japanese economy is also depicted by the upward shift of the price coordinate of the D-S intersection points on Figure 4 as we move to the right.

In this diagram we also suppose that effective fiscal and monetary fine-tuning supplement where necessary the rising marginal efficiency of capital to keep the economy at full employment. (This assumption is generally accurate for Japan during the greater part of its developmental history.) This feature is indicated by the location of D-S intersection points where S becomes vertical.

The moot question left unanswered thus far is why aggregate supply S

moves over time to S' rather than only to S'', given a rise in aggregate demand from D to D'. Or to put the matter differently, why can the Japanese economy, say, support aggregate demand at D' with no greater inflation than some other economy of equivalent size that can support only the smaller rise to D''?

The most obvious difference, given Japan's famous resource paucity, is the quality and quantity of the Japanese labor force. Attention has focused upon the man-hour (average) productivity of Japanese labor, which has apparently been increasing by 8 to 12 percent per year, approximately doubling corresponding rates for Western Europe and North America. To some extent, this difference itself reflects the higher rate of capital investment in Japan—marginal efficiency at second remove!—given the complementarity of inputs. But there is obviously more to it than that, although the mathematics is unwieldy even under

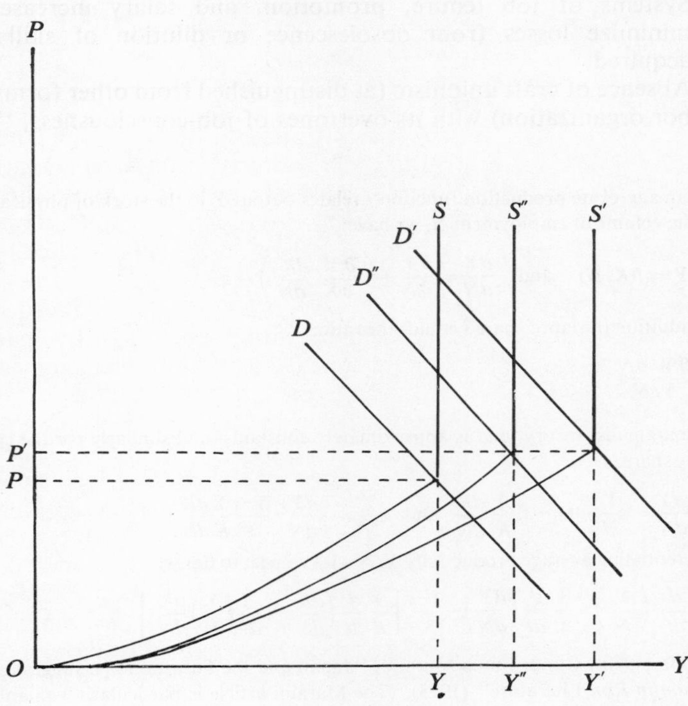

Figure 4

simplifying assumptions.[14]

Once agreed that increasing Japanese labor productivity as a factor in increasing the present marginal efficiency of Japanese capital investment is itself more than a spinoff from high past investment and high past marginal efficiency, immediate questions arise as to what the "something more" may be and from whence it may spring. It is then tempting to relapse into "national character" explanations, as Japanologists so diverse as Herman Kahn and Fosco Maraini have done,[15] but tautology-shyness steers me away into lesser degrees of abstraction. The principal components—marginal efficiency aside—of high Japanese labor productivity seem then to be the following quartet:

1. A well-educated and trained labor force, both blue- and white-collar, capable of learning new skills and disciplines well into adult life.
2. Availability of on-the-job training and retraining, comparable in the larger companies to what exists in modern military establishments.
3. Systems of job tenure, promotion, and salary increases that minimize losses from obsolescence or dilution of skills once acquired.
4. Absence of craft unionism (as distinguished from other forms of labor organization) with its overtones of job-consciousness, "lump-

[14] If an aggregate production function f relates output Y to the stock of physical capital K and the volume of employment N, we have:

$$Y = f(K, N) \quad \text{and} \quad \frac{dY}{dN} = \frac{\partial Y}{\partial N} + \left(\frac{\partial Y}{\partial K} \frac{dK}{dN} \right).$$

If in addition the labor share s equals the ratio:

$$\frac{\partial Y / \partial N}{Y/N}$$

as per marginalist theory, and is approximately constant—and similarly for the capital or property share $(1-s)$:

$$\frac{dY}{dN} = s\frac{Y}{N} + (1-s)\frac{Y}{K}\frac{dK}{dN} \quad \text{and} \quad \frac{Y}{N} = \frac{1}{s}\frac{dY}{dN} - \frac{1-s}{s}\frac{Y}{K}\frac{dK}{dN}.$$

Differentiating average productivity Y/N with respect to time t:

$$\frac{d}{dt}\left(\frac{Y}{N}\right) = \frac{1}{s}\frac{d}{dt}\left(\frac{dY}{dN}\right) - \frac{1-s}{s}\left[\frac{Y}{K}\frac{d}{dt}\left(\frac{dK}{dN}\right) + \frac{d}{dt}\left(\frac{Y}{K}\right)\frac{dK}{dN}\right].$$

[15] Kahn (1970), Ch. 2; Fosco Maraini, "Japan and the Future: Some suggestions from the *Nihonjin-Ron* Literature" (1975). (The Maraini article is particularly valuable for its treatment of the Japanese-language literature.) For a futurological slant, *see* also Robert Frager and Thomas P. Rohlen, "The Future of a Tradition: Japanese Spirit in the 1980s" (1976), Ch. 9.

of-work" theory, and reluctance to poach on "the other fellow's" specialty.

American white-collar folk, recalling negotiational disasters with construction tradesmen (whose job-conscious craft-unionism tradition is at its strongest) may ascribe near-exclusive responsibility to the last point cited. Such emphasis is probably exaggerated, after one realizes that our four points as listed are highly complementary and mutually intercorrelated. At any rate, and this controversy aside, the nature and significance of our developmental *tanto* or dagger should be obvious.

VI. THE DEVELOPMENTAL ARTS

The foregoing pages have been unduly mechanical. Not only do they permit but they positively support the implausible inference that a country can increase the marginal efficiency of its capital and the measured growth rate of its economy without bound, by strengthening guarantees (cushioning risks) for prominent firms, by accelerating inflation, and by training, multiplying skilled workers. Any such implications are exaggerated and overblown; this section proposes to indicate their limitations and assess Japanese success in the developmental arts of avoiding these same limitations.

As regards risk reduction: It is apparently important that conditions be implied if not imposed relative to product quality and technological alertness. Japan has done better than most rivals in this branch of developmental art. The case of passenger automobiles is an example: When the American Occupation ended (1952) the Japanese car was still an overpriced "pile of junk" imitating Western styling of a few years back. Its major technical defect was poor durability over Japanese roads. The Ministry of International Trade and Industry (MITI) guarantee system encouraged, permitted, and may have required the major Japanese companies to overcome such deficiencies, and they did so in less than a decade.

As regards inflation: The chronic danger is hyperinflation. In a hyperinflation consumers hoard as consumers' capital significant amounts of intermediate and final products, which belong in production pipelines and business inventories, and whose diversion slows down growth and employment. In addition, the effective unit of circulation and account ceases in hyperinflation to be the national currency. At least, the national currency must share its position increasingly with something else whose supply the monetary authorities cannot expand—gold, cigarettes, foreign currency, etc. Under such circumstances, further inflation is again counterproductive, as it diverts resources out of production and consumption into the provision of liquidity. Japan has avoided hyperinflation, except during the 1940s when it could be blamed rightly or wrongly on "Acts of God and the King's Enemies"—first the Imperial Armed Forces during the first half of the decade and the unreasonable demands of the Supreme Commander of the Allied Powers (SCAP) during the second half.

As regards labor productivity and transferability between jobs: The jack-of-all-trades remains master of none, Renaissance men to the contrary notwithstanding. Flexibility and variety, training and retraining, can both be overdone. The professional linguist should not shift his emphasis from Japanese to Chinese or from French to German until his Japanese (or his French) has had time to penetrate his semiconscious if not his unconscious. Nor should the professional musician shift from the piano to the violin in similar circumstances. Flexibility and training can indeed be overdone, but Japan has thus far refrained from overdoing them—perhaps because capital instruments are too scarce and specialized to permit indiscriminate fad-following.

VII. CONCLUSION

Modifying the "one-sword" approach to Japanese economic development, we have suggested the marginal-efficiency schedule as raised by risk-removal to be the major sword. This sword we have supplemented by such additional aids to the marginal-efficiency schedule as secular inflation and the high productivity of a flexible labor force. Finally, we have eschewed the mechanical by suggesting as "developmental art" the importance of not carrying any of these devices so far as to become involved in self-defeating activities or negative returns, and we have given Japan high marks for avoiding such excesses.

REFERENCES

Bronfenbrenner, Martin. 1965. "Economic Miracles and Japan's Income-Doubling Plan," in *The State and Economic Enterprise in Japan.* Edited by William W. Lockwood. Princeton, N.J.: Princeton University Press, Chap. 11, pp. 523–53.

———. 1970. "The Japanese Growth Path: Equilibrium or Disequilibrium?" *Keizai Kenkyu,* May.

———. 1971. *Income Distribution Theory.* Chicago: Aldine.

Burmeister, Edwin and Dobell, A. Rodney. 1970. *Mathematical Theories of Economic Growth.* New York: Macmillan.

Frager, Robert and Rohlen, Thomas P. 1976. "The Future of a Tradition: Japanese Spirit in the 1980s," in *Japan: The Paradox of Progress.* Edited by Lewis Austin. New Haven: Yale University Press, Chap. 9.

Kahn, Herman. 1970. *The Emerging Japanese Superstate: Challenge and Response.* Englewood Cliffs, N.J.: Prentice-Hall.

Kanamori, Hisao. 1970. "Nihon no Seicho-Ritsu wa Naze Takai ka?" *Ekonomisuto,* 24 November, pp. 26–31; translated as "What Makes Japan's Economic Growth Rate So High?" *Japanese Economic Studies,* vol. 1, Fall, 1972, pp. 31–48.

Kelley, Allen C. and Williamson, Jeffrey G. 1974. *Lessons from Japanese Development: An Analytical Economic History.* Chicago: University of Chicago Press.

Keynes, John Maynard. 1936. *General Theory of Employment, Interest and Money.* New York: Harcourt Brace.

Knight, Frank H. 1944. "Diminishing Returns from Investment," *Journal of Political Economy,* vol. 52, March, pp. 26–47.

Maraini, Fosco. 1975. "Japan and the Future: Some suggestions from the *Nihonjin-Ron* Literature," in *Social Structures and Economic Dynamics in Japan up to 1980.* Edited by Gianni Fodella. Milan: Luigi Bocconi University, Institute of Economic and Social Studies for East Asia.

Ohlin, Bertil G. 1937. "Some Notes on the Stockholm Theory of Savings and Investment," *Economic Journal,* Pts. I–II, vol. 47, March and June, pp. 53–69, 221–40.

Samuelson, Paul A. 1939. "Interactions Between the Multiplier Analysis and the Principle of Acceleration," *Review of Economic Statistics,* vol. 21, May, pp. 75–78; reprinted in *Readings in Business Cycle Theory.* Edited by Gottfried Haberler. 1944. Philadelphia: Blakiston.

Tobin, James. 1955. "A Dynamic Aggregative Model," *Journal of Political Economy,* vol. 63, April, pp. 103–15.

———. 1965. "Money and Economic Growth," *Econometrica,* vol. 33, October, pp. 671–84.

Comment

Tuvia Blumenthal (Israel)

After rejecting the "laundry list" approach to explain long-term Japanese economic development, the paper presents a modified version of the "one sword" theory. Here we encounter a long sword (marginal efficiency of capital), a short sword (inflation) and a dagger (labor productivity). Instead of *itto* (one stroke) we have what might be called *Sanshu no Jingi* (three holy treasures).

As a prologue to the main body of the paper, there is a contention that the Japanese economy has a basically unstable growth pattern, "stabilized" by the government. The supporting argument seems, however, to include a logical *non-sequitur*: Professor Bronfenbrenner argues, quite rightly, that if we observe a "mixed" economy and find it to be stable, this does not mean that the underlying "private" economy is stable. However, it does also not necessarily mean that the latter is unstable. It may as well be stable with a lower rate of growth.

Let me turn now and inspect the three treasures. Nobody could dispute the role of the high marginal efficiency of investment in the growth process; the interesting question is, what are the factors which cause it to be so high. The paper singles out one such factor, the risk-averting function of government, which increased the *ex-ante* marginal efficiency of investment for private firms and pushed them to undertake further investment. This policy will be growth promoting only if the *ex-post* fruits of the investment live up to the *ex-ante* expectations. Thus, a necessary condition of such a policy to be successful is a sound assessment by government of the various investment alternatives and a correct guidance of the private sector. Risk insurance by the government leading to undertakings such as Pertamina or Concord may hinder rather than promote growth. Moreover, guarantee against failure is only one part of the government's policy to encourage investment. The other, and in my opinion not less important, is the change in the relative prices of capital and labor through a "dual structure" in both the labor and capital markets. This had the effect of increasing the use of capital-intensive technologies and the pace of capital formation.

The second sword is inflation which has, according to the paper, two aspects: forced saving and diversion of saving from cash balances to physical and human capital. Nothing is said in the paper about the role of voluntary saving, and on the effect of inflation on these non-forced savings. Here it is interesting to note that contrary to expectations, inflation did not have a depressing effect on voluntary savings. The ratio of saving out of disposable income for wage earners increased from 12.8% in 1965 to 17.2% in 1972 and kept climbing to 19.2% in 1974, a year when consumers' prices increased by 24.3%. The paper does not give any empirical evidence to show that forced savings through inflation were more important than voluntary savings.

The effect of inflation on the composition of savings is also not clear-cut in the Japanese case. The share of saving deposits out of total saving showed an increasing trend up to 1972 when it reached a peak of 82.9%. Only the high inflation of 1973 and 1974 reduced the share to the 60% level.

As to the dagger of high labor productivity, Professor Bronfenbrenner points to the difficulty of disentangling it from the large sword of capital productivity, and gives a small "laundry list" to account for it. This list includes quality of labor, on-the-job training, tenure and promotion as well as the Japanese union system. I have no quarrel with these but would like to point out that they represent a dangerous retreat into the "multiple factor" explanation.

The most important deficiency of the paper is, to my mind, the lack of any shred of empirical evidence to back up the hypotheses presented in it. Why is capital the large sword, while labor is only a dagger? Was this ranking uniform during the last hundred years? Were there other important factors that operated during a part of this period? What role did the introduction of borrowed technology, foreign trade, war and armament play in the growth process? Only a quantitative analysis which attempts to answer these questions can show whether the hypotheses presented in the paper can be accepted.

U.S. and Japanese Economic Growth 1952-1973: An International Comparison*

Dale W. Jorgenson *Harvard University*

Mieko Nishimizu *Princeton University*

In this paper we analyze differences between levels of U.S. and Japanese output and sources of economic growth in the U.S. and Japan during the period 1952–73. Our major conclusion is that the gap between Japanese and U.S. technology was eliminated during this period. Capital intensity of production, defined as capital input per unit of labor input, was twice as great for the United States as for Japan in 1973. This difference in capital intensity fully accounts for the differences in output per unit of labor input for the two countries in 1973.

* Financial support of this research by the RANN Program, National Science Foundation, is gratefully acknowledged.

I. INTRODUCTION

The purpose of this paper is to provide an international comparison of aggregate economic growth in the United States and Japan during the period 1952–73. Throughout this period the United States has maintained its position as the world's largest economy and as the leader among industrialized countries in output per capita. During the period Japan has risen to its current position as the world's third largest economy behind the United States and the Soviet Union, but a substantial gap between U.S. and Japanese levels of output per capita remains.[1]

Our initial objective is to analyze differences between U.S. and Japanese levels of output and to allocate these differences between differences in factor input in the two countries and differences in levels of technology. We divide differences between levels of factor input between differences in capital and labor input. Our second objective is to analyze the sources of economic growth in the United States and Japan during the period 1952–73 and to allocate growth of output between growth in factor input and growth in total factor productivity.[2] We further divide the growth of factor input between growth in capital input and growth in labor input.

Our methodology for measuring output, factor input, and total factor productivity is based on the economic theory of production, beginning with a production function that gives output as a function of capital input, labor input, a dummy variable equal to one for the United States and zero for Japan, and time. From this theory we derive index numbers for output, factor input, and total factor productivity for the United States relative to Japan and for each country at different points of time. We also derive purchasing power parities between the yen and the dollar for output and factor input.

Using index numbers for the United States and Japan, we carry out an international comparison between the two countries and an analysis of economic growth for each country over the period 1952–73. Japanese output grew rapidly relative to U.S. output during the period 1952–60, but most of the gain was due to a substantial increase in Japanese labor input relative to U.S. labor input. Relative levels of capital input and technology in the two countries remained almost unchanged until the end of the period. Capital intensity, defined as capital input per unit of labor input, rose in the United States but remained unchanged in Japan.

The last half of the 1950s and the early 1960s marked a transition in

[1] A recent survey of the Japanese economy in English is presented by Hugh Patrick and Henry Rosovsky (1976). Historical perspective on Japanese economic growth is provided by Kazushi Ohkawa and Henry Rosovsky (1973).

[2] A summary and comparison of studies of aggregate economic growth of Japan is given by Mieko Nishimizu and C.R. Hulten (1976).

the relative growth of the U.S. and Japanese economies, involving four significant events: In 1955 the own rate of return on business capital in Japan, the marginal product of all capital in the business sector, surpassed that in the United States; in that same year the annual rate of growth of capital input in Japan overtook that in the United States; in 1957 the share of investment goods in the value of total output in Japan overtook that in the United States; in 1961 the ratio of U.S. investment goods output to the Japanese output of investment goods fell below the corresponding ratio for consumption goods output.

The years 1960–73 were a period of substantial economic growth in the United States and extremely rapid growth in Japan. During these years the average rate of growth of capital input in Japan was more than double that in the United States, running at three times the U.S. growth rate during the years 1969–72; the own rate of return on business capital in Japan exceeded that in the United States from 1960–73 and was more than double that in the United States in every year from 1967 to 1971. The ratio of U.S. output of investment goods to Japanese output fell from 7.2 in 1960 to 2.2 in 1973; the corresponding ratio for the output of consumption goods fell from 6.9 in 1960 to 4.3 in 1973.

The period 1960–73 was characterized by growth in labor input in the United States and Japan at roughly similar rates. Almost all of the narrowing of the gap between U.S. and Japanese levels of output during this period was due to the increase in the relative capital intensity of production in Japan and to the rapid rise in the level of technology in Japan relative to that in the United States. The Japanese level of technology remained at less than half the corresponding U.S. level during the period 1952–59. Beginning in 1960 the relative level of Japanese technology moved up rapidly, reaching 90 percent of the U.S. level by 1970. Between 1970 and 1973 the remaining gap between Japanese and U.S. technology was eliminated, so that the remaining difference in levels of output between the two countries is due to differences in levels of labor and capital input.

In 1973 the ratio of U.S. labor input to Japanese labor input was 2.4. The corresponding ratio of U.S. and Japanese capital input was 5.2, so that the capital intensity of production in the United States was more than double that in Japan in 1973. However, the year 1973 marked another milestone in the relationship of U.S. and Japanese economic growth: The ratio of investment goods output to labor input in Japan overtook that in the United States. This development suggests that the remaining gap between U.S. and Japanese output per capita can be closed, even in the absence of any change in relative levels of technology in the two countries.[3]

[3] A useful discussion of future Japanese growth prospects in English is given by Patrick (1977).

In *Section* II we present a brief discussion of our methodology. This methodology was developed by Laurits Christensen and Dale W. Jorgenson (1969, 1970, 1973a) and has been implemented by them for the United States for the period 1929–69. Mitsuo Ezaki and Jorgenson (1973) have implemented this approach for Japan for the period 1951–68. Christensen, D. Cummings, and Jorgenson (1976) have brought these estimates up to date through 1973 and have compared the results with corresponding estimates for Canada, France, Germany, Italy, Korea, the Netherlands, and the United Kingdom.

In *Section* III we present new measures of U.S. output relative to Japanese output, purchasing power parities between the yen and the dollar in terms of output, and growth rates of output in both countries for the period 1952–73. In *Section* IV we present corresponding measures of U.S. factor input relative to Japanese factor input, including purchasing power parities in terms of factor input and growth rates of factor input for both countries. In *Section* V we present measures of the aggregate level of technology in the United States relative to Japan and growth rates of total factor productivity in both countries.

II. METHODOLOGY

Our first objective is to allocate differences between levels of output for the United States and Japan between differences in factor input and differences in total factor productivity. Our second objective is to separate growth in factor input from growth in total factor productivity in accounting for growth in output over time in each country. For this purpose we require comparable measures of factor input, output, and total factor productivity for the two countries. To achieve comparability in measuring output, we employ purchasing power parities for the United States and Japan, based on the work of Irving B. Kravis and his associates (1975). To achieve comparability in measuring factor input, we develop corresponding purchasing power parities for capital and labor input for the two countries.

Our methodology is based on the economic theory of production. The point of departure for this theory is a production function giving output as a function of inputs, a dummy variable equal to one for the United States and zero for Japan, and time. We consider production under constant returns to scale, so that a proportional change in all inputs results in a proportional change in output. In analyzing differences in production patterns between the two countries and changes in production patterns over time for each country, we combine the production function with necessary conditions for producer equilibrium. We express these conditions as equalities between shares of each input in the value of output and the elasticity of output with respect to that input. The elasticities depend on inputs, the dummy variable for each country, and time, the variables that enter the production function. Under constant returns to scale the sum of the elasticities with respect to all inputs is equal to unity, so that the value shares also sum to unity.

To analyze differences in the pattern of production between countries, we consider the difference in output between countries, holding all inputs and time constant. Under constant returns to scale, the necessary conditions for producer equilibrium for both countries can be combined with differences between inputs and outputs to produce an index of the difference in the level of technology that depends only on the prices and quantities of inputs and outputs in the two countries. Similarly, to analyze changes in the pattern of production for each country with time, we consider the rate of technical change with respect to time for that country, defined as the rate of growth of output, holding all inputs and the country dummy variable constant. The necessary conditions for producer equilibrium for each country can be combined with growth rates of inputs and outputs for that country to provide an index of the rate of technical change for the country that depends only on the prices and quantities of its inputs and outputs.

Our methodology for productivity measurement is based on a specific form for the production function:

$$\begin{aligned} Y = \exp[&\alpha_0 + \alpha_K \ln K + \alpha_L \ln L + \alpha_D \cdot D + \alpha_T \cdot T \\ &+ \frac{1}{2}\beta_{KK}(\ln K)^2 + \beta_{KL} \ln K \ln L + \beta_{KD} \ln K \cdot D + \beta_{KT} \ln K \cdot T \\ &+ \frac{1}{2}\beta_{LL}(\ln L)^2 + \beta_{LD} \ln L \cdot D + \beta_{LT} \ln L \cdot T + \frac{1}{2}\beta_{DD} \cdot D^2 \\ &+ \beta_{DT} \cdot D \cdot T + \frac{1}{2}\beta_{TT} \cdot T^2], \end{aligned}$$

where Y is output, K is capital input, L is labor input, D is a dummy variable equal to one for the United States and zero for Japan, and T is time. For this production function, output is a transcendental or, more specifically, an exponential function of the logarithms of inputs. We refer to this form as *the transcendental logarithmic production function* or, more simply, the translog production function.[4] The translog production function is characterized by constant returns to scale if and only if the parameters satisfy the conditions:

$$\begin{aligned} \alpha_K + \alpha_L &= 1, \\ \beta_{KK} + \beta_{KL} &= 0, \\ \beta_{KL} + \beta_{LL} &= 0, \end{aligned}$$

[4] The translog production function was introduced by L.R. Christensen, D.W. Jorgenson, and L.J. Lau (1971, 1973). The treatment of technical change outlined below is due to W.E. Diewert (1977) and to Jorgenson and Lau (1977).

$$\beta_{KD} + \beta_{LD} = 0,$$
$$\beta_{KT} + \beta_{LT} = 0.$$

Denoting the price of output by q_Y, the price of capital input by p_K, and the price of labor input by p_L, we can define the shares of capital and labor input in the value of output, say v_K and v_L, by:

$$v_K = \frac{p_K K}{q_Y Y}, \quad v_L = \frac{p_L L}{q_Y Y}.$$

Necessary conditions for producer equilibrium are given by equalities between each value share and the elasticity of output with respect to the corresponding input:

$$v_K = \frac{\partial \ln Y}{\partial \ln K}(K, L, D, T),$$
$$= \alpha_K + \beta_{KK} \ln K + \beta_{KL} \ln L + \beta_{KD} \cdot D + \beta_{KT} \cdot T;$$
$$v_L = \frac{\partial \ln Y}{\partial \ln L}(K, L, D, T),$$
$$= \alpha_L + \beta_{KL} \ln K + \beta_{LL} \ln L + \beta_{LD} \cdot D + \beta_{LT} \cdot T.$$

Under constant returns to scale the elasticities and the value shares sum to unity.

We can define the difference in technology between the two countries, say v_D, as the logarithmic difference between levels of output between the countries, holding capital input, labor input, and time constant:

$$v_D = \frac{\partial \ln Y}{\partial D}(K, L, D, T),$$
$$= \alpha_D + \beta_{KD} \ln K + \beta_{LD} \ln L + \beta_{DD} \cdot D + \beta_{DT} \cdot T.$$

Finally, we can define the rate of technical change, say v_T, as the growth of output with respect to time, holding capital input, labor input, and the country dummy variable constant:

$$v_T = \frac{\partial \ln Y}{\partial T}(K, L, D, T),$$
$$= \alpha_T + \beta_{KT} \ln K + \beta_{LT} \ln L + \beta_{DT} \cdot D + \beta_{TT} \cdot T.$$

III. OUTPUT[5]

We can also consider specific forms for the functions defining aggregate output Y, capital input K, and labor input L. For example, the translog form for aggregate output as a function of its components is:

$$Y = \exp[\alpha_1 \ln Y_1 + \alpha_2 \ln Y_2 + \ldots + \alpha_m \ln Y_m + \frac{1}{2}\beta_{11} (\ln Y_1)^2 + \beta_{12} \ln Y_1 \ln Y_2 + \ldots + \frac{1}{2}\beta_{mm} (\ln Y_m)^2].$$

The translog output aggregate is characterized by constant returns to scale if and only if:

$$\alpha_1 + \alpha_2 + \ldots + \alpha_m = 1,$$
$$\beta_{11} + \beta_{12} + \ldots + \beta_{1m} = 0,$$
$$\ldots\ldots$$
$$\beta_{1m} + \beta_{2m} + \ldots + \beta_{mm} = 0.$$

The value shares of individual outputs $\{w_{Yi}\}$ can be expressed as:

$$w_{Yi} = \alpha_i + \beta_{1i} \ln Y_1 + \ldots + \beta_{im} \ln Y_m, \qquad (i = 1, 2, \ldots, m).$$

Considering data for the United States and Japan at a given point of time, the difference between logarithms of aggregate output for the two countries can be expressed as a weighted average of differences between logarithms of individual outputs with weights given by average value shares:

$$\ln Y(\text{US}) - \ln Y(\text{JAPAN}) = \Sigma\, \hat{w}_{Yi} [\ln Y_i(\text{US}) - \ln Y_i(\text{JAPAN})],$$

where:

$$\hat{w}_{Yi} = \frac{1}{2}[w_{Yi}(\text{US}) + w_{Yi}(\text{JAPAN})], \qquad (i = 1, 2, \ldots, m).$$

Similarly, considering data for a given country at two discrete points of time, the difference between successive logarithms of aggregate output can be expressed as a weighted average of differences between logarithms of individual outputs with weights given by average value shares:

$$\ln Y(T) - \ln Y(T-1) = \Sigma\, \overline{w}_{Yi} [\ln Y_i(T) - \ln Y_i(T-1)],$$

[5] The quantity indexes were introduced by Irving Fisher (1922) and discussed by L. Tornqvist (1936), Henri Theil (1965), and T. Kloek (1966). These indexes of output and input were first derived from the translog production function by Diewert (1976). The corresponding index of technical change was introduced by Christensen and Jorgenson (1970). The translog index of technical change was first derived from the form of the translog production function given above by Diewert (1977) and by Jorgenson and Lau (1977).

where:

$$\overline{w}_{Yi} = \frac{1}{2}[w_{Yi}(T) + w_{Yi}(T-1)], \qquad (i = 1, 2, \ldots, m).$$

We refer to these expressions for aggregate output Y as the *translog indexes of output*.

To define price indexes corresponding to the translog indexes of aggregate output, we employ the fact that the product of price and quantity indexes for the aggregate must equal the sum of the values of the components of each aggregate. The price index for aggregate output is defined as the ratio of the sum of the values of the individual outputs to the translog output index. For the United States and Japan at a given point of time the price index represents the purchasing power parity between the yen and the dollar in terms of aggregate output. If U.S. output is measured relative to Japanese output, the corresponding purchasing power parity represents the price of one dollar's worth of output in terms of yen. For a given country at different points of time the price index represents the price of aggregate output at these points of time, measured in the country's own currency. For the United States this price index provides a measure of the price of U.S. aggregate output in dollars, while for Japan it provides a measure of the price of Japanese output in yen.

The price indexes for aggregate output corresponding to translog quantity indexes can be determined solely from data on prices and quantities of the components of the aggregate. Although these price indexes do not have the form of translog index numbers, they are nonetheless well defined. By definition the product of price and quantity indexes for an aggregate is equal to the sum of the values of its components. Price indexes for the components of aggregate output are available from the national accounts of the United States and Japan for the period 1952–73. Purchasing power parities for these components are available for the year 1970 from the work of Kravis and his associates (1975).

The starting point for the construction of translog indexes of output for the United States and Japan for the period 1952–73 is the measurement of the value of total product and the value of total factor outlay for each country in current prices. The fundamental accounting identity for the production account is that the value of total product is equal to the value of total factor outlay for each country. The product and factor outlay accounts are linked through capital formation and the compensation of property. To make this link explicit, we divide total output between consumption and investment goods and total factor outlay between labor and property compensation. In analyzing productive activity, we have limited the scope of our production account to the private domestic sector of each country.

The production account in a complete system of national economic accounts includes the activities of the private sector, the government sector, and the rest of the world. Rest of the world production is excluded on the grounds that it can reflect a different physical and social environment for productive activity than the environment provided by the domestic sector; the government sector is also excluded from our private domestic production account. One unconventional aspect of our measure of total output is an imputation for the services of consumer durables. Our objective is to attain consistency in the treatment of owner-occupied residential structures and owner-utilized consumer durables. The services of consumer durables are included in consumption goods output, while investment in consumer durables is included in investment goods output.

Given the value of total output for the United States and Japan, the remaining task is to separate these data into price and quantity components. Total product is first divided between investment and consumption goods. The value shares of investment goods product for both countries for the period 1952–73 are presented in Table 1. Translog indexes of consumption and investment goods output in the United States and Japan, defined as ratios between these outputs in the two countries, and annual rates of growth of consumption and investment goods output for both countries are presented in Table 1. The quantity indexes of private domestic investment and consumption goods output can be combined into a translog quantity index of private domestic product. Translog indexes of private domestic output in the United States and Japan and annual rates of growth in both countries are presented in Table 1. Finally, purchasing power parities between the yen and the dollar in terms of investment goods, consumption goods, and total output are given in Table 1.

For the United States the share of investment goods output declined slightly from .321 to .306 over the period 1952–73. For Japan the share of investment goods output rose dramatically from a level below that in the United States for the years 1952–56, overtaking the share of investment goods output in the United States in 1957, and rising to .475, more than 50 percent above the U.S. share, by the end of the period in 1973. At the beginning of the period, the quantity of investment goods output in the United States was more than 16 times that in Japan; this ratio fell to 2.2 by the end of the period. The ratio of consumption goods output in the United States to that in Japan fell from 8.7 at the beginning of the period 1952–73 to 4.3 at the end of the period. The ratio of investment goods output in the United States to that in Japan dropped below the corresponding ratio for consumption goods output in 1961. Purchasing power parities between the yen and the dollar for investment goods exhibited a very slight upward trend over the period 1952–73. The corresponding purchasing power parities for consumption goods increased by 75 percent, but remained below those for investment goods throughout the period. Purchasing power parities between the yen and the dollar for total output began to rise steadily from 1958 onwards.

In 1952 total output in the United States was more than 10 times that in Japan. By 1960 the ratio of U.S. output to Japanese output had fallen to 7.0, and by 1969 the ratio had fallen to 4.0. By the end of the period in 1973, this ratio had fallen still further to 3.3. In both countries growth of total output accelerated substantially between the years 1952–60 and the years 1960–73. The average annual rate of growth of total output in the United States rose from 2.9 percent for the period 1952–60 to 4.3 percent for 1960–73. For Japan the rate of growth of output rose from 8.4 percent for 1952–60 to 10.5 percent for 1960–73. Our first conclusion is that the acceleration in the rate of growth of Japanese output was accompanied by a rapid rise in the share of investment goods in total output, while the corresponding acceleration in the rate of growth of U.S. output was unaccompanied by any substantial change in the investment goods share.

IV. INPUT

If aggregate capital and labor input are translog functions of their components, we can express the differences between logarithms of aggregate inputs for the two countries in the form:

$$\ln K(\mathrm{US}) - \ln K(\mathrm{JAPAN}) = \Sigma \hat{v}_{Kj} \left[\ln K_j (\mathrm{US}) - \ln K_j (\mathrm{JAPAN})\right],$$

$$\ln L(\mathrm{US}) - \ln L(\mathrm{JAPAN}) = \Sigma \hat{v}_{Lk} \left[\ln L_k (\mathrm{US}) - \ln L_k (\mathrm{JAPAN})\right],$$

where:

$$\hat{v}_{Kj} = \frac{1}{2}[v_{Kj}(\mathrm{US}) + v_{Kj}(\mathrm{JAPAN})], \qquad (j = 1, 2, \ldots, n),$$

$$\hat{v}_{Lk} = \frac{1}{2}[v_{Lk}(\mathrm{US}) + v_{Lk}(\mathrm{JAPAN})], \qquad (k = 1, 2, \ldots, p).$$

Similarly, considering data for a given country at two discrete points of time, we can express the difference between successive logarithms in the form:

$$\ln K(T) - \ln K(T-1) = \Sigma\, \bar{v}_{Kj} \left[\ln K_j (T) - \ln K_j (T-1)\right],$$

$$\ln L(T) - \ln L(T-1) = \Sigma\, \bar{v}_{Lk} \left[\ln L_k (T) - \ln L_k (T-1)\right],$$

where:

$$\bar{v}_{Kj} = \frac{1}{2}[v_{Kj}(T) + v_{Kj}(T-1)], \qquad (j = 1, 2, \ldots, n),$$

$$\bar{v}_{Lk} = \frac{1}{2}[v_{Lk}(T) + v_{Lk}(T-1)], \qquad (k = 1, 2, \ldots, p).$$

We refer to these expressions for aggregate capital input K and labor input L as *translog indexes of capital and labor input*.

Price indexes for capital and labor input can be defined in a manner that is strictly analogous to the definition of price indexes for output. For the United States and Japan at a given point of time, the price index for capital input represents the purchasing power parity between the yen and the dollar in terms of aggregate capital input. If U.S. capital input is measured relative to Japanese capital input, the corresponding purchasing power parity represents the price of one dollar's worth of capital input in terms of yen. Similarly, the price index for labor input represents the number of yen required to purchase one dollar's worth of labor input. For a given country at different points of time, the price indexes represent the prices of aggregate capital and labor input at these points of time, measured in the country's own currency. For the United States these price indexes provide measures of prices of U.S. aggregate capital and labor input in dollars, while for Japan they provide measures of the prices of Japanese aggregate inputs in yen.

To complete our comparison of differences in levels of output for the two countries and sources of economic growth for each country for the period 1952–73, we require price indexes for the components of capital and labor input for both countries for this period. Similarly, we require purchasing power parities for these components. The measurement of capital input begins with data on the stock of capital for each component of capital input. For each country the stock of capital, say $A(T)$, is the sum of past investments, say $I(T-\tau)$, each weighted by the relative efficiency of capital goods of each τ, say d_τ:

$$A(T)=\sum_{\tau=0}^{\infty} d_\tau\, I(T-\tau).$$

Similarly, the price of acquisition of new capital goods, say $p_I(T)$, is the discounted value of the future prices of capital input, say $p_K(T+\tau)$, each weighted by relative efficiency:

$$p_I(T)=\sum_{\tau=0}^{\infty} d_\tau \prod_{S=1}^{\tau} \frac{1}{1+r(T+S)}\, p_K(T+\tau+1),$$

where $r(T)$ is the *rate of return on capital* in period T and

$$\prod_{S=1}^{\tau} \frac{1}{1+r(T+S)}$$

is the discount factor in period T for future prices in period $T+S$.

Using data on decline in relative efficiency of capital goods with age for each country, estimates of capital stock for that country can be compiled from data on prices and quantities of investment in new capital goods at each point of time by means of the perpetual inventory

TABLE 1
OUTPUT

Year	*1.*	*2.*	*3.*	*4.*	*5.*	*6.*	*7.*
1952	.321	.286	8.7	148.	—	—	16.3
1953	.317	.254	8.6	177.	.051	.061	16.8
1954	.305	.255	8.2	164.	−.005	.038	14.3
1955	.337	.282	7.9	162.	.036	.074	14.0
1956	.334	.327	7.8	162.	.041	.053	12.3
1957	.326	.370	7.8	164.	.033	.035	10.4
1958	.297	.313	7.4	163.	.027	.081	9.9
1959	.318	.342	7.2	164.	.033	.061	9.3
1960	.300	.381	6.9	165.	.037	.084	7.2
1961	.294	.450	6.7	173.	.027	.043	5.4
1962	.305	.404	6.3	182.	.041	.106	5.8
1963	.309	.411	6.0	191.	.029	.073	5.2
1964	.310	.419	5.7	199.	.050	.113	4.6
1965	.317	.395	5.6	210.	.045	.062	4.9
1966	.317	.410	5.5	216.	.057	.073	4.6
1967	.304	.436	5.3	222.	.047	.074	3.7
1968	.306	.452	5.1	229.	.045	.082	3.2
1969	.303	.458	4.9	228.	.030	.073	2.9
1970	.291	.478	4.7	241.	.022	.057	2.3
1971	.296	.459	4.4	243.	.028	.088	2.2
1972	.303	.450	4.3	247.	.046	.084	2.3
1973	.306	.475	4.3	259.	.052	.054	2.2

Variables:

1. Value share of investment goods output, U.S.
2. Value share of investment goods output, Japan.
3. Ratio between U.S. and Japanese consumption goods output.
4. Purchasing power parity, consumption goods output, yen per dollar.
5. Annual rates of growth of consumption goods output, U.S.
6. Annual rates of growth of consumption goods output, Japan.
7. Ratio between U.S. and Japan investment goods output.

TABLE 1 (concluded)
OUTPUT

Year	8.	9.	10.	11.	12.	13.	14
1952	277.	—	—	10.6	196.	—	—
1953	283.	.033	.046	10.6	207.	.045	.057
1954	271.	−.019	.010	9.7	213.	−.009	.031
1955	260.	.145	.175	9.5	210.	.071	.101
1956	283.	−.009	.231	9.2	211.	.024	.108
1957	297.	−.020	.225	8.7	212.	.016	.101
1958	272.	−.059	−.116	8.1	209.	.000	.014
1959	274.	.113	.189	7.9	209.	.057	.103
1960	283.	−.013	.278	7.0	210	.021	.154
1961	306.	.013	.363	6.2	219	.023	.177
1962	302.	.091	−.026	6.1	225.	.056	.049
1963	300.	.058	.159	5.7	232	.038	.108
1964	302.	.059	.179	5.2	239	.053	.141
1965	300.	.092	.017	5.3	247.	.060	.043
1966	302.	.064	.126	5.1	251.	.059	.094
1967	308.	−.017	.228	4.6	257.	.026	.139
1968	300.	.044	.198	4.2	259.	.045	.134
1969	293.	.031	.137	4.0	254.	.031	.102
1970	295.	−.054	.182	3.6	261.	−.000	.116
1971	285.	.049	.059	3.4	258.	.034	.074
1972	286.	.100	.090	3.3	258.	.062	.087
1973	312.	.085	.146	3.3	274.	.062	.097

Variables:

8. Purchasing power parity, investment goods output, yen per dollar.
9. Annual rates of growth of investment goods output, U.S.
10. Annual rates of growth of investment goods output, Japan.
11. Ratio between U.S. and Japanese total output.
12. Purchasing power parity, total output, yen per dollar.
13. Annual rates of growth of total output, U.S.
14. Annual rates of growth of total output, Japan.

method.[6] We assume that relative efficiency of capital goods declines geometrically with age,

$$d_\tau = (1-\delta)^\tau, \qquad (\tau = 0, 1, \ldots).$$

Under this assumption capital stock at the end of each period is equal to investment during the period less a constant proportion δ of capital stock at the beginning of the period:

$$A(T) = I(T) - \delta A(T-1).$$

Similarly, the price of capital input is equal to the sum of the nominal return to capital $p_I(T-1)\, r(T)$ and depreciation $\delta\, p_I(T)$, less revaluation $p_I(T) - p_I(T-1)$:

$$p_K(T) = p_I(T-1)\, r(T) + \delta\, p_I(T) - [p_I(T) - p_I(T-1)].$$

We can also express the price of capital input as the sum of the price of investment in the preceding period $p_I(T-1)$ multiplied by the *own rate of return on capital*

$$r(T) - \frac{p_I(T) - p_I(T-1)}{p_I(T-1)}$$

and the current price of investment $p_I(T)$ multiplied by the rate of depreciation δ:

$$p_K(T) = p_I(T-1)\left(r(T) - \frac{p_I(T) - p_I(T-1)}{p_I(T-1)} \right) + \delta\, p_I(T).$$

Although we estimate the decline in efficiency of capital goods for each component of capital input separately for the United States and for Japan, we assume that the relative efficiency of new capital goods is the same in both countries. Accordingly, the appropriate purchasing power parity for new capital goods is the purchasing power parity for the corresponding component of investment goods output. To obtain the purchasing power parity for capital input, we first multiply the appropriate price index for investment goods in each country by the own rate of return on capital and the rate of depreciation and sum the two components to obtain the price index for capital input. We then multiply the purchasing power parity for investment goods in yen per dollar by the ratio of the price of capital services to the price of capital goods for Japan divided by the corresponding ratio for the United States:

[6] The perpetual inventory method has been employed by Raymond W. Goldsmith (1955). The dual to the perpetual inventory method, involving investment goods prices and capital input prices, was introduced by Christensen and Jorgenson (1969, 1973a). For further discussion of the underlying model of durable capital goods, *see* Jorgenson (1973).

$$\frac{p_K(\text{JAPAN})}{p_K(\text{US})} = \frac{p_I(\text{JAPAN})}{p_I(\text{US})} \quad \frac{p_K(\text{JAPAN}) \, / \, p_I(\text{JAPAN})}{p_K(\text{US}) \, / \, p_I(\text{US})}$$

The resulting price index represents the purchasing power parity between the yen and the dollar in terms of aggregate capital input.

Given the value of total factor outlay for the United States and Japan, the next task is to separate these data into price and quantity components. Total factor outlay is first divided between property and labor compensation. The value shares of property compensation for both countries for the period 1952–73 are presented in Table 2. The starting point for the computation of a translog quantity index of capital input is a perpetual inventory estimate of the stock of each type of capital, based on past investments in constant prices. We have compiled capital stock estimates for seven asset classes: consumer durables, nonresidential structures, producer durable equipment, residential structures, nonfarm inventories, farm inventories, and land. For each of these asset classes we derive perpetual inventory estimates of the stock as follows: First, we obtain a benchmark estimate of capital stock from data on national wealth for each country in constant prices. Second, we deflate the investment series from the national income and product accounts for each country to obtain investment in constant prices. Third, we choose an estimate of the rate of replacement from data on lifetimes of capital goods in each country.

To construct translog quantity indexes of capital input for the United States and for Japan, we require value shares of individual capital inputs in total property compensation and stocks of individual assets. Property compensation for each asset is equal to the product of the price of capital input and the quantity of capital stock at the beginning of the period; in the absence of taxation, property compensation is the sum of depreciation and the own rate of return on capital:

$$p_K(T) \cdot K(T) = \left[p_I(T-1)\left(r(T) - \frac{p_I(T) - p_I(T-1)}{p_I(T-1)} \right) + \delta \, p_I(T) \right] \cdot A(T-1).$$

Given property compensation, the stock of assets, the price of investment goods, and the rate of depreciation, we can determine the rate of return $r(T)$. In measuring the rate of return differences in the tax treatment of property, compensation must be taken into account. For tax purposes the private domestic sector in each country can be divided among corporate business, noncorporate business, and households and nonprofit institutions. Households and institutions are not subject to direct taxes on property and compensation. Noncorporate business is subject to personal income taxes on property compensation, while corporate business is subject to both corporate and personal income taxes. Both

TABLE 2
CAPITAL INPUT

Year	*1.*	*2.*	*3.*	*4.*	*5.*	*6.*	*7.*	*8.*
1952	.397	.377	.058	.066	18.9	362.	—	—
1953	.380	.323	.052	.039	20.5	335.	.036	−.001
1954	.399	.326	.052	.051	18.5	327.	.039	.026
1955	.402	.328	.060	.072	18.5	327.	.032	.033
1956	.388	.344	.049	.088	18.6	383.	.052	.045
1957	.384	.353	.047	.081	16.3	410.	.042	.069
1958	.403	.332	.048	.053	15.5	335.	.033	.105
1959	.398	.352	.051	.073	15.1	368.	.017	.068
1960	.401	.384	.046	.108	13.3	427.	.033	.084
1961	.406	.427	.048	.154	11.8	512.	.030	.120
1962	.410	.394	.058	.107	10.7	442.	.022	.165
1963	.413	.392	.060	.111	9.9	441.	.034	.125
1964	.416	.421	.062	.147	9.0	487.	.038	.123
1965	.426	.402	.070	.119	8.9	435.	.043	.133
1966	.429	.415	.076	.137	8.8	448.	.053	.100
1967	.421	.432	.067	.165	7.9	508.	.057	.101
1968	.415	.442	.061	.181	7.1	518.	.044	.130
1969	.412	.440	.053	.169	6.8	492.	.046	.146
1970	.396	.435	.045	.163	5.8	527.	.045	.146
1971	.406	.409	.047	.124	5.3	461.	.030	.153
1972	.415	.405	.054	.116	5.1	426.	.034	.123
1973	.420	.394	.057	.107	5.2	401.	.045	.117

Variables:

1. Value share of capital input, United States.
2. Value share of capital input, Japan.
3. Own rate of return on business capital, United States.
4. Own rate of return on business capital, Japan.
5. Ratio between U.S. and Japanese capital input.
6. Purchasing power parity, capital input, yen per dollar.
7. Annual rates of growth of capital input, U.S.
8. Annual rates of growth of capital input, Japan.

TABLE 3
LABOR AND TOTAL FACTOR INPUT

Year	1.	2.	3.	4.	5.	6.	7.	8.
1952	3.3	58.	—	—	6.5	116.	—	—
1953	3.1	62.	.017	.066	6.0	122.	.024	.042
1954	2.9	65.	−.033	.024	5.7	128.	−.005	.024
1955	2.9	65.	.035	.039	5.7	128.	.033	.037
1956	2.7	63.	.021	.083	5.5	134.	.033	.070
1957	2.6	64.	−.008	.049	5.1	141.	.011	.056
1958	2.4	65.	−.026	.032	4.8	133.	−.002	.057
1959	2.4	65.	.040	.044	4.8	137.	.031	.052
1960	2.3	69.	.013	.059	4.6	150.	.021	.068
1961	2.3	77.	−.004	.011	4.5	174.	.009	.056
1962	2.3	85.	.027	.021	4.3	170.	.025	.080
1963	2.2	93.	.015	.029	4.1	179.	.023	.067
1964	2.2	99.	.021	.028	4.0	194.	.028	.066
1965	2.2	106.	.037	.042	4.0	189.	.039	.079
1966	2.2	111.	.038	.026	4.0	198.	.044	.056
1967	2.2	118.	.015	.040	3.8	217.	.033	.066
1968	2.2	125.	.024	.034	3.6	229.	.032	.076
1969	2.2	134.	.032	.014	3.6	233.	.038	.072
1970	2.2	151.	−.011	.001	3.3	256.	.011	.064
1971	2.2	168.	.005	−.003	3.2	253.	.015	.063
1972	2.3	183.	.038	.000	3.2	256.	.036	.050
1973	2.4	213.	.049	.019	3.3	278.	.047	.058

Variables:
1. Ratio between U.S. and Japanese labor input.
2. Purchasing power parity, labor input, yen per dollar.
3. Annual rates of growth of labor input, U.S.
4. Annual rates of growth of labor input, Japan.
5. Ratio between U.S. and Japanese total factor input.
6. Purchasing power parity, total factor input, yen per dollar.
7. Annual rates of growth of total factor input, U.S.
8. Annual rates of growth of total factor input, Japan.

households and business are subject to indirect taxes on property compensation through taxes levied in the value of property.

To allocate property compensation among individual capital inputs for the United States and Japan, we first measure property compensation for each component of the private domestic sector of each country—corporate and noncorporate business and households and nonprofit institutions. Second, we assume that the rate of return after taxes is the same on all assets within each sector; under this assumption we can allocate property compensation for each sector among the classes of assets utilized within the sector. We combine the price and quantity of capital for all classes of assets within each sector into a translog index of capital input for the sector. Finally, we combine these indexes for all sectors into an index of capital input for the country as a whole. We present own rates of return for the business sector, comprising corporate and noncorporate business, for the United States and Japan for the period 1952–73 in Table 2. Translog indexes of capital input in the two countries, defined as the difference between logarithms of these inputs in the two countries together with purchasing power parities between the yen and the dollar in terms of capital services, are also presented in Table 2. Finally, annual rates of growth of capital input for the United States and for Japan are given in Table 2.

For the United States the share of capital services input in the value of total factor outlay rose from .397 to .420 over the period 1952–73. For Japan the share of capital services also rose from .377 to .394 over the same period. The capital shares in the United States and Japan were remarkably similar, fluctuating around an average of two-fifths or 40 percent. The own rates of return to capital of the United States follow the business cycle over the period 1952–73. This rate of return reached a high of 7.6 percent in 1966 and a low of 4.5 percent in 1970. For the period 1952–54, the own rate of return to capital in the business sector for Japan was comparable to or slightly below that in the United States. Beginning in 1960 and continuing through the end of the period 1973, the own rate of return to capital in Japan was approximately double that in the United States. This remarkable increase occurred with virtually no change in the share of property compensation in total factor outlay. Reflecting the dramatic rise in own rates of return to capital in the business sector in Japan, beginning in 1960, purchasing power parities between the yen and the dollar in terms of capital input also rose sharply over the period 1952–73, reaching a peak of 527 yen to the dollar in 1970. These parities also reflect a much more modest increase in purchasing power parities in terms of investment goods.

The quantity of capital input in the United States was nearly nineteen times that in Japan at the beginning of the period 1952–73. By the end of the period in 1973, this ratio had dropped to 5.2. The cyclical movement of the U.S. economy is again reflected in the annual rates of growth of capital input in the United States for this period. Relatively high rates of growth characterize the beginning of the period, 1952–54; the short-lived investment boom of 1956–57; the more prolonged boom of 1964–70; and the final year of the period, 1973. Average annual growth rates of capital input rose modestly from 3.6 percent for the period 1952–60 to 4.1 percent for the period 1960–73. For Japan, the annual rates of growth of capital input are well below the corresponding U.S. rates of growth for the period 1952–54. Beginning in 1955 annual rates of growth of capital input in Japan rose above U.S. levels, running at more than three times the corresponding U.S. levels during the remarkable period 1969–72. Annual growth rates of capital input increased dramatically from an average of 5.4 percent for the period 1952–60 to 13.0 percent for 1960–73.

To construct translog quantity indexes of labor input for the United States and Japan, it would be desirable to develop data for each country disaggregated by level of education, sex, age, occupation, and so on. Following Dale W. Jorgenson and Zvi Griliches (1967), we have limited consideration to a breakdown of labor input by educational attainment. This results in indexes of the quantity of labor input for the private domestic sector of both countries. The quantity indexes of private domestic capital and labor input can be combined into a translog quantity index of private domestic factor input. Translog indexes of labor input and total factor input in the two countries, defined as the difference between logarithms of these inputs in the two countries, together with purchasing power parities between the yen and the dollar in terms of labor input and total factor input are presented in Table 3. Finally, annual rates of growth of labor input and total factor input for the United States and Japan are given in Table 3.

The quantity of labor input in the United States was only 3.3 times that in Japan at the beginning of the period 1952–73. This ratio fell to a low of 2.2 in 1963 and rose toward the end of the period to a level of 2.4 in 1973. The ratio of labor input in the United States to that in Japan began at a level less than one-fifth the corresponding ratio of capital input in the two countries in 1952. The ratio of labor input in the United States to that in Japan remained at less than one-half the corresponding ratio of capital input in the two countries in 1973. Annual rates of growth of labor input in Japan fell from an average of 5.0 percent for the period 1952–60 to 2.0 percent for the period 1960–73. The corresponding averages for the United States are 0.7 percent for the period 1952–60 and 2.2 percent for the period 1960–73. Purchasing power parities between the yen and the dollar for labor input almost quadrupled over the period 1952–73, rising from 58 to 213 yen to the dollar. These purchasing power parities remained at less than half those for capital input until 1973, the final year of the period. Most of this dramatic rise in the purchasing power parities of labor input took place during the period 1960–73. Purchasing power parities between the yen and the dollar for total factor input rose much more substantially than the corresponding parities for total output.

In 1952 total factor input in the United States was 6.5 times that in Japan. By 1960 the ratio of U.S. to Japanese factor input had fallen to 4.6, and by 1969 this ratio had fallen to 3.6. By the end of the period in 1973, this ratio had fallen still further to 3.3, almost precisely the same level as the corresponding ratio of U.S. to Japanese output. In both countries growth of total factor input accelerated substantially between the years 1952–60 and the years 1960–73. The average annual rate of growth of total factor input in the United States rose from 1.9 percent for the period 1952–60 to 3.0 percent for 1960–73. For Japan, the rate of growth of factor input rose from 5.1 percent for 1952–60 to 6.6 percent for 1960–73. Our second conclusion is that acceleration in the rate of growth of Japanese factor input was accomplished with a substantial decrease in the rate of growth of labor input, by a truly astonishing increase in the rate of growth of capital input, while the corresponding acceleration in the rate of growth of U.S. factor input was associated with a substantial increase in the rate of growth of labor input and a much more modest increase in the rate of growth of capital input.

V. PRODUCTIVITY

If we consider data for both countries at a given point of time, the average difference in technology can be expressed as the difference between logarithms of output for the two countries less a weighted average of the differences between logarithms of capital and labor input for the two countries with weights given by the average value shares:

$$\ln Y(\mathrm{US}) - \ln Y(\mathrm{JAPAN}) = \hat{v}_K\,[\ln K(\mathrm{US}) - \ln K(\mathrm{JAPAN})] + \hat{v}_L\,[\ln L(\mathrm{US}) - \ln L\,(\mathrm{JAPAN})] + \hat{v}_D,$$

where:

$$\hat{v}_K = \frac{1}{2}[v_K(\mathrm{US}) + v_K(\mathrm{JAPAN})],$$

$$\hat{v}_L = \frac{1}{2}[v_L(\mathrm{US}) + v_L(\mathrm{JAPAN})],$$

$$\hat{v}_D = \frac{1}{2}[v_D(\mathrm{US}) + v_D(\mathrm{JAPAN})].$$

We refer to this expression for the average difference in technology as the *translog index of difference in technology*.

Similarly, if we consider data at any two discrete points of time for a given country, say T and $T-1$, the average rate of technical change can be expressed as the difference between successive logarithms of capital and labor input with weights given by average value shares:

$$\ln Y(T) - \ln Y(T-1) = \bar{v}_K\,[\ln K(T) - \ln K(T-1)] + \bar{v}_L\,[\ln L(T) - \ln L(T-1)] + \bar{v}_T,$$

where:

$$\bar{v}_K = \frac{1}{2}[v_K(T) + v_K(T-1)],$$

$$\bar{v}_L = \frac{1}{2}[v_L(T) + v_L(T-1)],$$

$$\bar{v}_T = \frac{1}{2}[v_T(T) + v_T(T-1)].$$

We refer to this expression for the average rate of technical change $\bar{v}_T$ as the *translog index of technical change.*

Given translog indexes of total output and total factor input, we can construct translog indexes of difference in technology and technical change. The translog index of difference in technology between the United States and Japan is given in Table 4. We also present the contributions of differences between U.S. and Japanese technology, U.S. and Japanese capital input, and U.S. and Japanese labor input in explaining differences between U.S. and Japanese output. Each of these contributions is defined as the ratio of the corresponding difference between logarithms for the U.S. and Japan to the difference between the logarithms of U.S. and Japanese output. Annual rates of technical change are given for the United States and for Japan in Table 4. Finally, we present the contributions of the rate of technical change and rates of growth of capital and labor input to the growth of total output in each country.

The results presented in Table 4 describe a very remarkable closing of the gap in technology between the United States and Japan over the period 1952–73. For the period 1952–59 the difference between U.S and Japanese technology was nearly stationary with the Japanese level remaining at less than half the U.S. level. Beginning in 1960 the level of Japanese technology moved up sharply relative to that in the United States, reaching 90 percent of the U.S. level by 1970. Between 1970 and 1973 the level of Japanese technology actually overtook that in the United States, so that by 1973 the level of technology at the aggregate level in Japan was very slightly ahead of that in the United States. Our third conclusion is that none of the remaining difference between U.S. and Japanese aggregate output in 1973 was due to a difference in levels of technology.

By analyzing the contributions of differences in the level of technology, capital input, and labor input to differences between U.S. and Japanese levels of output, we can obtain a different perspective on the disappearance of the gap in technology between the United States and Japan. For the period 1952–59, all of the reduction in the difference between U.S. and Japanese total output was due to the increase in Japanese labor input relative to U.S. labor input. Relative levels of capital input and total factor productivity in the two countries remained almost unchanged. During this period capital intensity, defined as capital input per unit of labor input, increased more rapidly in the United States than in

TABLE 4

PRODUCTIVITY

Year	*1.*	*2.*	*3.*	*4.*	*5.*	*6.*
1952	.494	.208	.481	.309	—	—
1953	.554	.234	.450	.315	.020	.014
1954	.523	.229	.465	.304	−.004	.006
1955	.504	.223	.471	.304	.037	.064
1956	.498	.224	.483	.292	−.009	.037
1957	.523	.241	.476	.281	.004	.045
1958	.518	.246	.480	.272	.003	−.043
1959	.485	.234	.492	.272	.026	.050
1960	.409	.210	.522	.267	.000	.086
1961	.311	.170	.562	.267	.013	.120
1962	.358	.196	.525	.277	.030	−.031
1963	.330	.188	.528	.283	.015	.041
1964	.264	.158	.553	.287	.024	.074
1965	.288	.171	.541	.286	.020	−.036
1966	.240	.146	.561	.292	.014	.037
1967	.193	.125	.574	.299	−.006	.072
1968	.157	.108	.579	.312	.012	.057
1969	.103	.074	.589	.335	−.007	.030
1970	.079	.061	.572	.365	−.012	.051
1971	.070	.056	.555	.387	.018	.011
1972	.051	.041	.548	.409	.025	.037
1973	−.004	−.003	.567	.435	.015	.038

Variables:

1. Difference between U.S. and Japanese technology.
2. Contribution of differences between U.S. and Japanese technology to differences between U.S. and Japanese output.
3. Contribution of differences between U.S. and Japanese capital input to differences between U.S. and Japanese output.
4. Contribution of differences between U.S. and Japanese labor input to differences between U.S. and Japanese output.
5. Annual rates of technical change, U.S.
6. Annual rates of technical change, Japan.

Japan; capital intensity in Japan was almost unchanged during the period.

For the period 1960–73, the dramatic reduction in the difference between U.S. and Japanese total output was due to the increase in Japanese capital input relative to U.S. capital input and to the closing of the gap between Japanese and U.S. technology. Japanese and U.S. labor input grew at almost the same rate during this period. While the gap between U.S. and Japanese technology had closed by 1973, there still remains a substantial gap between U.S. and Japanese capital intensity of production. Our fourth conclusion is that all of the remaining difference between U.S. and Japanese output per unit of labor input is due to differences in the capital intensity of production in the two countries.

There is a slight upward trend in annual rates of growth of total factor productivity for the United States over the period 1952–73. The average for the period 1952–60 is 1.0 percent per year; the average for the period 1960–73 is 1.3 percent per year. For Japan, the average rate of growth of total factor productivity also rose modestly from 3.3 percent per year for the period 1952–60 to 3.9 percent per year for the period 1960–73. The contribution of growth in real labor input to U.S. growth in total output rose from .157 during the earlier period to .306 during the later period, while the contribution of real capital input fell from .495 during the earlier period to .393 in the later period. For Japan, the contribution of real capital input rose from .220 during the earlier period to .515 in the later period, while the contribution of real labor input fell from .390 to .114. In both countries the contribution of total factor productivity was essentially unchanged between the two periods, falling from .348 to .301 in the United States and from .389 to .371 in Japan. Our final conclusion is that the great acceleration of growth in Japan during the period 1960–73 was due to an acceleration in the growth of capital input relative to labor input.

Our results suggest some fruitful directions for future research. In view of the central role of capital input in our comparison of the United States and Japan, first priority must be given to further analysis of the growth of capital. This will require a complete accounting system, consisting of the production account we have presented in this paper, an income and expenditure account that includes saving, an accumulation account that includes capital formation, and a wealth account that includes capital stock as a component of wealth. Christensen and Jorgenson (1969, 1970, 1973a, 1973b) have developed a system that is well adapted to the analysis of aggregate economic growth and have implemented this system for the United States for the period 1929–69. Ezaki (1977) has implemented this system for Japan for the period 1952–1971.

Another fruitful direction for future research is the disaggregation of the production account to the level of individual industrial sectors. This will make it possible to analyze relative output, input, and productivity at the level of individual industries.[7] F. Gollop and Jorgenson (1977) have completed a study of U.S. economic growth for 51 industrial sectors for the period 1947–73. Nishimizu (1975) and Nishimizu and Hulten (1976) have completed a study of Japanese economic growth for 10 industrial sectors for the period 1955–71. We are currently engaged in a comparative study of U.S. and Japanese economic growth for 30 industrial sectors. This study will include a comparison of levels of output, input, and productivity by industrial sector for the two countries and an analysis of industrial growth by sector in each country.

REFERENCES

Christensen, Laurits R.; Cummings, D.; and Jorgenson, Dale W. 1977. "Economic Growth, 1947–1973: An International Comparison," in *New Developments in Productivity Measurement*. Edited by J.W. Kendrick and B. Vaccara. NBER, Studies in Income and Wealth, Vol. 41. New York: Columbia University Press.

——— and Jorgenson, Dale W. 1969. "The Measurement of U.S. Real Capital Input, 1929–1967," *Review of Income and Wealth*, Series 15, December, pp. 293–320.

——— and Jorgenson, Dale W. 1970. "U.S. Real Product and Real Factor Input, 1929–1967," *Review of Income and Wealth*, Series 16, March, pp. 19–50.

——— and Jorgenson, Dale W. 1973a. "Measuring Economic Performance in the Private Sector," in *The Measurement of Economic and Social Performance.* Edited by Milton Moss. New York: National Bureau of Economic Research, pp. 233–338.

——— and Jorgenson, Dale W. 1973b. "U.S. Income, Saving, and Wealth, 1929–1969," *Review of Income and Wealth*, Series 19, December, pp. 329–38.

———; Jorgenson, Dale W.; and Lau, Lawrence J. 1971. "Conjugate Duality and the Transcendental Logarithmic Production Function," *Econometrica*, vol. 39, July, pp. 255-56.

———; Jorgenson, Dale W.; and Lau, Lawrence J. 1973. "Transcendental Logarithmic Production Frontiers," *Review of Economics and Statistics*, vol. 55, February, pp. 28–45.

Diewert, W. E. 1976. "Exact and Superlative Index Numbers," *Journal of Econometrics*, vol. 4, May, pp.115–46.

———. 1977. "Aggregation Problems in the Measurement of Capital, " Discussion Paper No. 77-09, Department of Economics, University of British Columbia.

Ezaki, Mitsuo. 1977. *Nihon Keizai no Moderu Bunseki: Kokumin Keizai Keisan Kara no Sekkin [An Analysis of Japanese Economy: An Approach from the System of National Accounts]*. Tokyo: Sobun-sha.

——— and Jorgenson, Dale W. 1973. "The Measurement of Macroeconomic Performance in Japan, 1951–1968," in *Economic Growth: The Japanese Experience Since the Meiji Era*. Edited by Kazushi Ohkawa and Yujiro Hayami. Vol. 1. Tokyo: Japan Economic Research Center, pp. 286–361.

Fisher, Irving. 1922. *The Making of Index Numbers*. Boston: Houghton Mifflin.

[7] Detailed discussion of the development of Japanese technology at the level of individual industries is provided by Ozawa (1974) and by Merton J. Peck and Shuji Tamura (1976).

Goldsmith, Raymond W. 1955. *A Study of Saving in the United States*. 3 Vols. Princeton, N.J.: Princeton University Press.

Gollop, F. and Jorgenson, Dale W. 1977. "U.S. Productivity Growth by Industry, 1947–1973," in *New Developments in Productivity Measurement*. Edited by J. W. Kendrick and B. Vaccara. NBER, Studies in Income and Wealth, Vol. 41. New York: Columbia University Press.

Jorgenson, Dale W. 1973. "The Economic Theory of Replacement and Depreciation," in *Econometrics and Economic Theory*. Edited by Willy Sellekaerts. New York: Macmillan, pp. 189–221.

——— and Griliches, Zvi. 1967. "The Explanation of Productivity Change," *Review of Economic Studies*, vol. 34, July, pp. 249–83.

——— and Lau, Lawrence J. 1977. *Duality and Technology*. Amsterdam: North-Holland.

Kloek, Trun. 1966. *Indexcijfers: enige methodologisch aspecten*. The Hague: Pasmans.

Kravis, Irving B., *et al*. 1975. *A System of International Comparisons of Gross Product and Purchasing Power*. Baltimore: Johns Hopkins University Press for the World Bank.

Nishimizu, Mieko. 1975. *Total Factor Productivity Analysis: A Disaggregated Study of the Post-War Japanese Economy with Explicit Consideration of Intermediate Inputs, and Comparison with the United States*. Ph.D. Dissertation, Johns Hopkins University.

——— and Hulten, C. R. 1976. "The Sources of Japanese Economic Growth: 1955–1971," Research Memo No. 200, Econometric Research Program, Princeton University.

Ohkawa, Kazushi and Rosovsky, Henry. 1973. *Japanese Economic Growth: Trend Acceleration in the Twentieth Century*. Stanford, Calif.: Stanford University Press.

Ozawa, Terutomo. 1974. *Japan's Technological Challenge to the West, 1950–1974: Motivation and Accomplishment*. Cambridge, Mass.: M.I.T. Press.

Patrick, Hugh. 1977. "The Future of the Japanese Economy: Output and Labor Productivity," *The Journal of Japanese Studies*, vol. 13, Summer, pp. 219–49.

———and Rosovsky, Henry, eds. 1976. *Asia's New Giant: How the Japanese Economy Works*. Washington, D.C.: Brookings Institution.

Peck, Merton J. with the collaboration of Shuji Tamura. 1976. "Technology," in *Asia's New Giant*. Edited by H. Patrick and H. Rosovsky. Washington, D.C.: Brookings Institution, pp. 525–86.

Theil, Henri. 1965. "The Information Approach to Demand Analysis," *Econometrica*, vol. 33, January, pp. 67–87.

Tornqvist, L. 1936. "The Bank of Finland's Consumption Price Index," *Bank of Finland Monthly Bulletin*, no. 10, pp. 1–8.

Relative Productivity of Labor in American and Japanese Industry and Its Change, 1958-1972

Kenzo Yukizawa *Kyoto University*

The ratio of American to Japanese physical productivity of labor is measured for 60 products based on the U.S. and Japanese Censuses of Manufactures. On average, it was 3.1 in 1958, 2.9 in 1963, 2.0 in 1967, and 1.5 in 1972. The gap has narrowed rapidly but still remains. There exist correlations between relative productivity and market size, between relative productivity and relative exports, and between productivity growth and export growth. Next the international level of Japanese efficiency is considered. As a check, unit labor cost including intermediate labor input is calculated using American and Japanese input-output tables.

I. THE CHARACTERISTICS OF THE APPROACH OF THIS PAPER

The research work, the results of which form the main foundation of my argument here, consists of comparing physical productivity of labor in the American and Japanese manufacturing industries, branch by branch, along the lines followed by Lászlő Rostás (1948) and Marvin Frankel (1957).

There are several concepts of productivity and, corresponding to them, several ways of measurement, the results of which are complementary to each other. Among them is the concept of *physical* output of *labor*, which I adhere to in my research work. A justification for using the concept of *labor* productivity instead of total factor productivity may be found in the following passage in Irving B. Kravis's recent survey article (1976, p. 40):

> In principle, it is desirable that total factor productivity rather than merely labour productivity should be compared, but until there is substantial progress in international comparisons of capital inputs, it is probably preferable to continue to stress labour productivity in international comparisons because of the large errors likely to be introduced in the measurement of inputs other than labour.

Productivity of labor can be regarded as a measure of the joint effect of efficiency of many factors of production.

Then, on what reasoning do I base my adherence to the concept of *physical* output of labor? Nowadays, in contrast to this concept, the so-called "net output value method" seems to have become a more popular method of comparing labor productivity. Each way of measuring output, needless to say, has its own particular significance. For instance, Deborah Paige's and Gottfried Bombach's joint work observes that the net output value method takes account of differences in input purchased from other sectors and, thus, has the advantage of relating the results to real costs (1959, p. 58); whereas Rostás's (1948) and Frankel's (1957) investigations aim to derive, from global data, results approximating more closely an efficiency concept of labor productivity. I have been more interested in facts related to the latter concept.

Besides, if we compare *real* net output internationally, difficulty of measurement is greater than by my method, as will be explained later in relation to the Geary formula. Otherwise, we could not help having recourse to the expediency of converting the value of output or materials in terms of one local currency into another and hence of enlarging the room for arbitration in choosing the rate of conversion. If a project of comparing purchasing power parity of currencies such as ICP (United Nations International Comparison Project), the results of which appeared in Kravis *et al.* (1975), were carried out for the other years and in more detail, the difficulty involved in choosing a suitable rate of conversion would be mitigated.

The second characteristic of my research work may be found, as it were, in its time horizon. As a matter of fact, my comparison has been carried out at four separate points in time, 1958, 1963, 1967, and 1972, the recent four years for which reports of *American Census of Manufactures* have been made. This fact not only permits checking the results of one year against another, but also makes it possible to analyze actuality of productivity from the standpoint of growth and of change of comparative productivity.

The third characteristic of the present work consists of the countries compared. There is a special significance in making a comparison with the productivity of a specific nation, namely *that of the United States.* In particular, we may safely assume that the attainable level of productivity efficiency at the present time is found, in most cases, to have materialized in the actual level of productivity in American industry.

On the other hand, the choice of *Japan,* as the other party of comparison might be more significant than the simple fact that it is my home country. Japanese industry has attained a very high rate of growth in productivity in the ten years following 1962. This high rate may be said to be historically exceptional (Denison and Chung, 1976, p. 1), and my comparison of productivity happens to cover this period, which may reveal interesting evidence that cannot be found in more usual years.

Fourthly, the above reasons are supplemented by observations on the international level of Japanese economy as a whole. In addition, the results of another way of productivity comparison are referred to as a check. Unit labor cost, including indirect labor, has been estimated and compared making use of input-output tables.

II. RELATIVE PRODUCTIVITY IN AMERICAN AND JAPANESE INDUSTRY

II-1 What is Measured

Now, taking the physical output of product $i\,(i = 1, 2, \ldots, n)$ to be represented by q^i, and the labor input to be represented by l^i, the physical productivity per worker p^i can be measured in terms of q^i/l^i. Then, the following formula can be obtained as an *individual* index of productivity: p^i_{AJ} representing the level of productivity with respect to each product of country A on the basis of country J:

$$p^i_{AJ} = \frac{q^i_A}{l^i_A} \bigg/ \frac{q^i_J}{l^i_J} = (p^i_A / p^i_J)$$

What has been pursued in the present research work is, fundamentally speaking, none other than the quantitative approach toward this concept. The next step is, aggregating these individual indices in conformity with the formula set forth later, to make an evaluation of the aggregate indices of labor productivity P_{AJ} by which means we can conjecture the relative level of efficiency of material production and indirectly, the relative level of real income per capita in each country as a whole. The indices also show how overall productivity is a result of industrial structure and of productivity in individual industries. Figure 1 shows in advance, some of the results of such an aggregation. This figure may suggest that during the 1960s the gap in manufacturing productivity between the two countries narrowed considerably, and around 1972 the productivity of the U.S. was, on the average, 1.5 times as much as Japan.

Let us turn to the choice of the products ($i=1,\ldots,n$). Only particular products of such nature that would not lead to any serious errors in making comparisons of *physical* volume of output were picked out, and then out of them only products of such nature that would be least affected by any possible error in assessing labor input were selected.[1] Through this screening process, there finally remained sixty products of the manufacturing industry.[2] The relative weight exercised in the whole manufacturing industry by the observed branches to which the said products belong, *i.e.,* the coverage, is shown in Table 1.

Next, as to the *labor input* needed to produce these products, its measurement was made in the following manner. First, the labor

TABLE 1
COVERAGE OF THE SELECTED PRODUCTS

Year (Symbol of each year)	*Japan*		*U.S.A.*	
	Number of Employees	*Value Added (gross)*	*Number of Employees*	*Value Added (gross)*
1958–59 (*a*)	27	39	23	26
1963 (*b*)	27	46*	21	27
1967 (*c*)	24	32*	21	25
1972 (*d*)	21	26*	20	24

* Net value added.

[1] The problem we face here is related to a gap between *industry* and *product* statistics. The details of it are explained in Yukizawa (1975, pp. 26–32).

[2] The definition of products with the code number of the Industrial Classification of respective countries is shown in Yukizawa (1975), Appendix Table A2, pp. 34–35.

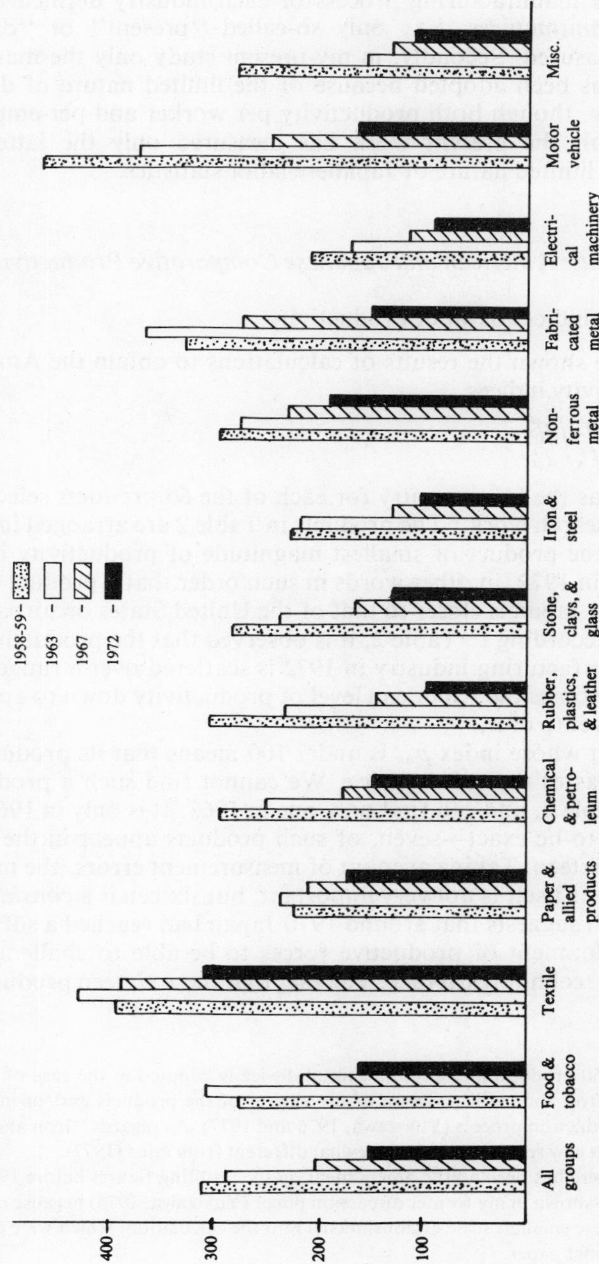

Fig. 1 Output per head comparisons in Japanese and American manufacturing industry, averages by major industry group: Japan = 100.

Source: indices (C) of Table 3.

engaged in the manufacturing process of each industry defined by the *Census of Manufactures, i.e.,* only so-called "present" or "direct" labor, was measured. Secondly, in my present study only the man-year comparison has been adopted because of the limited nature of data in Japan. Thirdly, though both productivity per worker and per employee are meaningful, the present work has measured only the latter, on account of the limited nature of Japanese labor statistics.

II-2 The Results: American and Japanese Comparative Productivity

(a) Individual indices of labor productivity

In Table 2 are shown the results of calculations to obtain the American labor productivity indices

$$p^i_{AJ} = \frac{q^i_A}{l^i_A} \bigg/ \frac{q^i_J}{l^i_J}$$

taking Japan as the base country for each of the 60 products selected in the present research work.[3] The products in Table 2 are arranged in order starting with the product of smallest magnitude of productivity indices *per employee* in 1972, in other words in such order that a product whose productivity in Japan is closer to that of the United States or surpasses it comes first. According to Table 2, it is observed that the productivity of Japanese manufacturing industry in 1972 is scattered over a range varying from three times the American level of productivity down to approximately one-seventh of it.

The product whose index p_{AJ} is under 100 means that its productivity in Japan surpasses that of America. We cannot find such a product in 1958–59 in Table 2, and can find only one in 1963. It is only in 1967 that several cases, to be exact—seven, of such products appear in the table, and in 1972, sixteen. Taking account of measurement errors, the number of such products itself is not very important, but sixteen is a *considerable* number, which suggests that around 1970 Japan had reached a sufficient stage of development of productive forces to be able to challenge any country in the technological race. The fact that these sixteen products be-

[3] Among the 60 products measured, a special device is adopted in the case of "Motor vehicles" and "Iron and steel" because of the variety of the products and/or integrated nature of the production process (Yukizawa, 1976 and 1977). As regards "Iron and steel," Mark calculated a new result, which is somewhat different from mine (1977).

By the way, there has been a little amendment on the resulting figures before 1967 from those which were shown in my former discussion paper (Yukizawa, 1975) because of the inclusion of Japanese smallest scale establishments into the calculation, which were excluded from it in the former paper.

long to various groups within the manufacturing industry might support the above observations.

(b) Aggregate indices of labor productivity

There are several formulae for calculating the level of productivity of an industrial group or of manufacturing industry as a whole, using aggregate productivity indices. In the present study only the number of employees has been adopted as a weight in aggregating the individual indices, resulting in the following two kinds of aggregate indices: (A) aggregate indices weighted by the number of American employees l^i_A and (B) aggregate indices weighted by the number of Japanese employees l^i_J: each of these having its own particular meaning.

In other words, this aggregate index (A) can be taken to be an indicator showing how many times more labor would be required in Japan relative to America if both countries' physical composition and scale of production were the same as those actually realized in the United States, and the aggregate index (B) indicates how many times more labor input Japan requires, assuming that each country is to produce each item in just the same amounts as actual production in Japan.[4]

Table 3 shows the results of such aggregations with respect to all the products and each industrial group (two-digit code) for the years 1958–59, 1963, 1967, and 1972. It must be kept in mind throughout Table 3 that the calculated values listed there are concerned only with the products selected as objects of the present study, out of many other products belonging to the respective groups of industry, and consequently whether or not the values can be regarded as reflecting the real situation of one whole industrial group depends upon how much the products selected represent the general circumstances in their own industries. The fact that there are differences between the measured values (A) and (B) is due to differences in the physical composition of production between the two countries, as already clarified. Therefore, another result obtained from the aggregate indices (C), which could in one sense be interpreted to indicate an average of (A) and (B), is added. Letting r mean unit labor requirement l/q, the method of calculating it runs as follows:

$$\text{Aggregate index (C)} = \frac{\sum_i r^i_J (q^i_J + q^i_A)}{\sum_i r^i_A (q^i_J + q^i_A)}$$

therein signifying the ratio of the total labor input required, for each country assuming that, with respect to each product, both countries are

[4] In more detail, refer to Yukizawa (1975, pp. 14–15).

TABLE 2

INDIVIDUAL PRODUCTIVITY INDICES, U.S. VERSUS JAPAN

Product title	Productivity indices $\left(P_{AJ} = \frac{q_A}{l_A} \Big/ \frac{q_J}{l_J}\right)$				Rank order according to the indices			
	(a) 1958–59	(b) 1963	(c) 1967	(d) 1972	(a) 58–59	(b) 63	(c) 67	(d) 72
Radio and TV receiving type electron tubes	112	74	50	35	2	1	1	1
Canned seafood	133	145	97	67	4	5	6	2
Piano	392	176	84	67	44	13	4	3
Sheet glass, except tinted	105	106	67	71	1	3	2	4
Storage batteries	633	438	115	76	54	51	12	5
Cement, hydraulic	166	128	98	78	10	4	7	6
Tires	319	247	117	82	38	29	14	7
Steel rolling and finishing	220	195	110	84	16	17	11	8
Carded and combed cotton yarn	161	171	129	86	9	11	15	9
Acetate yarn	148	179	108	89	7	14	8	10
Home-type television sets	134	159	109	90	5	7	9	11
Pencil, nonmechanical	118	98	138	93	3	2	16	12
Petroleum products	239	185	116	95	21	16	13	13
Brass, bronze, copper castings	210	222	189	95	13	25	32	14
Leather gloves	295	221	173	96	36	23	26	15
Steel castings	137	161	96	97	6	8	5	16
Plastic materials	289	237	143	100	33	27	17	17
Iron and steel forgings	160	257	77	102	8	31	3	18
Rayon yarn	220	222	148	105	18	24	19	19
Aluminum castings	231	208	143	107	19	20	18	20
Bolts, nuts, and rivets	215	240	173	112	15	28	27	21
Men's shoes	284	209	155	113	29	22	21	22
Household refrigerators	266	170	110	114	24	9	10	23
Refined cane sugar	242	365	218	117	22	42	35	24
Wool yarn, including carpet and rug yarn	287	248	177	121	31	30	28	25

Malleable iron castings	318	279	186	127	37	35	31	26
Tobacco	287	316	233	128	32	40	39	27
Printing ink	278	203	198	133	28	18	33	28
Watches	274	150	148	133	26	6	20	29
Zinc slab, including remelt zinc	526	302	159	134	49	37	24	30
Paper	172	181	183	135	11	15	30	31
Paperboard	193	226	180	149	12	26	29	32
Motor vehicles and equipment	458	367	232	160	46	44	37	33
Fatty acid	272	398	340	166	25	46	51	34
Refind unalloyed aluminum	392	311	247	178	44	39	43	35
Metal cans	361	423	348	179	41	49	53	36
Beer and ale	220	208	165	179	17	21	25	37
Lime	600	405	247	181	53	48	44	38
Synthetic organic fibers except cellulosic	261	203	155	181	23	19	22	39
Reclaimed rubber	366	366	321	185	42	43	47	40
Manufactured ice	276	263	211	188	27	33	34	41
Gray iron castings	293	326	233	196	34	41	38	42
Copper rolling and drawing	213	263	305	203	14	32	46	43
Wheat flour	328	305	230	204	39	38	36	44
Inorganic color pigments	675	399	240	211	55	47	42	45
Fertilizers	238	267	158	214	20	34	23	46
Aluminum rolling and drawing	343	299	236	215	40	36	41	47
Woven carpets and rugs	(284)	(174)	346	222	(30)	(12)	52	48
Brick	475	433	325	255	47	50	48	49
Steel spring	550	558	234	236	50	53	40	50
Cotton broad woven fabrics	385	382	340	240	43	45	50	51
Starch	510	482	330	254	48	52	49	52
Adhesives and gelatin	1,040	916	282	258	59	58	45	53
Compressed and liquefied gas	946	1,144	1,080	281	58	59	58	54
Industrial explosive	561	593	493	284	51	54	54	55
Wool fabrics	573	868	700	349	52	57	57	56
Wine and brandy	1,295	1,995	1,126	467	60	60	59	57
Wood pulp	776	698	530	591	57	55	55	58
Woven fabrics, man made fiber, and silk	710	793	685	710	56	56	56	59
Matches	294	170	–	–	35	10	–	–

TABLE 3

AGGREGATE INDICES OF LABOR PRODUCTIVITY BETWEEN U.S. AND JAPAN

Industrial Groups and U.S. Census Code		*1958–59 year a*			*1963 year b*			*1967 year c*			*1972 year d*		
		Indices			*Indices*			*Indices*			*Indices*		
		(A)	*(B)*	*(C)*	*(A)*	*(B)*	*(C)*	*(A)*	*(B)*	*(C)*	*(A)*	*(B)*	*(C)*
All groups	(20~39)	316	267	308	294	252	286	205	174	197	161	127	147
Food and tobacco	(20,21)	281	247	275	311	272	303	225	183	214	172	125	158
Textile mill products	(22)	409	319	386	451	337	422	401	317	382	318	180	309
Paper and allied products	(26)	220	226	221	232	235	232	206	211	207	165	166	165
Chemicals and petroleum products	(28,29)	298	258	292	256	233	251	176	156	171	141	134	139
Rubber, plastics and leather products	(30,31)	301	308	301	228	232	229	140	139	140	93	93	93
Stone, clay and glass products	(32)	291	212	276	253	162	230	187	109	165	148	87	128
Iron and steel	(331,332)	225	226	226	217	211	216	128	122	126	106	94	101
Nonferrous metal	(333,335,336)	296	259	292	274	259	272	238	213	234	183	151	174
Fabricated metal products	(34)	337	276	326	386	301	367	293	208	268	159	137	151
Electrical machinery	(36)	218	157	204	180	149	169	103	100	102	89	87	88
Motor vehicles and equipment	(371)	458	458	458	367	367	367	232	232	232	160	160	160
Miscellaneous	(38,39)	284	255	275	156	151	153	127	123	125	105	108	106

$$\text{Index (A)} = \frac{\sum (p_A/p_J)\, l_A}{\sum l_A}, \quad \text{Index (B)} = \frac{\sum l_J}{\sum (p_J/p_A)\, l_J}, \quad \text{Index (C)} = \frac{\sum r_J(q_J+q_A)}{\sum r_A(q_J+q_A)}.$$

$p = q/l,\ r = l/q$

TABLE 4

GROWTH RATE OF PRODUCTIVITY BY INDUSTRIAL GROUPS OF EACH COUNTRY*

		Japan			U.S.		
Industrial groups and U.S. Census Code		*1963/59 (b/a)*	*1967/63 (c/b)*	*1972/67 (d/c)*	*1963/58 (b/a)*	*1967/63 (c/b)*	*1972/67 (d/c)*
All groups	(20~39)	(9.9) 146	(13.0) 163	(10.7) 166	(5.3) 130	(2.4) 110	(4.4) 124
Food and tobacco	(20,21)	(3.6) 115	(8.2) 137	(9.6) 158	(4.1) 122	(2.4) 110	(3.9) 121
Textile mill products	(22)	(3.6) 115	(6.4) 128	(7.9) 146	(4.1) 122	(2.1) 109	(3.4) 118
Paper and allied products	(26)	(5.1) 122	(6.6) 129	(11.7) 174	(4.4) 124	(3.6) 115	(6.2) 135
Chemicals and petroleum products	(28,29)	(13.7) 167	(20.0) 207	(12.8) 183	(6.8) 139	(7.8) 135	(8.6) 151
Rubber, plastics and leather products	(30,31)	(14.5) 172	(16.5) 184	(11.5) 172	(6.2) 135	(1.2) 105	(3.0) 116
Stone, clay and glass products	(32)	(12.1) 158	(11.8) 156	(10.4) 164	(3.9) 121	(2.4) 110	(5.4) 130
Iron and steel	(331,332)	(7.3) 133	(16.5) 184	(8.0) 147	(4.6) 125	(1.7) 107	(2.7) 114
Nonferrous metal	(333,335,336)	(9.5) 144	(9.7) 145	(11.3) 171	(5.7) 132	(5.7) 125	(5.5) 131
Fabricated metal products	(34)	(1.0) 104	(11.8) 156	(10.2) 163	(1.6) 108	(2.6) 111	(−0.4) 98
Electrical machinery	(36)	(9.0) 141	(17.1) 188	(15.1) 202	(6.2) 135	(2.4) 110	(8.9) 153
Motor vehicles and equipment	(371)	(15.0) 175	(11.6) 155	(12.7) 182	(7.0) 140	(−0.5) 98	(4.6) 125
Miscellaneous	(38,39)	(13.8) 168	(13.7) 167	(4.2) 123	(−0.2) 99	(8.0) 136	(1.4) 107

Growth rate $(b/a) = \dfrac{\sum (p_b/p_a)\, l \cdot d}{\sum l \cdot d}$, $(c/b) = \dfrac{\sum (p_c/p_b)\, l \cdot d}{\sum l \cdot d}$, $(d/c) = \dfrac{\sum (p_d/p_c)\, l \cdot d}{\sum l \cdot d}$.

a: 1958/59, *b*: 1963, *c*: 1967, *d*: 1972.

* Figures in parentheses are annual averages.

to produce an amount that is equal to the sum of the volume actually produced by the two countries.

Judging from the indices calculated according to the respective methods described above, it has been disclosed, as far as all the products selected as objects of the present study are concerned, that labor productivity in American manufacturing industry was 2.7–3.2 times greater than that in Japan in the period 1958–59, 2.5–2.9 times greater in the year 1963, 1.7–2.1 times in the year 1967 and 1.3–1.6 times in the year 1972. *This reveals both the existence of a productivity gap between the two countries and the considerable speed by which the gap has narrowed during the four points of time compared.*

Most industrial groups have shown the same type of catching up by Japan as that shown by the figures for "all groups" described above, but in a few cases such as "Food and tobacco" and "Textile mill products," the productivity gap between the two countries widened between 1958–59 and 1963, but then narrowed considerably. Finally, if we compare the aggregate indices of 1972 with those of 1958–59 group by group, we find that the gap has narrowed in *all* of the groups.

II-3 Growth Rate of Productivity in the Respective Countries

These movements of productivity gaps by groups of industry are, in their turn, the product of different growth rates of productivity between the two countries. Let us now examine this.

Table 4 shows the growth rate of productivity by industrial groups of each country. This aggregate rate of growth has been attained using the figures of employment for 1972 as a weight. Anyhow, according to the table, in both countries almost all the groups have shown a more or less positive growth rate during the three periods; between 1958–59 and 1963, between 1963 and 1967, and between 1967 and 1972. The exceptions are American "Miscellaneous" in the first, American "Motor vehicles" in the second, and American "Fabricated metal products" in the third period, although all three groups have shown a positive growth rate over the periods as a whole.

Secondly, Japanese industrial groups, in general, have shown a higher growth rate of productivity than those of the United States, and this is especially conspicuous in the second and the third periods observed.

The productivity growth rate of American manufacturing industry was 30 percent in the first period, a considerably high rate. During this period there were a few groups of industry whose growth rates of productivity surpassed those of Japan, and, as a result, as pointed out above, the productivity gap between the two countries widened in these groups. But in the second and the third periods, the growth rates of all the Japanese groups examined surpassed those of the corresponding groups in

America and, thus, the same thing can be said for the whole period between 1958–59 and 1972. This may be said to be the spectacular example of Hicks's "uniform improvement in productivity" (Hicks, 1953).

III. PRODUCTIVITY AND MARKET

III-1 Productivity and Market Size

As already pointed out, the individual productivity indices for both countries vary over a wide range from product to product. Now, in order to do research on those factors that may cause such variations in labor productivity, it is necessary, as pointed out by L. Rostás (1948), to make a detailed survey with respect not merely to common factors, such as the size of market and factory and standardization, but also to specific factors arising from individual industries.

However, it must also be noted that, because even the "common" factors are after all subject to different technical conditions as required by the different industrial groups, it becomes more or less necessary to study the bias of each product by all methods. Although this has not yet been worked out in my present survey, it can be pointed out that, as far as the results of this paper are concerned, the scale of production and consequently the size of market can be taken as fairly justifiable factors in accounting for differences in productivity.[5] In other words, they are correlated with the latter. Let us consider this.

Concerning the correlation between relative *productivity* and relative *volume of production*, it is conceivable both interspatially and intertemporally.

First, international *relative productivity* between America and Japan might be correlated with their *relative volume of production.* The higher the relative productivity of a certain product is, the higher its relative volume of production might be. Because relative productivity has been measured for four points of time, the above test of correlation should be made for each of these years.

Next, we can test the correlation between *growth rate of productivity* and that of *volume of production* in each country. The higher the growth rate of productivity of a certain product in Japan is, the higher its growth

[5] Concerning this problem Professor Kravis observes that among other variables, such as the amount of capital per man and scale, only the total size of the market emerges from the studies with any degree of consistency as being associated with the industry-to-industry differences in relative productivity, and there the line of causation probably runs both ways (1976, p. 41).

My results can be taken as additional evidence for his observation, and on the other hand, the latter may authorize my results in the general context of the problem.

rate of production might be. The same thing may be true in America, too. As we have measured productivity for four points of time, we can divide the whole range of time into three periods, 1958–63, 1963–67, and 1967–72. The growth rate has been measured for these three periods for each country, and the correlation has been tested.

Because productivity is obtainable by dividing volume of production by labor input, the existence of rank correlation is regarded to be more suitable for testing than that of linear correlation. We have, therefore, used Kendall's formula, and the resulting coefficients prove the existence of some positive correlation within 0.1 percent level of confidence, for *all* the ten pairs treated, both interspatial and intertemporal.

The main reason for this *intertemporal* correlation, if it exists in any degree, may be found in the fact that the higher the growth rate of an economy, the more opportunity can be given for the replenishment of machinery. The existence of the *interspatial* correlation might also be a result of this fact.

III-2 Labor Productivity and Exports

(a) Relative labor productivity and relative exports

Since Sir Donald MacDougall, connecting Rostás's results concerning relative productivity in British and American industry around 1935 with relative exports of both countries in the same period, pointed out the existence of a close correlation between the two indices (MacDougall, 1951), a few similar affirmative results have been reported for different countries and different points of time, for instance by Bela Balassa (1963).

Yoshiaki Yanagida, connecting my results for 1963, 1967, and 1972 concerning relative productivity in American and Japanese industry with relative exports of the corresponding year, pointed out that the rank correlation between the two indices may be said to be significant within 0.1 percent level of confidence, for all three years tested.

From among 60 products chosen for my productivity measurement, he has tried to take up those products whose export statistics can be found in the U.N.'s *World Trade Annual.* As a result, some 45 products are chosen, each year, for his test of correlationship. To give an example, Table 5 presents a list of products chosen for the 1972 test arranged according to the level of American relative productivity with Japan as the base country and the figures of relative exports of the respective products are given in the second column.

These results would suggest, for the period in question, that labor productivity or relative labor productivity has been a major factor of international competitive power. Next let us pay attention to the dynamic phase of the situation, by comparing the change in relative productivity over a period of time.

(b) Growth of productivity and the structure of exports

In *Section* II-3 of this paper, it was pointed out that the growth rate of productivity in Japan has been higher in *all* the branches compared, and in this sense it can be characterized as a case of Hicks's "uniform improvement." We may call our case more exactly "overall" instead of "uniform" because the difference in growth rates of productivity between the two countries has been uneven or varied from branch to branch. The disparate increase in productivity between Japan and America for the period in question, therefore, may be characterized as

TABLE 5
RELATIVE PRODUCTIVITY AND RELATIVE EXPORTS, 1972
(percent)

Product	*Relative productivity** *(U.S./Japan)*	*Relative exports** *(U.S./Japan)*
Wood pulp	591	3021
Wine & brandy	467	228
Industrial explosives	284	404
Adhesives & gelatin	258	261
Brick	255	∞**
Cotton board woven fabrics	240	88
Steel spring	236	123
Woven carpets & rugs	222	155
Rolled & drawn aluminum	215	408
Fertilizers	214	1096
Inorganic pigments	211	126
Wheat flour	204	3685
Rolled & drawn copper	203	62
Gray iron castings	196	158
Reclaimed rubber	185	∞**
Synthetic organic fibers	182	37
Iron & steel forgings	181	385
Beer & ale	180	129
Metal cans	179	2666
Refined unalloyed aluminum	178	1206
Fatty acids	166	1090
Motor vehicles & equipment	160	76
Paperboard	149	1372
Paper	135	226
Zinc slab	134	5

TABLE 5 *(Continued)*

Product	*Relative productivity* (U.S./Japan)*	*Relative exports* (U.S./Japan)*
Watches	133	1
Printing ink	132	175
Malleable iron castings	127	8299
Wool yarn	121	2
Refined cane sugar	117	15
Household refrigerators	114	209
Men's shoes	113	600
Bolts, nuts, & rivets	112	54
Iron & steel forgings	102	48
Plastic materials	100	122
Steel castings	97	700
Pencils	93	49
Hometype TV sets	90	19
Cotton yarn	86	182
Iron & steel	84	20
Tires	82	33
Cement, hydraulic	78	16
Storage batteries	76	208
Sheet glass	71	7
Piano	68	0
Canned seafood	67	17
Radio & TV receiving-type electron tubes	35	35

$n = 47$ $T = 3.604$ Sig.001

*In order to make the rank exact, some indexes are one point more or less than its actual rounded value.

**Japanese exports are zero.

having been not only "overall" but "uneven." This disparate increase in productivity must have changed the structure of comparative productivity and hence the structure of exports during the period, and this change might have been introduced mainly by Japanese initiative.

Concerning *the "overall" phase of this disparate increase in productivity*, it can be pointed out that as the level of American productivity relative to that of Japan has narrowed, that of relative exports has also narrowed. This has been true for most of the products under consideration, though not all. As a whole, the total exports of the United States were 4.2 times as great as Japan's in 1963, and the ratio fell to 1.7 times in 1972.

By the way, we may safely conclude that such an "overall" phase of disparate increase in productivity between the United States and Japan, by changing their relative competitive power of manufactured goods in

TABLE 6

RELATIVE RATIO OF PRODUCTIVITY AND THOSE OF WAGES, THE U.S. VERSUS JAPAN

(Japan = 1)

	1958–59	*1963*	*1967*	*1972*
Physical productivity of labor (Manufacturing industry)	3.1	2.9	2.0	1.5
Nominal wages (Manufacturing industry)	6.0[a]	5.3[b]	4.1[b]	2.3[c]

[a] *Census of Manufactures* of the U.S. and Japan. Converted with the official exchange rate.

[b] Extrapolated from the results of a joint research committee of the U.S. and Japanese Governments for 1964 (5:1).

[c] 2.7 when extrapolated in the same way as (b) and 2.3 when taking into account the Smithsonian rate of \$1 = ¥308, instead of \$1 = ¥360 before summer in 1971.

the world market, may have been the cause of the tendency of a relative rise in the market rate of exchange of yen to dollars since 1971, though it was reversed temporarily by the so-called oil shock.

Speaking of international competitive power in the world market in relation to relative labor productivity, we should not forget the important *role played by the relative wage level.*

According to Table 6, the wage differential between the two countries narrowed from 5.3 in 1963 to 2.3 in 1972 as the result of the rapid increase in the Japanese wage level during those years and the Smithsonian agreement about the exchange rate. Thus, the differential in money wages decreased a little more rapidly than the productivity differential, which narrowed from 2.9 in 1963 to 1.5 in 1972. If we assume that the increase in the Japanese wage level reduces international price competitiveness proportionally, the narrowing differential must have offset the unbalanced growth of productivity. I have not yet solved this question to my satisfaction, but one reason for this might be that the differential in wage levels was so much greater than the differential in productivity already in 1963 that the gap between the two differentials still remains in spite of the narrowing process of the gap. In addition to this, the time-lag proper to exports of manufactured goods and Japanese efforts at exporting with an improving market research and a network of after-care in the importing area must have contributed to the growth of exports more or less.

Another reason is related to the *"uneven" phase of the disparate increase of productivity.* Concerning this phase, my conclusion is that

TABLE 7

INDICES OF LABOR PRODUCTIVITY IN JAPANESE INDUSTRIES (1970/1965)

	Industry	*Index*
	Total industry	196
	Manufacturing industry	197
(1)	Iron and steel[a]	230
(2)	Metal fixtures	208
(3)	Iron construction materials and metal cable materials	204
(4)	Metal tableware	240
(5)	General machinery	283
(6)	Electrical machinery	258
(7)	Transportation machinery[b]	215
(8)	Precision instruments	209
(9)	Organic chemicals	264
(10)	Plastics	250
(11)	Synthetic fibers	236
(12)	Synthetic rubber	233

Source:
Prepared from Japan Productivity Center, *Kikan Seisansei Tokei [Quarterly Productivity Statistics]*.

Notes:
[a] Excluding pig iron castings.
[b] Excluding railroad rolling stock and ships.

those branches of manufacturing industry in Japan whose growth rate of labor productivity was above its average, surmounting the above offsetting factor, were the branches whose growth rate of exports was most rapid. I have come to this conclusion based on the following facts of American and Japanese trade in the latter half of the 1960s.

In this period, the Japanese-American productivity differential in manufacturing industry as a whole was rapidly reduced, with the remarkable Japanese productivity growth of 14.2 percent per annum against the American growth of 1.4 percent. But this is the story about the average. There were, of course, those branches of the Japanese manufacturing industry, whose growth rate of labor productivity was above the average. Table 7 enumerates those branches.

Next let us shift our attention from productivity growth to *export growth* and examine what Table 8 reveals. Table 8 enumerates those commodities in Japan whose exports to the United States for the same period grew above the average and to one side of the various com-

TABLE 8

INDICES OF JAPANESE EXPORTS TO THE U.S.

(percent)

Commodity	*Corresponding industry*[a]	*Index (1970/1965)*
All commodities		240
Light industrial goods		155
Heavy and chemical industrial goods		297
Others		155
High-growth commodities		
[Metal goods]		187
Tin plate sheets	(1)	269
Iron and steel structural materials and construction materials	(2) (3)	1,500
Hand tools and implements		269
Table knives, forks, and spoons	(4)	286
[Machinery and equipment]		402
General machinery	(5)	418
Electrical machinery	(6)	352
Television sets		425
Radio receivers		381
Transportation machinery	(7)	580
Passenger cars		1,751
Precision instruments	(8)	233
Camera lenses		406
[Chemical goods]		348
Organic chemicals and compounds	(9)	614
Artifical plastics	(10)	380
[Light industrial goods]		
Synthetic textile fabrics	(11)	332
Buttons	(11)	333
Tires and tubes	(12)	455

Source:
Prepared from *Tsusho Hakusho [White Paper on International Trade]*.

Notes:
[a] The number in parentheses corresponds to the industry number in Table 7.

modities there is provided a number in parentheses, which is the same as the number attached to each branch of the manufacturing industry enumerated in Table 7. We can find almost perfect correspondence between the former commodities and the latter branches. In other words, commodities whose exports expanded particularly can be found in those branches whose productivity growth was above the average, and it was these branches that played a dynamic role to change the structure of trade between the two countries.[6]

After all, as a result of "overall" and "uneven" development of labor productivity in Japan, the improvement of competitive power was realized in almost all branches of manufacturing industry and, in some cases, the strengthening of this power was especially spectacular. Generally speaking, the latter phenomenon can take place only when the growth rate of productivity is very high.

And it seems that the condition of such a high rate of productivity growth for Japan is being lost in the 1970s. So far, the Japanese economy has enjoyed a fairly high growth rate by importing technology, but since the U.S.–Japan productivity differential was reduced to 1.5 in 1972, there seems little leeway left for growth through imports of technology. America's stock of technology relative to Japan is now being depleted; Japan has to depend on its own research and development from now on, but it seems to me that Japan has not yet prepared itself for this.[7] From the standpoint of market, Japan will have to, and had better, depend more on her home market expansion and on her own foreign aid than before.

IV. THE INTERNATIONAL LEVEL OF EFFICIENCY OF THE JAPANESE ECONOMY

IV-1 What I Could Not Compare

Thus far, my reasoning has been based on the facts which *I could compare* and, accordingly, are not enough to estimate the international level of Japanese economy as a whole. For the sake of doing so, we should examine what *I could not compare.*

First of all, taking only the sector of manufacturing industry in the national economy, there are important branches of it that were not compared here mainly because Japan has none or has barely established them yet. Among them, atom and aircraft are the two branches which, in some sense, concentrate the essence of the new technology after the war. In

[6] I have reached such a conclusion in an earlier article (1973). The recent event in Spring-Summer 1977 looks like the continuation of the same situation with slower steps.

[7] Kazuo Sato's paper in this volume discusses this in detail.

addition to this, there may be a gap accountable to a decade of technology in space development.

Turning to sectors other than manufacturing, a much wider gap exists in labor productivity in American and Japanese agriculture than in manufacturing. And this is also true for most branches of mining such as coal and oil.

After all, *in the area of the traditional manufacturing industry*, producing consumer goods of daily life, durable and nondurable, and capital goods related to them, *Japan has narrowed the gap with the United States.* And, on the other hand, on account of *what I could not compare,* there needs to be some reservation when we estimate the international level of the Japanese economy based on the results of may comparison. Then to what degree should this reservation be held? The next section will make a suggestion concerning this.

IV-2 Comparison Based on Input-Output Calculations

Input-output tables of the two countries permit another disaggregated calculation of relative labor productivity and, besides, cover a whole industrial sector of national economy.

Such tables are available for 1960–65–70 (Japan) and for 1967 (U.S.). Based on the tables, labor input that is *directly and indirectly* necessary to produce a value unit (one million yen for Japan and one million dollars for the U.S.) of product j ($j=1, .., 59$ for Japan and $j=1, ..., 65$ for the U.S.[8]) has been computed.[9] Mathematically expressed, we have solved the following multilateral equations (1) to get the value of t_j.

$$t_j = \sum_{i=1}^{n} a_{ij} t_i + \tau_j \quad (j = 1, 2, \ldots, n) \tag{1}$$

Where t_j denotes an amount of labor input *directly and indirectly* necessary to produce a value unit of product j,

τ_j denotes an amount of labor input *directly* necessary to produce a value unit of product j,

a_{ij} denotes an amount of intermediate input of product i to produce a value unit of product j,

[8] 1967 input-output tables of the U.S. consists of 71 branches, but here they are integrated to 65 in relation with employment statistics.

[9] Helped by Tadashi Inoue (Ph.D. Minnesota)., We are to prepare another discussion paper on the details of calculation and analysis.

thus,

$$\sum_{i=1}^{n} a_{ij}t_i$$

is an amount of labor *indirectly* necessary to produce a unit (in value terms) of product *j*.

Table 9 shows our results for Japan grouped into three industrial divisions.[10] In Part (a) of the table, labor input, direct and indirect together, *per million yen* of final output is shown for primary, secondary, and tertiary industries, respectively, for the years 1960, 1965, and 1970 *in-constant 1970 prices.* Recalling that a sort of purchasing power parity rate of ICP elaborated by Irving B. Kravis, Zoltan Kenessey, Alan Heston, and Robert Summers (1975) is computed for the year *1970,* we get man-year input per *million dollars* (in 1970 prices) for Japan as shown in Part (b) of Table 9 for the years 1960, 1965, and 1970 and, by extrapolation, for 1967.[11]

As regards the United States, Table 10 (a) shows the resulting figures for man-year input *per million dollars* for the year 1967 by industry, *in 1967 prices.* Deflating the figures with the wholesale price index, we get the same man-year input per million dollars *in 1970 prices.* Now we can compare our results for the two countries as shown in Table 11. Column (a) of Table 11 shows the figures for labor productivity by industry of the two countries obtained by way of input-output calculation, and column (b) shows relative productivity resulting from them. According to this, the ratio of labor productivity of the United States over Japan in 1967 is 2.3 times for the national economy as a whole and, seen at the industry level, 3.9 in primary, 2.6 in secondary, and 1.7 in tertiary industry.

Thus, the productivity gap between the two countries is much wider in primary than secondary industry, to be exact, just 50 percent more in the primary sector. In contrast, the gap in tertiary industry is the narrowest and this brings the gap in national economy as a whole (2.3) to a little

[10] The formula of the aggregation is as follows:

$$T_i = \sum_j \omega_j t_j$$

where

$$\omega_j = \frac{X_j}{\sum_j X_j}$$ X_j: final output value of product *j* in industry *i* ($i = 1, 2, 3$).

[11] Professor D. J. Daly of York University gave me some suitable advice that I had better adopt different conversion rates by industry group instead of a common rate of conversion. It is especially true concerning cereals and dairy products, the prices of which are supported at a very high level by governmental measures.

TABLE 9

MAN-YEAR INPUT PER FINAL OUTPUT UNIT BY INDUSTRY IN JAPAN, 1960, 1965, AND 1970

(a) Man-year per million yen (1970 prices)

	1960	*1965*	*1970*
Primary industry	2.07	1.64	1.40
Secondary industry	1.64	1.06	0.61
Tertiary industry	1.09	0.87	0.59
Whole industry	1.48	1.03	0.63

(b) Man-year per million dollars (1970 prices converted with ICP binary ideal rate*)

	1960	*1965*	*1970*	*1967***
Primary industry	509	403	344	379
Secondary industry	403	261	150	209
Tertiary industry	268	214	145	182
Whole industry	364	253	155	209

Source:

Government of Japan, *1960–1965–1970 Link Input-Output Tables,* Data Report (1), Feb. 1975.

Notes:

* $1 = ¥246 (Kravis *et al.*, 1975, p.177)

** Intrapolation with annual average rate of change between 1965 and 1970.

lower level than that in secondary industry (2.6). This observation will serve as supplemental information for what I did not measure in my physical productivity calculation and will give a more comprehensive idea about the international level of the Japanese economy as a whole.

IV-3 Differences in the Results among Different Measurements

Now, let us examine the gap between the two kinds of results reported here. Secondary industry contains mining and construction, and if we separate manufacturing industry from it, the productivity gap between the two countries is as high as 2.8 times, exclusively because of higher value productivity of manufacturing industry relative to other groups in the United States, as shown in Table 11. Whereas, the corresponding figure of physical productivity of labor in 1967 is 2.0, according to index

TABLE 10

MAN-YEAR INPUT PER FINAL OUTPUT UNIT BY INDUSTRY IN THE U.S., 1967

Man-year per million dollars in terms of 1967 prices and 1970 prices

	1967	
	(a) 1967 prices	*(b) 1970 prices**
Primary industry	108	98
Secondary industry	88	80
Tertiary industry	119	107
Whole industry	102	93

Source:
Input-Output Structure of the U.S. Economy: 1967, *Survey of Current Business,* Feb. 1974. Employment Figures from *Monthly Labor Review*, Jan. 1969 and Dec. 1976.

Notes:
* (*a*) ÷ 1.104 Wholesale price increase in America between 1967 and 1970 = 1.104.

TABLE 11

RELATIVE LABOR PRODUCTIVITY BY WAY OF INPUT-OUTPUT CALCULATION, 1967

	(a) Final output value per man-year in 1967 (Unit: 1,000 dollars in 1970 prices)		*(b) Relative Productivity U.S./Japan (%)*
	(A) U.S.	*(J) Japan*	*[(A)/(J) × 100]*
Primary industry	10.22	2.64	387
Secondary industry	12.50	4.79	261
(Manufactures)	(13.33)	(4.79)	(278)
Tertiary industry	9.32	5.46	171
Whole industry	10.81	4.81	225

(A) from Table 10 (b); *(J)* from Table 9 (b).

(C) in Table 3.

Professor D.J. Daly, examining my results (Yukizawa, 1975) expressed concern that my method of measurement would involve the possibility of overstating Japan's level of productivity relative to that of North America's (Daly, 1976, pp. 4–6). Mr. J. A. Mark also joins in this concern (1977, pp. 21–21a). Professor Daly points out, first, that a number of studies for other countries suggest that the differences are less in gross output measures (such as mine) than net output and, secondly, that the low coverage of my measurement (shown in Table 1) suggests the possibility that Japanese data could involve an overpresentation of high productivity industries in Japan.

Concerning the first point, I would like to try to measure relative productivity by the so-called net output method, but I am afraid that it would be much more difficult than by my method. For instance, when applying *strictly* the Geary formula,[12] a formula that can be said to be most fitted to the concept of relative net output productivity, we have to compare *physically* not only output itself, like that of my case, but also intermediate input goods and, preferably, the depreciated part of fixed capital (q' and q'' in footnote 12). Therefore, if we apply the Geary formula strictly, I am afraid that the coverage will be lower than that of my case.

Before this trial, the above results of input-output calculation could serve as a kind of substitute for the net output method because that calculation takes into account intermediate labor input instead of intermediate goods. Actually, the resulting indices of American productivity relative to Japan are higher than those of physical gross output productivity. Among possible factors of this measurement gap, the following fact could be regarded as related to Daly's first point, though it belongs to a specific factor rather than a common factor explaining Daly's first point. Examining the results of the above input-output calculation, we can find that the ratio of direct labor input over total labor input (τ/t) in America is about 20 percent higher than that in Japan, so far as the manufacturing sector is concerned. In other words, indirect labor input ($t-\tau$) in America is lower relatively to total labor input (t) than that in Japan. This fact suggests either input of intermediate goods per product

12
$$\left(\frac{\sum q_A p - \sum q'_A p' - \sum q''_A p''}{\sum q_B p - \sum q'_B p' - \sum q''_B p''}\right) \div \frac{\sum m_A}{\sum m_b}$$

where q and p denote the quantities and prices of gross output,

q' and p' denote the quantities and prices of intermediate consumption,

q'' and p'' denote the quantities and prices of consumption of fixed capital,

m denotes labor input, and subscripts A and B refer to the two countries compared. (United Nations Statistical Commission and Economic Commission for Europe, 1971, p. 5).

is less in American manufactures, or the relative labor productivity of those intermediate goods is higher than that of other manufactured goods.

Anyhow, this fact makes relative productivity of total labor input $(O_A/t_A)/(O_J/t_J)$ higher than that of direct labor input $(O_A/\tau_A)/(O_J/\tau_J)$ by about 20 percent.[13] The concept of the latter belongs to the same category as relative physical productivity of labor $(q_A/\tau_A)/(q_J/\tau_J)$, which I have tried to measure.

Regarding Daly's second points, I share his concern. As already expressed in *Section* IV-1 of this paper, there are those branches of manufacturing industry that I could not compare because they have not been established in Japan, and accordingly, the technological gap of these is not reflected in the measured gap.[14]

On the other hand, there is still room to examine whether the gap of 2.0 and 2.8 appearing between the two kinds of measurements concerning relative productivity in American and Japanese manufacturing industry in 1967 can be explained mostly by this second point. For instance, we have adopted as the conversion rate, in our input-output calculation, ICP binary rate (\$1 = ¥246), but if we adopt IMF fixed rate at that time (\$1 = ¥360), the resulting productivity of Japan is 46 percent lower, relatively speaking.

All these observations suggest that a little amendment is required to assess the international level of efficiency of Japanese economy based on the results of my calculation of physical productivity of labor and that this amendment consists of the correction of *overvaluation* of Japanese efficiency, but the degree of amendment is difficult to make precise.

Lastly, let us compare my results with Jorgenson-Nishimizu's and K. Sato's conclusions in their papers read at this Congress in Tokyo. According to mine, American relative productivity of labor over Japan in 1972 is 1.5 times, and I regard it as a measure of the gap in technology still remaining between the two countries. On the other hand, from the

[13] According to our calculation concerning manufacturing industry, $(\tau_A/t_A)>(\tau_J/t_J)$, therefore $(t_A/t_J)<(\tau_A/\tau_J)$, then,

$$\left(\frac{O_A}{t_A}\bigg/\frac{O_J}{t_J}\right)=\left(\frac{O_A}{O_J}\bigg/\frac{t_A}{t_J}\right)>\left(\frac{O_A}{O_J}\bigg/\frac{\tau_A}{\tau_J}\right).$$

[14] In some sense, this concern could be extended to Jorgenson-Nishimizu's results. Though they derive the ratio between U.S. and Japanese total output by aggregating individual outputs with weights given by average value shares, relative backwardness in such branches of Japanese industry as atomic, aircraft, large-sized computer and medical machines could not be suitably reflected in that ratio *so long as* their disaggregation of industrial sectors is insufficient. In fact, they, themselves, make it as one of their directions for future research to disaggregate of the production account to the level of individual industrial sectors.

point of view of Jorgenson-Nishimizu, this would not necessarily contradict their conclusion that between 1970 and 1973 the remaining gap between Japanese and U.S. technology was eliminated because, since then, all of the remaining difference between U.S. and Japanese output per unit of labor input is due to differences in the capital intensity of production in the two countries. As a matter of fact, their results show that the ratio between U.S. and Japanese total output is 3.3 times in 1972 and that of labor input is 2.3 in the same year. This may suggest the gap of labor productivity is about 1.4 times, which is very near to my results. Professor Sato, based on my former results (Yukizawa, 1973), says that the productivity gap of around 1.3 in 1970 between the two countries is the result of calculation of all establishments, and if the comparison is limited to large firms, we might reason that the productivity gap disappeared around 1970. Thus, concerning a technology gap between the two countries, I am most pessimistic in my reservation, and Professor Sato is pessimistic about Japan's possibility as an innovator of technology. I wonder if Professors Jorgenson and Nishimizu would agree with these conclusions. Rather, there is still room for further factual study before we can push our reasoning any further.[15]

REFERENCES

Balassa, Bela. 1963. "An Empirical Demonstration of Classical Comparative Cost Theory," *Review of Economics and Statistics,* vol. 45, August, pp. 23–38.

Daly, D. J. 1976. *Estimates of Manufacturing Levels, United States, Canada and Japan.* York University, July, 9 page mimeo.

Denison, Edward F. and Chung, William K. 1976. *How Japan's Economy Grew so Fast.* Washington, D.C.: Brookings Institution.

Frankel, Marvin. 1957. *British and American Manufacturing Productivity: A Comparison and Interpretation.* University of Illinois Bulletin No. 81. Urbana: University of Illinois.

Hicks, John R. 1953. "The Long-Term Dollar Problem, An Inaugural Lecture," *Oxford Economic Papers,* vol. 5, June, pp. 117–35.

Kravis, Irving B. 1976. "A Survey of International Comparisons of Productivity," *Economic Journal,* vol. 86, March, pp. 1–44.

———; Kenessey, Zoltan; Heston, Alan; and Summers, Robert. 1975. *A System of International Comparisons of Gross Product and Purchasing Power.* United Nations International Comparison Project: Phase One. Baltimore: Johns Hopkins University Press for the World Bank.

MacDougall, Donald. 1951. "British and American Exports: A Study Suggested by the Theory of Comparative Costs," Part I, *Economic Journal,* vol. 61, December, pp. 697–724.

[15] Professor Masaru Saito has supplemented my reasoning, in his comment, by stating the relationship of Japanese productivity growth with the "dual structure" of its economy, which I could not investigate so far.

Mark, J. A. 1977. *Comparative Growth in Manufacturing Productivity and Labor Costs in Selected Industrialized Countries.* Mimeo, to be forthcoming as a report of the BLS bulletin.

Paige, Deborah and Bombach, Gottfried. 1959. *A Comparison of National Output and Productivity of the United Kingdom and the United States.* Joint study by the OEEC and the Department of Applied Economics, University of Cambridge. Paris: OEEC.

Rostás, László. 1948. *Comparative Productivity in British and American Industry.* NIESR Occasional Papers, No. XIII. Cambridge: Cambridge University Press.

United Nations Statistical Commission and Economic Commission for Europe. 1971. *Methodological Problems of International Comparison of Levels of Labor Productivity in Industry.* Conference of European Statisticians, Statistical Standards and Studies, No. 21, New York: U.N.

Yukizawa, Kenzo. 1973. "The Narrowing Japanese-United States Productivity Gap—as Related to the Yen Revaluation," *Japanese Economic Studies,* Summer.

——— . 1975. *Japanese and American Manufacturing Productivity: An International Comparison of Physical Output per Head.* KIER Discussion Paper, No. 087. Kyoto: Kyoto University, Institute of Economic Research, March.

——— . 1976. *Comparative Productivity in the Iron and Steel Industry; the United States and Japan, 1958–1972.* KIER Discussion Paper, No. 092. Kyoto: Kyoto University, Institute of Economic Research, June.

——— . 1977. *Comparative Productivity in the Automobile Industry; the United States and Japan, 1958–1972.* KIER Discussion Paper, No. 105. Kyoto: Kyoto University, Institute of Economic Research, March.

Comments

Masaru Saito (Japan)

Professor Yukizawa suggests the following concluding points in his paper, that is, (1) speedy catching-up of Japan to the United States in the labor productivity of manufacturing industry, (2) Japan had the biggest growth rate of labor productivity in the middle of the 1960s, 1963–67, in the period, (3) Japan has had bigger rate of growth of labor productivity in the heavy and chemical industries, and (4) in the latter period, the gaps in the productivity growth rate between the United States and Japan have been equalizing among sectors.

It is necessary to make clear the process of structural transformation for the analysis of the structural characteristics. Certainly, we can understand the relationship between the labor productivity on the one hand and market, export, the scale of production, etc. on the other, by focusing upon the fact that the relative productivity change brings about the change of the comparative advantage between the United States and Japan.

So, at first, I would like to take up the problem of catching-up mechanism from the viewpoint of the relationship between both countries in the productivity growth. Especially in this period, we can find the intimate relation of labor productivity between both countries through the technology transfer from the United States. For example, 56 percent of total number of cases of imported technologies (important technologies, type A) was from the United States in this period; and generally, the more successful the industries in the catching-up process of the productivity were, the more they imported technologies. In terms of monetary values, 83 percent of total imports was from the United States in 1972.

The second problem is how the industrial structure has been transformed by improving the productivity. The period for which Professor Yukizawa selects includes the turning point of Japanese economy. The growth mechanism of Japanese economy changed in the middle of the 1960s. Until the middle of the 1960s, the basic character of Japanese industrial structure or growth mechanism was the dual structure. The dual economic structure of the growth mechanism means here that there are two sectors, an advanced sector and a backward sector in the national economy. Almost all the large-scale companies belong to the former, and almost all the small-scale companies belong to the latter; and in the Japanese-type of dual economy, the advanced sector grows by the support of backward sector, while the backward sector shares indirectly in the benefit of development of the advanced sector through total economic growth, such a relationship between both sectors being useful for economic growth. In Japan, small scale companies have played an important role in economic revival and growth after World War II. They have accelerated the development of large-scale companies, and pushed up the ceiling of the balance of payments against economic growth. After the turning point, the leading sectors of economic growth and exports shifted from textile and light industries to chemical and heavy industries. The dual economic structure has been disorganized, and the gaps between advanced and backward sectors have been decreased. If Professor Yukizawa does the comparative research between large- and small-scale companies by dividing into two sectors, I think he would be able to make clear the disorganizing process of dual industrial structure and new mechanism of industrial transformation and development. For example, when the productivity level of large-scale companies is 100, that of small-scale companies is about 80 in the United States in terms of value added, whereas in Japan, that of small-scale companies is 50 in 1958, 44 in 1959, 57 in 1963, 55 in 1967, and 50 in 1972.

Industrial Organization and Economic Growth in Japan*

Ken'ichi Imai *Hitotsubashi University*
Masu Uekusa *Tokyo University*

The base of Japan's postwar industrial organization stems from the large-scale deconcentration measures during the occupation period. It was the competitive and centrifugal forces set in motion by the measures that performed the innovation in both technological and social spheres and promoted economic growth in Japan. The centrifugal forces have been gradually weakened by the centripetal policy of the Japanese government and ministries, favoring fast-growing big business and administratively controlling market competition. Subsequently, distortion of allocative efficiency and stagnation of economic growth have begun to appear.

* We wish to express our thanks for the valuable comments of the discussant, Professor Soichiro Giga.

INTRODUCTION

The authors will point out in this paper that the large-scale deconcentration measures imposed on Japan during the occupation period formed a relatively competitive market structure of Japanese industry, which has been a driving force for rapid growth in the Japanese economy; at the same time, rapid macroeconomic growth has had various kinds of effects on microeconomic activities and thus industrial organization in Japan. It should be noted, however, that the centrifugal forces set in motion by the competition policy have been gradually weakened by the centripetal policy of the Japanese government and ministries, favoring fast-growing producers and administratively controlled market competition. This has begun to generate distortion of allocative efficiency, administered inflation, and stagnation of capital formation.

I. THE SHAPING OF THE MARKET MECHANISM AND ORGANIZATION FRAMEWORK

1. The Deconcentration Measures during the Occupation Period

The economic democratization policy imposed on Japan during the occupation period (1945–51), including the deconcentration measures, revolution in agricultural landholding, and liberalization of labor movements, was one of the most important determinants of economic growth in postwar Japan. Especially, the deconcentration measures represented probably the greatest use of government power in modern times to reform industrial market structure and had a crucial influence on economic growth in postwar Japan. The measures went through a complex sequence, but they broadly include: dissolution of *zaibatsu* organizations; reorganization (division, divorcement, and divesture) of private enterprises; and dissolution of cartel organizations. The outline of these measures is described in this section.

Dissolution of Zaibatsu. The *zaibatsu* were a major target of the occupation policy. The ultimate control of the *zaibatsu* organizations reposed in a few wealthy families. Each *zaibatsu* family held a top holding company, with that company controlling the operation companies, which in turn controlled many other affiliates through subsidiaries and fractional shareholdings. This pyramidal form of control was solidified by interlocking directorates and the absolute loyalty of the company managements to the *zaibatsu* families. The *zaibatsu* thus achieved monolithic unity under the centralized direction of the top holding company and controlled principal sectors of the Japanese economy. There were the big four *zaibatsu*, the six smaller and incomplete *zaibatsu*, and the seventeen local ones.

The *zaibatsu* organizations were dissolved by the Japanese government under the direction of the occupation powers as follows: (a) prohibition—confiscation—of stock ownership of the *zaibatsu* families; (b) dissolution of top holding companies; (c) severance of vertical and horizontal ties of ownership of the *zaibatsu* group companies; (d) severance of personal ties through interlocking directorates, and a purge of officers of major companies within the *zaibatsu* organizations; (e) dissolution of the two big trading companies (Mitsui and Mitsubishi); and (f) prohibition of further use of *zaibatsu* trademark. An ambitious program of projected dissolution of "*zaibatsu* combines" was almost completely accomplished, leaving the banks and financial intermediaries in the *zaibatsu* organizations untouched.[1]

Reorganization of Private Enterprises. Reorganization of private enterprises had the aim to "bust the trusts," and included the three measures as follows: (a) division, divorcement, and divesture of big businesses; (b) restriction of intercorporate shareholdings; and (c) prohibition of interlocking directorates and economic purge of officers of big businesses.

The program to bust the trusts was first of all enforced by the Excessive Economic Power Deconcentration Law. Three hundred twenty-five firms were designated for reorganization, but only 18 companies were finally subject to antitrust actions because of the shift of occupation policy toward the reconstruction of Japan as an economic ally of the United States. It should be noted, however, that division of private enterprises was enforced on a far larger scale under the Enterprise and Financial Institution Reconstruction and Reorganization Laws. Of the 325 firms designated by the Deconcentration Law, 60 big companies were subject to reorganization measures under the Reconstruction and Reorganization Laws.[2]

The prohibition of intercorporate shareholdings was enforced under the Antimonopoly Law enacted in 1947. Then, 4,044 companies were deprived of shareholdings in their subsidiaries in order to reduce their vertical and conglomerate economic power.[3]

Evidence revealed that the Antimonopoly Law prohibited interlocking directorates between a large number of big corporations and their subsidiaries.[4] In addition to the prohibition of interlocking directorates, 1,535 officers of 245 big companies were purged through the measures for

[1] *See* Eleanor M. Hadley, *Antitrust in Japan* (1970) and Richard E. Caves and Masu Uekusa, *Industrial Organization in Japan* (1976).

[2] Masu Uekusa, "Shuchu Haijo Seisaku ni tsuiteno Atarashii Shiten" ["New Look into the Deconcentration Measures"] (1977), pp. 6–19.

[3] Fair Trade Commission of Japan, *Dokusen Kinshi Seisaku Niju-nenshi [Twenty-year History of Antimonopoly Policy]* (1968), p. 83.

[4] Fair Trade Commission of Japan (1968), p. 83.

eliminating war-cooperators and replaced by younger managements.

Although the program of deconcentration of private enterprises was undermined and partially dropped, it should not be overlooked that the Reconstruction and Reorganization Laws and the Antimonopoly Law supplemented the aim of the Deconcentration Law.

Dissolution of Cartel Organizations. There were a large number of cartel organizations in prewar Japan. The roots of collusive behavior in Japanese business run back to the pre-Meiji associations. Modern cartel organizations developed through the price collapse of the depression after World War I and the Great Depression at the beginning of the 1930s, and rapidly spread as compulsory cartels during the military control period. After World War II the Japanese government was empowered—and required—to remove the wartime economic control systems. One of the measures was to sweep away compulsory cartel organizations. Cartel organizations of 1,022 industries were dissolved and 69 national policy companies were closed under the Close Institution Law.

The Effects of the Deconcentration Measures. The deconcentration policy was an event of some significance from the point of view of the growth and development of postwar industry in Japan.

First of all, the dissolution of *zaibatsu* increased the decision-making autonomy of operating companies and other subsidiaries, which were released from the centralized direction of the top holding companies. The reorganization of big businesses reduced their market power and created a large number of independent firms. The dissolution of cartel organizations formed independent firm behavior. In short, these deconcentration measures had a great role in increasing independency and autonomy of business.

Secondly, the purge of top executives was also an important component in enhancing the independent decision-making capability of Japanese firms because those who replaced the former officers were younger men who had relatively little loyalty to the *zaibatsu* families. In addition, the events described meant that within Japanese industry after the war, there took place the so-called separation of ownership and control. In effect, at a stroke, the era of management control had arrived.

Thirdly, the dissolution of *zaibatsu* and the reorganization of big businesses directly decreased the level of market concentration for a number of industries and indirectly—in the long run—decreased it through reduction of barriers to entry into industries, as shown in the following section. Summing up, the companies' new autonomy under the competitive market structure was a factor causing them to import foreign technologies, helping create a large number of new industries, and causing them to compete actively in expanding capital investment outlays. Therefore, the deconcentration measures during the occupation period had a great impact on Japanese industries not only in the short run but also over the long run.

2. *Public Policy toward Industry after the Occupation*

Although postwar Japanese industries started under relatively competitive conditions, the Japanese government and ministries adopted an economic control system rather than the utilization of market mechanisms under competitive market structure after the occupation ended. It might have been unavoidable that, in the postwar economic revival period, the so-called priority production system of controlled allocation of raw materials was adopted on the basis of the Temporary Materials Control Law and through the medium of the Reconstruction Finance Bank. Long after the reconstruction period is recognized to have ended, however, Japan still utilized a system of mobilization of resources based on government planning for individual industries, which is given effect through the practice that has come to be known as "administrative guidance." What produced this policy was a latent consciousness that in the Japan of that time, market forces alone were not capable of providing direction for the economy. Certainly, among government officials there existed a firm belief that new products should be planned, new technologies imported, and industries generally organized from the point of view of the national advantages. They in no way subscribed to the idea that market forces should be allowed to dictate the direction of resources.

As a result, until the latter half of the 1950s the development of Japan's industry was dictated by the "Five-Year Production Plan for Mining and Manufacturing Industry" under the Economic Stabilization Board and by individual industrial plans under the Ministry of International Trade and Industry (MITI), which was established in 1949. MITI controlled many industries by means of a system of government licensing, foreign exchange allocation, and validation of induction of foreign technology.

Since 1960 after the reconstruction period ended, the Japanese economy transferred from the closed economy to an open economic system under the organization of free international trade, centered on the GATT and IMF system. The transformation into the free international trade system had the following two impacts on Japanese industries. On the one hand, Japanese industries enthusiastically subscribed to the merit of free international trade through free import of raw materials and advanced technologies from abroad. Japan benefited especially from one of the major technological advances of the postwar years, the great reduction in the cost of marine transportation, which rendered it positively advantageous for a maritime nation like Japan not to possess domestic sources of raw materials. On the other hand, Japanese industries experienced a ballooning of mergers, which resulted from the policy to increase international export competitiveness. The annual average number of mergers more than doubled from the 1950s to the 1960s. Most of the combinations have been among relatively small companies, but mergers among very large companies became increasingly common during the 1960s.

MITI played the most important role in increasing international competitiveness. MITI has promoted the movement of resources toward certain favored industries—to the heavy and chemical industries in which Japan seemed to enjoy a comparative advantage in international markets. The ministry has also undertaken a policy to enlarge the scales of plant and firm in certain industries. This goal has led to a considerable enthusiasm for mergers and the restriction of new entry into industries of interest to MITI. The ministry has promoted industry-wide cooperation in its use, and it has sought to protect Japanese markets from the intrusion of foreign competition through imports or foreign investment. MITI's control over industrial structure and behavior through "administrative guidance" was the most active during the 1960s.

The principal "bogey" raised to justify "administrative guidance" has been that of "excessive competition." It is a fact that the competitive market structure has produced unnecessary duplication. It should be noted, however, that "administrative guidance" involving some forms of allocation has produced a special type of "excessive competition" among businesses.

After all, MITI's justification of "administrative guidance" stemmed from the formation of a system favoring the fast-growing producer. But it is a fact that market competition has gotten over a complicated element of control by MITI. Hence, one of the most striking features of industrial organization in Japan has been the conflict between the centrifugal forces set in motion by the competition policy during the occupation period and the centripetal policy of the Japanese government and ministries.

Antimonopoly Policy. The Antimonopoly Law enacted in 1947 was highly systematic: (1) private monopolization was prohibited absolutely; (2) the law provided authority to break up a dominant firm; (3) unreasonable restraint of trade (cartel and other agreement) was "per-se illegal"; (4) international agreement in restraint of trade was strictly prohibited; (5) establishment of trade associations was restricted; (6) formation of holding companies was prohibited; (7) mergers, interlocking directorates, and intercorporate shareholding to restrict competition substantially might be restricted; (8) six categories of unfair trade practices were prohibited. This law was too ideal in the Japan of that time and represented a barrier to reconstruction of the postwar Japanese economy. The law was extensively weakened in 1953, immediately after the occupation ended. Also the enforcement of the amended law was

TABLE 1

THE NUMBER OF VIOLATIONS OF ANTIMONOPOLY LAW

	1947–1951	*1952–1959*	*1960–1969*	*1970–1975*	*Total Violations, 1947–1975*
Private monopolization	3	2	0	1	6
Unreasonable restraint of trade	57	29	137	224	447
International agreement	22	1	1	6	30
Merger and other combination	28	1	2	2	33
Unfair trade practices	23	14	26	9	72
Other	0	0	2	19	21
Total	133 (26.6)	47 (5.9)	168 (16.8)	261 (43.5)	609 (21.0)

Sources:

Fair Trade Commission, *Annual Report—1975,* pp. 24–25.

gradually eroded by the passing of separate laws authorizing "administrative guidance."

Table 1 shows the number of violations of the Antimonopoly Law found by the Fair Trade Commission of Japan in each period. The annual average number of violations declined from about 27 during the occupation period (1947–51) to about 6 during the reconstruction period (1952–59). These data thus do show that competition policy was weakened by the development of "industrial policy." The number fairly increased during the rapid growth period (1960–69) and rapidly expanded during the stable growth period (1970–75). Table 1 shows that, of the 609 total number of illegal cases, 447 (73.4 percent) were cartel cases and that there were very few cases of private monopolizations, mergers, and other combinations. We could say that the increased mergers, shown in Table 2, have escaped serious restraint from the Antimonopoly Law.

II. CHANGES IN INDUSTRIAL ORGANIZATION

1. Changes in Market Structure

Rapid Growth and Change in Interindustry Structure. Within the institutional framework of the Japanese industrial organization described above, Japanese enterprises expanded rapidly, in the process introducing many innovations, broadening markets, and extending the division of labor and generally positively pursuing a competitive policy. It was indeed the sort of process referred to by Joseph A. Schumpeter as "con-

TABLE 2
NUMBER OF RECORDED JAPANESE MERGERS BY SIZE OF CAPITAL OF COMBINED COMPANIES, 1950–70

(billions of yen)

Year	*Under 0.1*	*0.1-1*	*1-10*	*Over 10*	*Total*
1950	413	7	0	0	420
1951	317	14	0	0	331
1952	359	23	3	0	385
1953	315	24	5	0	344
1954	293	28	4	0	325
1955	311	19	8	0	338
1956	359	15	7	0	381
1957	367	22	9	0	398
1958	348	25	8	0	381
1959	372	37	8	0	417
1960	381	49	9	1	440
1961	519	54	16	2	591
1962	585	101	26	3	715
1963	821	131	33	12	997
1964	730	104	21	9	864
1965	771	109	11	3	894
1966	763	78	25	5	871
1967	852	103	33	7	995
1968	876	116	21	7	1,020
1969	965	162	34	2	1,163
1970	925	179	37	6	1,147

Source:
Japan, Fair Trade Commission, *Nihon no Kigyo Shuchu [Corporate Mergers in Japan]* (1971), p. 171.

structive destruction." It was a process of innovation that can be described statistically in many different ways. Table 3 describes changes in industrial structure in terms of a statistical coefficient of change. It gives an international comparison of coefficients of changes and appeared in a discussion of industrial flexibility in the 1964 *Economic White Paper*. Even at that time, Japan's coefficient of change was noted to be relatively high, but it rose even higher during 1960–70. The Japanese economy can thus be seen to have undergone considerable change in industrial structure and to have exercised a high degree of flexibility over an extended period.

The process of change had a large impact on Japan's industrial structure. Table 4 is intended to give a bird's-eye view of the structure of Japan's industry by form of market structure. Based on statistics of domestic product for 450 industries in 1970, it separates out monopolistic and public industries to leave an area of competitive industry. The monopolistic sectors occupy 21.6 percent of the total domestic product, the public utilities 13.1 percent, and the competitive sectors 65.2 percent. Although some of the competitive sectors are under restriction of competition exempted from the Antimonopoly Law, it is not incorrect to say that about 70 percent of Japanese industries was under competitive market structure in 1970.

Looking into market structure of manufacturing industries, one can find that there are many oligopolistic industries: products for the mono-

TABLE 3

COEFFICIENTS OF CHANGE IN INDUSTRIAL STRUCTURE

	1954–61	*1960–70*
Japan	18.4	19.3
U.S.	5.6	10.9
U.K.	9.1	12.0
W. Germany	7.0	10.6
France	5.4	12.9
Italy	14.3	11.9

Sources:

1954–61, *Economic White Paper, 1964,* p. 54. 1960–70 based on material from United Nations: *Yearbook of National Accounts Statistics 1973.*

Note:

Coefficient of change in industrial structure (three sectors) is obtained as follows. In both the base year and the end year the industrial composition of gross national product is calculated and points given for differences between two years. The coefficient of change is the sum total of such points.

TABLE 4
PRODUCTION BY INDUSTRY AND BY FORM OF MARKET STRUCTURE

(Unit: one billion yen)

Industry	(1) Competitive	(2) Monopolistic	(3) Public Utilities	(4) Domestic Product Total	(5) Degree of Monopoly	(6) Proportion Public & Utilities
1 Agriculture & fishing	5,206.9	—	1,906.7	7,113.6	—%	26.8%
2 Mining	808.3	150.9	—	959.2	15.7	—
3 Food products	6,856.6	1,947.5	816.5	9,620.6	20.2	8.5
4 Textiles and clothing	5,892.2	242.3	—	6,134.5	4.0	—
5 Wood manufactures and furniture	3,399.2	—	—	3,399.2	—	—
6 Pulp, paper, printing, publishing	3,450.3	1,023.2	—	4,473.5	22.9	—
7 Leather and rubber products	639.9	289.2	—	929.1	31.1	—
8 Chemicals	4,416.5	1,694.5	—	6,111.0	27.7	—
9 Petroleum and coal products	646.0	2,373.7	—	3,018.7	78.6	—
10 Clay, stone	1,742.2	927.7	—	2,669.9	34.7	—
11 Iron and steel	4,042.5	7,242.7	—	11,285.3	64.2	—
12 Nonferrous and ferrous products	4,433.2	1,200.0	—	5,633.2	21.3	—
13 General machinery	6,319.1	2,004.6	—	8,323.7	24.1	—
14 Electrical machinery	4,175.0	3,457.4	—	7,623.4	45.3	—
15 Transportation machinery	2,633.5	4,990.4	—	7,623.9	65.5	—
16 Precision instruments	634.7	468.8	—	1,103.5	42.5	—
17 Other manufacturing	2,088.6	131.1	—	2,219.7	5.9	—
18 Construction and civil engineering	10,392.9	—	5,865.8	16,258.7	—	36.1

19 Electricity, gas, water	—	—	2,627.9	2,627.9	—	100.0
20 Wholesale and retail real estate	14,126.1	—	163.6	14,289.7	—	1.1
21 Finance, insurance	—	4,906.9	—	4,906.9	100.0	—
22 Real estate	5,907.7	—	—	5,907.7	—	—
23 Transport and communications	1,442.3	946.0	5,055.6	7,443.9	12.7	67.9
24 Services	12,611.7	944.3	4,775.8	18,331.8	5.2	26.1
25 Other	3,499.9	—	—	3,499.9	—	—
Total	105,365.3	34,940.4	21,212.0	161,517.8		
	65.2%		13.1%	100%		

Sources:

From material in 1970 Industrial Structure-Statistics (I) 1974–Administrative Management Agency (Gyosei Kanri-cho), etc.; Ken'ichi Imai, *Gendai Sangyo Soshiki [Contemporary Industrial Organization]* (1976), p. 17.

Notes:

In the above an industry is defined as monopolistic if 50% of the industry is accounted for by the top six or less concerns or if 12 or less concerns account for more than 75% of output. This follows the method of A. D. Chandler ("The Structure of American Industry in the Twentieth Century: A Historical Overview," 1969) where he discusses the weight of monopoly in 100 years of U.S. business history—thus allowing of comparison with that work. Finance and Insurance have been tentatively classified as monopolistic because of their degree of concentration and the existence of artificial limits on competition. The monopolistic element in service industries is advertising. In agriculture, production of rice and wheat has been placed in the public area; their distribution has been similarly separated from the commerce industry.

polistic markets as a percentage of the total 2-digit products are 78.6 percent for petroleum and coal products; 65.5 for transportation machinery; and 64.2 for steel industry. It should be noted that these industries have been affected by the policy favoring fast-growing producers. Imai found, however, that the level of seller concentration in Japanese industries is not higher than that for the United States.[5] Richard E. Caves and Uekusa also found that concentration in Japanese manufacturing industries is closely related to concentration in their U.S. counterparts and that the level of concentration in the two countries is about the same.[6]

Changes in Seller Concentration. In America a vast amount of study has been directed at industrial concentration. Alfred D. Chandler's study, "A 100-Year's Analysis of U.S. Business History,"[7] divides industries into the following three types, in terms of degree of concentration in the long term. Although understandably somewhat lacking in detailed statistics, this work abounds in significant analysis worthy of a well-known business-history researcher.

Group A represents industries exhibiting a low degree of concentration and includes leather, printing and publishing, timber and wood manufactures, furniture, and clothing. When we regard these industries in detail we find few signs of concentration (see notes to table), and similarly in the long period since 1909 there appears almost no tendency toward increasing concentration.

Group B represents industries with some inclination toward monopoly and includes textiles, paper, paper products, metal manufactures and general machinery, and food products. A detailed analysis shows 25–50 percent of the industries in the group to be monopolistic, and since 1940 there has been some trend toward further concentration.

Group C consists of typically monopolistic industries such as ceramics, quarrying, chemicals, petroleum, rubber, basic metals, electrical machinery, transportation machinery, precision instruments, and cigarette manufacture. In Group C the level of technological competence required dictates to some extent the degree of concentration, which is also influenced by the degree to which individual enterprises diversify their activities, there being not a few areas where diversification on the part of the large concerns has reduced the degree of concentration. Since the industries in Group C represent the driving force in economic development and employ virtually all the labor employed in research and development, it is evident that the activities of enterprises in this group

[5] Ken'ichi Imai, *Gendai Sangyo Soshiki [Contemporary Industrial Organization]* (1976), p. 18.

[6] Caves and Uekusa (1976), pp. 19–26.

[7] Alfred D. Chandler, "The Structure of American Industry in the Twentieth Century: A Historical Overview" (1969).

TABLE 5

INDICES OF CONCENTRATION IN THE JAPANESE MANUFACTURING SECTOR, 1950–74

Concentration Level	*Number of Industries*	*1950*	*1955*	*1960*	*1965*	*1970*	*1974*
3 firms	43	100.0	93.5	91.5			
	170			100.0	97.8	104.0	
	163				100.0	102.9	103.8
10 firms	43	100.0	96.4	95.5			
	170			100.0	100.4	102.8	
	161				100.0	101.2	101.5
Herfindahl	170			100.0	96.8	110.1	
	163				100.0	107.6	110.6

Sources:

Japan, Fair Trade Commission, *Nihon no Sangyo Shuchu: 1963–1966 [Industrial Concentration in Japan: 1963-1966]* (1969), pp. 58-59; *Syuyo Sangyo ni okeru Seisan Shuchudo:* 1955-1970 *[Concentration Ratios of Production in the Main Industries: 1955-1970]* (1973), p. 2; and *Syuyo Sangyo ni okeru Seisan Shuchudo Chosa [Survey of Concentration Ratios of Production in the Main Industries]* (1976), p. 2.

have a decisive influence on the performance of industry as a whole.

In Japan, continuous statistical data on concentration in industry is only available from 1950, but over the 20 years it is already possible to detect a pattern of industrial concentration, which tends to resemble the American picture described above. Table 4 gives basic data on monopoly in Japan. In Group A industries—leather, printing and publishing, timber, wood manufactures, furniture, and clothing—the only concern classifiable as monopolistic is the daily newspaper; the balance of the market is competitive in organization.

Table 4 also shows clearly that in Group B industries, which include textiles, paper, paper products, metal manufactures and general machinery, and food products, with the exception of textiles, the degree of monopoly is about 20 percent in each case. The Group C industries of chemicals, base metals, electrical machinery, and transportation machinery, just as in America, have been the leaders in postwar growth. Also, due to their technological content (particularly economies of scale), these industries are by their very nature fairly highly concentrated.

What trend in the average level of seller concentration in manufacturing industries can be seen in Japan? Table 5 shows the annual indexes of the unweighted average share of the top three and ten firms, and the Herfindahl Index for selected manufacturing industries from 1950 to 1974.

The indexes of seller concentration indicate an obvious decline from 1950 until at least 1960. In the 1960–65 period one sees decreases in the top three indexes and in the Herfindahl Indexes, but an increase in the top ten firms' concentration indexes. As the Herfindahl Index is an overall indicator of seller concentration, although it gives a heavy weight to inequality among the leading firms, the aggregate market concentration seems to have decreased in the 1960–65 period. In the 1965–74 period, however, the indexes point to a definite increase. We conclude that the trend of declining concentration gave way in 1965 to an upward trend.

Some evidence suggests that the high rate of growth of demand and affluent opportunities for import of technology favored entry of new firms into an industry, which in turn reduced seller concentration during the 1950–65 period. And it should not be overlooked that the base of active entry behavior of Japanese firms did lie in the new found independent decision-making ability of Japanese businesses and the fresh management outlook of the younger executives who took over during the economic purge during the occupation period. Hence, the trend of declining concentration was a product of the deconcentration measures during the occupation period and indicates how effectively the centrifugal forces set in motion by the measures worked.

How can we explain the reasons of the increasing seller concentration after 1965? The previous studies can be summarized as follows: (1) competition in construction of large-scale plants in the heavy and chemical industries has gradually lifted the shares of oligopolistic firms within the industries; (2) their possession of large-scale plants has served to enlarge their control over raw materials and distributive channels, which in turn has increased their market power; (3) product differentiation through advertising developed since the 1960s has raised concentration in consumer-goods industries; (4) because about half of the large mergers during the 1960s were horizontal, the merger movement has been an important component of the increase in seller concentration; and (5) the construction of large-scale plants, backward and forward integration, and product differentiation have gradually raised barriers to entry into the relevant industries and the slowdown of macroeconomic growth after 1970 has definitely reduced the expectation ratio of investment of Japanese firms. Of the five reasons, the first four are related to the policy favoring a fast-growing producer and strengthening international competitiveness of Japanese business. We can conclude that the fundamental determinant of increasing concentration after 1965 has been the industrial policy of the Japanese government and ministries, recognizing the fact that the policy has played a great role in attaining the high rates of growth of the nation's economy. Thus, the centrifugal forces set in motion by the competition policy during the occupation period has been gradually held back by the centripetal policy of the Japanese government and ministries.

TABLE 6
DIVERSIFICATION RATIO OF 124 LARGE CORPORATIONS IN JAPAN

	Industry Classification		
	2-digit level	*4-digit level*	*Commodity level*
1963	0.489	0.400	0.266
1964	0.486	0.399	0.260
1965	0.484	0.393	0:254
1966	0.479	0.393	0.251
1967	0.482	0.399	0.254
1968	0.481	0.401	0.254
1969	0.483	0.405	0.257
1970	0.485	0.414	0.259
1971	0.484	0.410	0.263
1972	0.492	0.420	0.273

Sources:
Ken'ichi Imai; Akira Goto; and Kei Ishiguro, "Kigyo no Tayoka ni kansuru Jissho Bunseki" ["An Analysis of Diversification of Enterprises"] (1975).

Note:
Diversification ratio = (1 − specialization ratio).

2. *Market Conduct of Japanese Enterprises*

We shall draw attention to the two aspects of market conduct of Japanese enterprises: diversification of enterprises as an aspect of development of Japanese firms and collusive practices as an aspect of monopolistic conduct of Japanese enterprises.

Diversification of Enterprises. Japanese enterprises have expanded not only through construction of larger-scale plants but also diversification of products. Diversification activities of enterprises means new entry into a particular market and, therefore, can have an influence on the market structure. The trend of decline in seller concentration in postwar Japan has a strong relation to such diversification activities of large enterprises. Although there is little data on diversification activities of Japanese enterprises from 1950 to the present, Table 6 shows the annual change in the average diversification ratio of 124 big corporations from 1963 to 1972. Whichever level of industry classification is chosen, the 124 corporations show a tendency toward diversification. In the broad industrial classification the average number of products per enterprise was 3.24 in 1963 and 3.64 in 1972.

Examining the degree of diversification by industry, we find that in industries producing raw material products (steel, glass, stone, petroleum and coal, pulp and paper, and so on), the degree of diversification of the enterprises within those industries was low and declining. In such other industries as chemicals, general machinery, electrical machinery and precision machinery, the degree of diversification was high and increasing.

If we examine the areas into which enterprises are diversifying, we find great numbers entering the fields of general machinery and chemical products, followed by electrical machinery, transportation machinery, and the service industries, where technological advance is proceeding apace.

In examining the type of diversification, we followed the classification produced by the Business Policy Study Group of Harvard University, which divides enterprises into the four types of "Single Business," "Dominant Business," "Related Business," and "Unrelated Business." The 124 Japanese corporations comprised 2.4 percent single business, 22.6 percent dominant business, 33.1 percent related business and 41.9 percent unrelated business.

Performance analysis of the enterprises in these categories suggests that "single industry" enterprises have the worst performance in terms of both profitability and growth rate, while those in the "dominant industry diversified" category have the best results.

The above observations allow us to draw the following conclusion. In the raw material industries diversification has not taken place but rather appears to be declining in many instances. If this continues, these industries will find it difficult to adapt themselves to change and will concentrate on trying to realize monopoly profit under their stagnant demand. If we, in addition, bear in mind that enterprises in these industries are exerting themselves to succeed in vertical integration, their ownership in other companies, and directing their own purchasing preferentially to other enterprises within their group, we have to recognize a strong trend toward stagnant oligopoly.

Collusive Practices. The Antimonopoly Law has been weakened by both general amendment and specific statutory exemptions or administrative interventions by MITI and other ministries. As a result, a large number of cartel agreements exempted from the Antimonopoly Law have been approved since the 1953 amendment of the law, as shown in Table 7. The total number of exempted cartels rose from 248 in 1956 to 595 in 1960, then to its peak in 1966. From that time the number has not changed greatly.

The importance of these exemptions is seen in the prevalence of officially sanctioned cartels in Japanese industry. Imai tabulated the incidence of statutory and administrative cartels in manufacturing industries at the six-digit level of disaggregation in 1970. The shipments of products subject to exempted cartels accounted for 30.7 percent of the total manufacturing shipments; 74.2 percent for lumber and wood products, 69.1 percent for textile mill products, 67.0 percent for apparel products, and

58.8 percent for iron and steel. Thus, officially sanctioned cartels have spread their roots widely throughout the industrial structure.

The prevalence of statutory and administrative cartels and the weak enforcement of cartel sections of the Antimonopoly Law—especially, the weak elimination measures—have evidently left ample scope for the operation of illegal cartels. A recent feature of cartel formation is the increasing number of cartels organized by large enterprises in the producers' goods industries, as shown in Table 8. Yokokura also found that cartels have been organized in the more concentrated and higher barriers-to-entry industries with high variability of demand.[8]

Many oligopolistic industries in Japan use not only the formal collusion but also the same informal, tacit devices that are known in the United States. Price leadership has been common in such highly concentrated sectors as the production of beer, film, flat glass, aluminium ingot, synthetic fibers, metal cans, newsprint, wire and cable, and many steel products. Especially in the steel industry, formation of a dominant firm, Shin Nihon Steel, through a merger between the two largest makers in 1969 has made it easy to form a dominant price leadership.

And the development of control of oligopolistic firms over distributive channels has exercised vertical agreements or resale price maintenances. The Fair Trade Commission has recently attempted to prohibit their behavior by blocking the erosion of prices structure from the retail end as unfair trade practices—Matsushita case, Nihon Gakki case, France Bed case, and Hakugen case. Matsushita Electric Company took the lead in a retrenchment designed to tighten the company's link with the distributive channels. The company granted exclusive territories to two hundred of its sales subsidiaries, supporting them with credit and service organizations. In the process it achieved tight control over resale prices and the structure of trade discounts.

The recent development of collusive practices described above is a product of monopolization in the market structure for the principal industries after 1965. Such collusive arrangements will distort the allocation of resources among industries, give rise to administered inflation, and stagnate investment activities.

3. *Economic Performance of Japanese Industry*

Allocative Efficiency. The allocative efficiency of Japanese industry can be investigated by statistical analysis of the influences on industry profit rates. Previous studies of Japanese manufacturing industries have found two interesting facts.

[8] Takashi Yokokura, "Cartels no Jissho Kenkyu" ["A Quantitative Analysis of Cartels"] (1977).

TABLE 7
CARTEL AGREEMENTS EXEMPTED FROM THE ANTIMONOPOLY LAW BY THE FAIR TRADE COMMISSION OF JAPAN, BY YEAR AND EXEMPTING STATUTE; 1964–73[a]

Statutory Basis for Exemption	*1964*	*1965*	*1966*	*1967*	*1968*	*1969*	*1970*	*1971*	*1972*	*1973*
Depression cartels	2	2	16	1	0	0	0	0	9	2
Rationalization cartels	14	14	14	13	13	12	10	13	10	10
Export cartels	201	208	211	206	213	217	214	192	175	180
Import cartels	1	2	3	4	3	4	4	3	2	2
Cartels under Medium and Small Enterprises Organization Act	588	587	652	634	582	522	469	439	604	607
Cartels under Environment Sanitation Act	106	122	123	123	123	123	123	123	123	123
Cartels under Coastal Shipping Association Act	15	14	16	15	22	22	22	21	19	19
Cartels under other statutes	43	50	44	44	47	48	56	53	34	42
Total	970	999	1,079	1,040	1,003	948	898	844	976	985

Sources:
FTCJ, Staff Office, *The Antimonopoly Act of Japan* (1973), p. 27.
[a] Number in force in March of each year.

TABLE 8

CARTEL VIOLATIONS BY TYPE OF PRODUCT AND SCALE OF ENTERPRISE

	1965	*'66*	*'67*	*'68*	*'69*	*'70*	*'71*	*'72*	*'73*	*Total*
Products of large enterprises		2	1	3	4	6	13	11	33	73
Products of medium/small enterprises	3	2	1	5	2	8	4	3	14	42
Producer Goods	1	4	2	4	4	11	14	14	41	95
Consumer Goods	2	0	0	4	2	3	3	0	6	20

Sources:
Kenji Sanekata; Masu Uekusa; and Joji Atsuya, "Cartels no Tetteiteki Bunseki" ["A Thorough Study of Cartels"] (1975).

Note:
Products of large enterprises are where the product represens more than 50 percent of the shipments of an enterprise with capital in excess of ¥50 million.

The first includes the significant influence of seller concentration on profits in periods when Japan's economy has grown with normal rapidity, but not when it has proceeded at an explosive pace. However, concentration has remained significant in the periods after 1965 when the trend of declining concentration gave way to an upward trend. The other finding is the relatively weak performance of statistical variables that present the barriers to entry of new firms in the periods before at least 1965 and the significant influence of entry barriers based on product differentiation in the periods after 1965. These facts suggest that industry profits were affected by macroeconomic growth before 1965, while they have been influenced by the market structure elements featuring oligopoly after 1965. Therefore, the recent increase in seller concentration and the development of collusive arrangements have gradually enlarged the distortion of the allocation of resources among industries.

Administered Inflation. The Weiss-Dalton type of administered inflation hypothesis has been tested in Japan. Previous studies have found the strong negative relation of concentration to annual price increases. It is a fact that most of concentrated industries have attained a high rate of growth of productivity, which has in turn sustained stability of wholesale prices. The most recent research reveals seller concentration's marginal positive influence on annual price increases and a strong tendency toward incidence of increases in wage and material costs on prices.[9] This finding suggests that administered inflation has begun to appear in Japan.

Stagnation of Capital Formation. Seller concentration affects capital formation as follows: concentrated industry might sometimes invest at an uneven pace over time because oligopoly rivalry entails close imitation of each other's actions, or because the same scale economies that cause concentration also force a lumpy course of investment in efficient-size additions to capacity; on the other hand, high concentration might cease to inflate investment because tight-knit oligopolists do not imitate each other's increases in capacity through their possession of excessive capacity, or because formation of high barriers to entry do not evoke investment by new entrants.

There is casual evidence to support the former hypothesis in Japanese industry. Iwasaki found a strong positive relation between the instability of investment over 1961–70 and seller concentration in Japanese manufacturing industries, after allowing for the significant influence of the instability of demand and the size of the market.[10] Capital formation in

[9] Koji Shinjo, "Business Pricing Policies and Inflation: the Japanese Case" (1975) and Teruhiro Tomita, "Kigyo no Kakaku Seisaku to Kanri Kakaku Inflation" ["Corporate Pricing Policy and Administered Inflation"] (1975).

[10] Akira Iwasaki, "Market Structure and Stability of Investment in Japanese Manufacturing Industries " (1976).

Japanese industry, however, has been very stagnant after 1970, through slowdown of macroeconomic growth, environmental factors, and probably collusive conduct in investment under the shift of concentration and barriers to entry in the principal industries. Especially, in such highly concentrated industry as the production of steel, petrochemicals, and so on, the oligopolistic firms tend to collude in investment in order to avert the instability of capital formation.

III. CONCLUDING REFLECTIONS

The authors have pointed out that the large-scale competition policy imposed on Japan during the occupation period has affected the Japanese economy over the long period, forming and promoting the relatively competitive market structure of Japanese industry, which has indeed been a principal factor in realizing rapid economic growth. The industrial policy of the Japanese government and ministries favoring a fast-growing producer and administratively controlled market competition has contributed in realizing macroeconomic growth, while at the same time it has gradually increased centripetalization of economic power into the hands of a few big businesses. This increased seller concentration in principal industries, generated undesirable economic performance, distorted allocative efficiency, and was accompanied by symptoms of administered inflation and stagnation of capital formation. From the point of view of a researcher of industrial organization, the Japanese economy would appear to stand at a crossroads.

We have to reconsider the significance of competition policy. The enforcement of the recently revised Antimonopoly Law, the elimination of a large number of the particular industrial laws restricting competition, and improvement of imperfection of factor markets—especially in the capital market—should be the major goals of public policy toward Japanese industry.

Finally, competition policy has become increasingly significant for all developed countries because each suffers from the trend of increasing seller concentration, and subsequently from the distortion of allocative efficiency and other undesirable economic performance. International cooperation and mutual understanding are necessary for the improvement of economic performance. An important new field of cooperation for the competition policy authorities lies ahead: such as new attempts to define those restrictive practices that are harmful for the economies of developing countries; to determine a Code of Conduct for transnational corporations; and to formulate the rules to solve a clash between the export policy of one country and the competition policy of the other. For such attempts, a thorny path lies ahead, but they should be very welcome.

REFERENCES

Caves, Richard E. and Uekusa, Masu. 1976. *Industrial Organization in Japan.* Washington, D.C.: Brookings Institution.

Chandler, Alfred D., Jr. 1969. "The Structure of American Industry in the Twentieth Century: A Historical Overview," *Business History Review,* vol. 43, Autumn, pp. 255–98.

Hadley, Eleanor M. 1970. *Antitrust in Japan.* Princeton, N.J.: Princeton University Press.

Imai, Ken'ichi. 1976. *Gendai Sangyo Soshiki [Contemporary Industrial Organization].* Tokyo: Iwanami Shoten.

———; Goto, Akira; and Ishiguro, Kei. 1975. "Kigyo no Tayoka ni kansuru Jissho Bunseki" ["An Analysis of Diversification of Enterprises"], Nihon Keizai Data Kaihatsu Center.

Iwasaki, Akira. 1976. "Market Structure and the Stability of Investment in Japanese Manufacturing Industries," *Economic Studies Quarterly,* December.

Japan, Fair Trade Commission. 1968. *Dokusen Kinshi Seisaku Niju-nenshi [Twenty-year History of Antimonopoly Policy].* Tokyo: Ministry of Finance.

———. 1969. *Nihon no Sangyo Shuchu: 1963–1966 [Industrial Concentration in Japan: 1963–1966].* Tokyo: Toyo Keizai Shinpo-sha.

———. 1971. *Nihon no Kigyo Shuchu [Corporate Mergers in Japan].* Tokyo: Ministry of Finance.

———. 1973. *Syuyo Sangyo ni okeru Seisan Shuchudo: 1955–1970 [Concentration Ratios of Production in the Main Industries: 1955–1970].* Tokyo: Fair Trade Institute.

———. 1973. *The Antimonopoly Act of Japan.* Fair Trade Commission Staff Office.

———. 1975. *Annual Report—1975.* Tokyo: Author.

———. 1976. *Sangyo ni okeru Seisan Shuchudo Chosa [Survey of Concentration Ratios of Production in the Main Industries].* Tokyo: Fair Trade Institute.

Japanese Government. 1964. *Economic White Paper.* Tokyo: Author.

Sanekata, Kenji; Uekusa, Masu; and Atsuya, Joji. 1975. "Cartels no Tetteiteki Bunseki" ["A Thorough Study of Cartels"], *Chuo Koron.*

Shinjo, Koji. 1975. "Business Pricing Policies and Inflation: The Japanese Case." Kobe: Kobe University.

Tomita, Teruhiro. 1975. "Kigyo no Kakaku Seisaku to Kanri Kakaku Inflation" ["Corporate Pricing Policy and Administered Inflation"], *Denryoku Keizai Kenkyu,* September.

Uekusa, Masu. 1977. "Shuchu Haijo Seisaku ni tsuiteno Atarashii Shiten' "New Look into the Deconcentration Measures"], *Keizai Hyoron,* February.

United Nations. 1973. *Yearbook of National Accounts: Statistics 1973.* New k: Author.

Yokokura, Takashi. 1977. "Cartels no Jissho Kenkyu" ["A Quantitati nalysis of Cartels"], *Keizai Hyoron,* April and May.

Comment

Soichiro Giga (Japan)

I. THE DRIVING FORCE OF RAPID ECONOMIC GROWTH

Ken'ichi Imai and Masu Uekusa state that a relatively competitive market structure of Japanese industry has been a driving force or a principal factor in realizing rapid economic growth (*See* p. 92). There are several questions concerning this statement.

(1) On the one hand, we can see the rapid economic growth in the highly concentrated industries (Petroleum and coal products, Iron and steel, Transportation machinery, Electrical machinery, etc. *See* Imai and Uekusa, Table 4). On the other hand, the industries that have the competitive market structure could not be the champion or the leader of rapid economic growth (*See* Table 4).

(2) The authors state: "The centripetal policy of the Japanese government and ministries...has begun to generate distortion of allocative efficiency, administered inflation, and stagnation of capital formation" (p. 92). I agree with this opinion. However, I wonder whether this centripetal policy has been a driving force of rapid economic growth or not.[1] I wonder whether the centrifugal forces never generate distortion of allocative efficiency and stagnation of capital formation or not. At least, according to the authors: "It is a fact that the competitive market structure has produced unnecessary duplication" (p. 96).

In addition, it should not be overlooked that one can find rapid economic growth in many socialist countries, where the centripetal policies are dominant in the conditions of the planned economy. In short, one should pay attention to the relative merits of capitalism and socialism in respect to this problem.

(3) We should consider economic growth within a multilateral context. It is important to investigate the features of economic growth in each industry and also the drastic changes in the industrial structure (*See* Imai and Uekusa, Table 3). I argue that we should not overlook the relations between large enterprises and smaller enterprises in postwar Japan. Since the end of the Korean War, large enterprises have organized many subsidiaries and affiliates. We can see some aspects on page 110. Therefore, we should consider exactly the monopolistic power of large enterprises including many smaller enterprises.

Using an extreme metaphor, Japan is a fatherland of smaller enterprises and the United States is a mother country of the large enterprises (*See* Tables 2 and 3 of these Comments). The smaller enterprises coexist with the large enterprises in many forms: independent or subordinate, growing or falling, etc. As a whole, the vitality of the smaller enterprises is remarkable. The large enterprises use the smaller enterprises so far as they can be useful. For instance, the largest 100 firms effectively control many firms by shareholding. The number of such firms

[1] *See* Shigeto Tsuru, *The Mainsprings of Japanese Growth: A Turning Point?* (1977).

TABLE 1

BREAKDOWN OF EMPLOYED PERSONS BY INDUSTRY ON BASIS OF POPULATION CENSUS

(As of Oct. 1, percent)

	1950	*1960*	*1965*	*1970*
Primary industry	48.3	32.6	24.7	19.3
Secondary industry	22.0	29.2	32.3	33.9
Tertiary industry	29.6	38.2	43.0	46.7
Unclassifiable	0.1	0.0	0.0	0.1
Total	100.0	100.0	100.0	100.0

Source:
Bureau of Statistics, Office of the Prime Minister.

Note:
Data for 1950 refer to 14 years old and over, for 1955–70 refer to 15 years old and over.

TABLE 2

BREAKDOWN OF EMPLOYED PERSONS BY STATUS IN SELECTED COUNTRIES (1973)

(Thousand persons and percentage in brackets)

	Japan	*U.S.*	*U.K.*	*W. Germany*	*Italy*
Total employed persons	52,330	84,409	24,641	26,202	18,310
	(100.0)	(100.0)	(100.0)	(100.0)	(100.0)
Self-employed workers	9,660	7,202	1,979	2,599	4,041
	(18.5)	(8.5)	(8.0)	(9.9)	(22.1)
Paid employes	35,950	76,249	22,662	22,054	13,049
	(68.7)	(90.4)	(92.0)	(84.2)	(71.2)
Unpaid family workers	6,630	958	...	1,549	1,220
	(12.7)	(1.1)	...	(5.9)	(6.7)

Source:
"Labor Force Statistics," OECD.

rose from 3,475 in 1960 to 7,612 in 1970 (percentage of shares owned by one of the 100 largest firms: 10–25 percent—1,731, 25–50 percent—3,603, over 50 percent—2,818, total 7,612).[2] The roles of subcontract are notably significant because the wide gaps in wages, fringe benefits, and working hours, etc., between the large firms and smaller firms are favorable for the large firms in Japan (*See* Table 4 of these Comments). Also we should not overlook that most of the laborers in small and medium enterprises as well as the many temporary laborers are not organized in trade unions (number of union members are 12.6 million persons, in 1975).

[2] Kosei Torihiki Iinkai [Fair Trade Commission], *Nihon no Kigyo Shuchu [The Concentration of Enterprise in Japan]* (1971), pp. 42-43, pp. 144-49.

TABLE 3

BREAKDOWN OF ENTERPRISES BY SCALE IN JAPAN AND THE UNITED STATES

(Percentage)

	Factory Scale	*Number of Factories*	*Number of Employed Persons*	*Shipment*
Japan	300 persons or more	0.6	30.8	48.8
	1–299	99.4	69.2	51.2
U.S.	250 persons or more	4.4	60.0	63.7
	1–249	95.6	40.0	36.3

Source:
Small and Medium Enterprises Agency, *Chushokigyo Hakusho [White Paper of Medium and Small Enterprises]* (1976), p. 91.

Notes:
U.S. Census of Manufactures, 1967; Japan, Kogyo Tokei-hyo, 1972.

TABLE 4

WAGE DIFFERENTIALS BY SCALE IN MANUFACTURING

(Wages in establishments with more than 1,000 employees = 100)

Scale (Number of Employees)	*Japan (1970)*	*France (1966)*	*Germany (1966)*	*Italy (1966)*	*U.K. (1954)*	*U.S. (1967)*
10– 19	60.3	83.5	87.0	67.5	79.3[a]	75.5
20– 49	66.6	81.7	86.1	70.8	80.3[b]	73.4
50– 99	66.9	79.1	87.0	74.2	80.9	73.4
100–199	70.5	80.0	87.0	79.2	82.0	74.3[c]
200–499	78.4	83.5	89.8	82.5	85.0	76.3[d]
500–999	84.9	91.3	92.6	88.3	89.3	81.7
Degree of differential	7.5 (7.6)	3.3 (2.7)	2.5 (3.9)	6.3 (11.4)	3.8 (4,7)	4.3 (2.7)

Source:
Andrea Boltho, *Japan: An Economic Survey 1953–1973* (1975), p. 28.

Notes:
Gross annual earnings of employees for Japan, the United Kingdom, and the United States; gross hourly earnings of workers for France, Germany and Italy.

[a] 10–24 employees.
[b] 25–49 employees.
[c] 100–249 employees.
[d] 250–499 employees.

Andrea Boltho has stated: "...a comparison made for the mid-1950s showed that Japanese wage differentials by size were then as large as, or even larger than, those recorded in several underdeveloped countries."[3] The large enterprises are the star performers in the drama of high economic growth. Now the role of the smaller enterprises is becoming more and more important in the new economic conditions.

In short, I argue that the problems of the industrial organization should be solved not only from the standpoint of the consumers or the enterprises that wish to enter some industry, but also from the standpoint of the smaller enterprises, the farmers, the laborers as well as the people living nearby the factories and highways, etc. We know the pollution and the other negative aspects of the regional developments, too.

The limitation of the deconcentration policies and the Antimonopoly Law is evident from this wide view.

II. THE LIMITATION OF THE ANTIMONOPOLY LAW AND THE ROLE OF THE SIX MAJOR ENTERPRISE GROUPS

(1) If the Antimonopoly Law did not exist, it would be more difficult for MITI, etc., to organize officially sanctioned cartels. Illegal cartels organized by large enterprises in the highly concentrated industries have escaped the restraint from the Antimonopoly Law. Illegal "cartels" of smaller enterprises have been the scapegoats and sometimes could not escape the restraint.[4]

The increased mergers were rather authorized by this law (*See* Imai and Uekusa, Table 2). Many important merger cases were planned by the six major finance capital groups. The effects of these mergers were to strengthen the cartels in relative industries and to accelerate the reorganization of the six major enterprise groups (Mitsubishi, Mitsui, Sumitomo, Fuji, Daiichi, Sanwa). The Antimonopoly Law has been able to do almost nothing about the reorganization of the six major enterprise groups. The Antimonopoly Law, in my opinion, has been a component of the so-called "Japan Inc." or the Japanese state monopoly capitalism. We must not overestimate the role of the Antimonopoly Law of Japan or the United States.

(2) Enterprise groupings (*keiretsu*) started in the early 1950s when the former *zaibatsu* executives were permitted to return, when the antitrust legislation was eased, and the former *zaibatsu* people began to make earnest plans for their future: the new presidents of the former three *zaibatsu* companies began to meet regularly; there was the Monday Club (*Getsuyokai*) for the Mitsui firms' presidents, the Friday Club (*Kinyokai*) for those of Mitsubishi, and the White Water Club (*Hakusuikai*) for the Sumitomo Group's presidents. Now there are Fuyo Club (*Fuyokai*) for the Fuji Group's presidents and Sansui Club (*Sansuikai*) for the Sanwa Group's presidents. It was essential for the emergence of the grouping phenomenon that the *zaibatsu* banks had been left untouched by the process of dissolution of the *zaibatsu*. They could thus take over the leadership role.[5] Apart

3. Andrea Boltho, *Japan: An Economic Survey 1953–1973* (1975), p. 28.

4. *See* Soichiro Giga, ed., *Gendai no Kigyo Keitai [Contemporary Forms of Enterprise]* (1966).

5. Johannes Hirschmeier and Tsunehiko Yui, *The Development of Japanese Business, 1600–1973* (1975), p. 264.

from the role of the banks, the trading companies began to exercise a unifying influence. In 1954 Mitsubishi Shoji reunited and became the largest general trading company (*sogoshosha*). In 1959 Mitsui Bussan reunited, and Sumitomo Shoji, Marubeni-Iida, Ito Chu, and Nissho-Iwai have been growing as the important components of the six major enterprise groups. The presidents clubs (*shachokai*) have worked systematically to strengthen the cohesive power by increasing mutual stock holdings and interlocking management. There the presidents discuss their joint problems, exchange information, and plan strategic moves.[6] Many important and huge joint companies in the new industries (atomic energy, petrochemistry, etc.) were the products of these presidents clubs. These groups connecting with foreign technique and capital are the top stars in rapid economic growth.

"It has been observed that the groupings have resulted in a 'one-set' syndrome whereby each group tries to re-create what the other does, and gain strength setting up a self-sufficient industrial and trading empire. This then leads to much unnecessary duplication of investment, and waste... Perhaps here is part of the explanation for the fierce competitive struggle and the strong investment thrust to enlarge the market share ahead of the rival."[7]

Therefore, first it is difficult for me to agree that the most significant effects of the deconcentration measures were the formation of the decision-making autonomy of operating companies of the *zaibatsu* and other big corporations. Secondly, we must not overestimate the role of the centrifugal forces in the process of rapid growth.

(3) In 1954, T.A. Bisson stated: "State ownership and operation of the basic enterprises controlled by the combines represented a second alternative to the *zaibatsu* system."[8] After the Second World War, we found many different cases of nationalization in the U.K., France, Italy, etc., and in many developing countries as well as socialist countries.

When we consider the problems of the industrial structure and industrial organization as a whole, we should analyze the relative merits of the private enterprises and the public enterprises including the mixed enterprises. Actually, the political and economic problems of the nationalization and the public enterprise have grown in importance in many countries, especially in the 1970s.

According to Kiyohiko Yoshitake: "In the 1970s public enterprise in Japan will have a bigger significance than in the 1960s. It will be analyzed from two entirely different viewpoints; one is the criticism against hasty and irrational establishment of public enterprises in the 1960s, the second is the reflection on the new role which public enterprise can play in the 1970s."[9] I agree with Yoshitake's opinion; I also cannot overlook the recent role of Japanese mixed enterprises abroad. Sometimes mixed enterprise means "creeping nationalization."

(4) According to Imai and Uekusa: "Since 1960 after the reconstruction period

[6] Hirschmeier and Yui (1975), pp. 265-66. For details, *see* Soichiro Giga, *Gendai Nihon no Dokusenkigyo [Monopoly Enterprise in Contemporary Japan]* (1962); Yoshikazu Miyazaki, *Sengo Nihon no Keizaikiko [Japanese Economic Structure after the Second World War]* (1966) and *Sengo Nihon no Kigyo Shuchu [Enterprise Groups in Postwar Japan]* (1976).

[7] Hirschmeier and Yui (1975), p. 267.

[8] Thomas A. Bisson, *Zaibatsu Dissolution in Japan* (1954), p. 47.

[9] Kiyohiko Yoshitake, *An Introduction to Public Enterprise in Japan* (1973), p. 69.

ended, the Japanese economy transferred from the closed economy to an open economic system..." (p. 95). Then I would like to know the authors' opinion of international industrial organization because when we consider the economic problems of oil, atomic energy, computer, airplane, textile, and so on, we cannot perfectly understand the features of the Japanese industrial organization without an investigation with respect to the influence of international cartels, international trusts, or so-called multinational enterprises, etc.[10] I wonder whether the Antimonopoly Law can adequately control the international monopolies or not.

REFERENCES

Ballon, Robert J. and Lee, Eugene H., eds. 1972. *Foreign Investment and Japan.* Tokyo: Sophia University Press.

Bisson, Thomas A. 1954. *Zaibatsu Dissolution in Japan.* Berkeley: University of California Press.

Boltho, Andrea. 1975. *Japan: An Economic Survey, 1953–1973.* London: Oxford University Press.

Frank, Isaiah, ed. 1975. *The Japanese Economy in International Perspective.* Baltimore and London. Johns-Hopkins University Press.

Giga, Soichiro. 1962. *Gendai Nihon no Dokusenkigyo [Monopoly Enterprise in Contemporary Japan].* Kyoto: Minerva Shobo.

———. 1976. "Foreign Enterprises in Japan and Japanese 'Multinational Enterprises'," *Keiei Kenkyu [Business Review],* July.

———. 1977. "Some Characteristics of the Japanese 'Multinational Enterprises'," *Keiei Kenkyu,* March.

———, ed. 1966. *Gendai no Kigyo Keitai [Contemporary Forms of Enterprise].* Tokyo: Sekai Shoin.

Henderson, Dan Fenno. 1973. *Foreign Enterprise in Japan.* Chapel Hill: University of North Carolina Press.

Hirschmeier, Johannes and Yui, Tsunehiko. 1975. *The Development of Japanese Business, 1600–1973.* London: Allen & Unwin.

Japan, Small and Medium Enterprises Agency. 1976. *Chushokigyo Hakusho [White Paper of Medium and Small Enterprises].* Tokyo: Printing Bureau, Ministry of Finance.

———, Fair Trade Commission. 1971. *Nihon no Kigyo Shuchu [The Concentration of Enterprise in Japan].* Tokyo: Printing Bureau, Ministry of Finance.

Miyazaki, Yoshikazu. 1966. *Sengo Nihon no Keizaikiko [Japanese Economic Structure after the Second World War].* Tokyo: Shin Hyoron.

———. 1976. *Sengo Nihon no Kigyo Shuchu [Enterprise Groups in Postwar Japan].* Tokyo: Nihon Keizai Shimbun.

Tsuru, Shigeto. 1977. *The Mainsprings of Japanese Growth: A Turning Point?* Paris: Atlantic Institute for International Affairs.

Yoshino, M.Y. 1975. "Emerging Japanese Multinational Enterprises," *Modern Japanese Organization and Decision-Making.* Edited by Ezra F. Vogel. Berkeley and London: University of California Press.

Yoshitake, Kiyohiko. 1973. *An Introduction to Public Enterprise in Japan.* Tokyo: Nippon Hyoron-sha.

[10] *See* S. Giga, "Foreign Enterprises in Japan and Japanese 'Multinational Enterprises'" (1976) and "Some Characteristics of the Japanese 'Multinational Enterprise'" (1977); Dan Fenno Henderson, *Foreign Enterprise in Japan* (1973); Robert J. Ballon and Eugene H. Lee, eds., Foreign Investment and Japan (1972); M.Y. Yoshino, "Emerging Japanese Multinational Enterprises" (1975); Isaiah Frank, ed., *The Japanese Economy in International Perspective* (1975).

Capital Accumulation and Economic Growth: A Comparison of Italy and Japan

Roberto Zaneletti *University of Genova*

This paper attempts to test, through a comparative analysis of Italy and Japan for the period 1960–73, some theoretical arguments on accumulation of capital and economic growth. An economic policy based on a high and sustained rate of qualified autonomous investments joined to controlled development of consumption may guarantee full employment as well as a high rate of increase of per capita gross national product (the case of Japan). An economic policy based mainly on consumption (private and public) and accompanied by huge deficit spending, strong reduction of profits as well as of financial means available for the private sector of the economy, and structural changes in the industrial relations may hinder the operation of the principle of acceleration. The results may be stagnation of investments, unemployment, and underemployment (the case of Italy).

I.

As we all know, economic growth is the result of many factors, human and material. Now and then one or another factor has been emphasized by economists as well as by politicians. At the end of the Second World War, because of the destruction, the main problem was the lack of equipment and inventories. Therefore, attention was concentrated on the capital factor. From the pioneering works of Roy Harrod and E.D. Domar sprang a series of growth models, at the heart of which was the accumulation of capital.

But around the end of the 1960s, the position of many economists began to become critical of the role of capital in the process of economic growth. Among those economists, we may quote Colin Clark, who in 1961 wrote a book entitled *Growthmanship: A Study in the Mythology of Investment.* Through empirical work, Clark tried to show that we can no longer be sure that increasing investment is the best way to increase the rate of economic growth. Clark states regarding estimates of gross domestic investment charted against GNP growth for 31 countries for the period 1950–56: "[they] might almost suggest that a high rate of investment goes with a low rate of economic growth."[1]

The main factors of development, according to Colin Clark, are human factors and not material factors. Therefore, economic growth cannot be accelerated through the acceleration of investment.

Unfortunately, these ideas have been accepted by some Italian economists and especially by some Italian politicians who have adopted a line of neglect towards the problem of capital accumulation. The Fifth World Congress of I.E.A. gives an opportunity to revisit the thesis of Colin Clark through empirical research referring to two countries—Japan and Italy.

Even though Japan's population is now double that of Italy, the two countries have similar economic structures. Both are developed countries, both have very few raw materials, both have concentrated their activity on transformation, and both are mixed economies, even if a little more mixed for Italy and a little less mixed for Japan.

Taking account of the cyclical fluctuations in both countries, I have considered the year 1960 the starting year and 1973 the ending year: that means the comparative research is on a medium-term period. I will later consider briefly the dynamics in the years 1974–76.

[1] Colin Clark, *Growthmanship: A Study in the Mythology of Investment,* Hobart Paper, No. 10 (London: Barrie and Rockliff for the Institute of Economic Affairs, 1961), p. 50.

II.

The usual indicators of capital accumulation are:

(a) The ratio between gross fixed investment and gross national product.
(b) The ratio between gross fixed investment in machinery and equipment and gross national product.
(c) The ratio between fixed capital and the number of dependent workers.[2] Unfortunately, in Italy we do not have statistics on fixed capital assets. Therefore, I have considered the ratio between gross fixed investment and the number of dependent workers.

The usual indicator of economic growth is the rate of increase of total and per capita gross national product.

III.

During the period 1960–73, Italian gross fixed investment increased at an average yearly rate of 4.2 percent; in Japan this rate has been three times as great, that is, 14 percent (both figures are based on constant prices).

The average yearly ratio between gross fixed investment and gross national product has been 21.5 percent for Italy and 31 percent for Japan. And we have to remember that at the beginning of the period—that is, in 1960—Japanese gross fixed investment was equal to 87 percent of Italian gross fixed investment.

During the same period, total gross national product increased at an average yearly rate of 5.1 percent in Italy and 10.3 percent in Japan. The corresponding per capita gross national product yearly rates are 4.4 percent for Italy and 9 percent for Japan.

IV.

A critical expert like Colin Clark could object at this point that the Japanese economic growth has been due not to the investment drive, but to the growth of the labor force. It must be underlined, as a matter of fact, that between 1960 and 1973 the number of dependent workers increased by about 12,250,000 persons in Japan, whereas in Italy it increased only by about 1,660,000 persons.

There is no doubt that Japanese economic growth is due to the physi-

[2] I define "dependent workers" as those who are dependent upon an employer for work and receive regularly a wage or salary.

cal productivity displayed by the new labor force, but would such productivity have been displayed without the support of a huge amount of investment? If we consider the data on capital accumulation per dependent worker, we notice that the average yearly rate of increase of this quantity was 3.2 percent in Italy and 10.4 percent in Japan. Japanese economic growth would have been different if there had been a different combination of capital and labor. Therefore, it seems hard to deny the existence of a direct causal relationship between capital accumulation and economic growth.

V.

This relationship becomes clearer if we look not only to the quantity but also and especially to the quality of investment.

During the period 1960–73, the relation between gross fixed investment in machinery and equipment and the gross national product was only 9 percent in Italy, whereas in Japan it was 18.1 percent, that is, almost two and one-half times. Saying it another way, out of the total gross fixed investment, Italy devoted 42 percent to investment in machinery and equipment, whereas in Japan that ratio was 66 percent. *Vice versa,* investment in dwellings was 30.3 percent of the total gross fixed investment in Italy and only 18.9 percent in Japan.

As a result of such a qualitative difference in the composition of investment, the marginal capital-output ratio of the two countries has been quite different. I have calculated such a ratio relating the gross fixed investment of one given year to the increase in gross national product of the following year (of course, at constant prices). During the period 1960–61/1972–73, the marginal capital-output ratio was 5.22 in Italy and 3.16 in Japan. Therefore, in Japan investment has been more productive.[3]

VI.

What are the reasons for such a different scale of capital accumulation in the two countries?

Economic theory—as we all know—gives many explanations. Besides the marginal theory, which is hard to submit to an empirical test, we have:

(a) The principle of acceleration, which tells us that investment is a function of the estimated future demand for goods and services,

[3] I owe the introduction of the marginal capital-output ratio to a suggestion made from the floor by Professor Martin Bronfenbrenner during the discussion of my paper.

given the degree of exploitation of productive capacity.

(b) The Keynesian theory, according to which investment is a function of the relationship between marginal efficiency of capital and the rate of interest.

(c) The theory that makes investment dependent upon the relative prices of capital and labor.

(d) The theory that makes investment dependent on the availability of financial means.

VII.

I will consider first the acceleration theories. In addition to autonomous investment, investment is derived from consumption and exports. But since it is impossible to distinguish between autonomous and derived investment, let us look at the first two types of demand.

Italian economic development during 1960–73 was based upon consumption, especially private consumption. The average yearly rate of increase of consumption for this period was 5.2 percent; consumption has always absorbed two-thirds of total demand. Italian exports increased during this period at an average yearly rate of 11 percent. The acceleration principle tells us that the fluctuations in the total demand have accelerating and decelerating effects on investment. In the Italian case, we might say that the decelerating effects appeared quickly, whereas the accelerating effects were quantitatively small. As I have already pointed out, the average yearly rate of increase of Italian gross fixed investment was 4.2 percent, less than the average yearly rate of increase of Italian consumption.

In Japan, the average yearly rate of increase of consumption was 8.4 percent, greater than the Italian rate; but we must remember that only in 1972 did the Japanese per capita final consumption become equal, in absolute terms, to the Italian one. Besides, the ratio between consumption and total demand decreased from 68 percent in 1960 to 52.5 percent in 1973. During the same period, Japanese exports increased yearly by 14.3 percent. Now, even taking account of the strong dynamics of exports, if we remember that the average yearly increase of Japanese gross fixed investment was 14 percent, much higher than the increase in consumption, we might say that in Japan the capital accumulation was to a large extent derived demand for capital goods.[4]

[4] The discussant of my paper, Professor Hisao Onoe, has emphasized that the pattern of Japanese production is export-oriented, and therefore exports have contributed to a certain extent to the Japanese investment drive.

VIII.

As far as the availability of financial means is concerned, the relation between net national saving and net national disposable income during 1960–73 was equal to 16.5 percent in Italy and 28.3 percent in Japan. There is no doubt that the higher propensity to save in Japan is a reflection of the controlled development of consumption.

But the point is that in Italy the savings of firms strongly decreased (from 11.1 percent of net national saving in 1961–70 to 4.7 percent in 1971–73), while the public administration changed from a position of positive saving to a position of negative saving. At about the end of the period, in the 1970s, more than 40 years after Keynes published his *General Theory*, the Italian government started a huge deficit spending policy. In 1973, the state budget had a deficit equal to 6.1 percent of the gross national product. In the same year, the Japanese state budget had a surplus of 1.8 percent of their gross national product. Needless to say, the negative saving of the Italian public administration reduced strongly the availability of financial means for the private sector of the economy.

IX.

There was a differential behavior in relative prices of labor and capital in the two countries.

During 1960–73 both countries experienced creeping inflation at the same average yearly rate (5.5 percent in Japan and 5.4 percent in Italy). But in Italy, prices of capital goods increased more than prices of consumer goods, whereas in Japan the opposite occurred, and this was an additional incentive to investment in favor of Japan.

It has been claimed for a long time, and not only in Italy, that Japan has largely benefited in its development from low salaries. As far as this point is concerned, I think it is necessary to distinguish absolute levels from dynamic movements. In 1960, the average income per dependent worker in Japan was 63 percent of the corresponding figure in Italy. Only in 1973—that means at the end of the period I have considered—did the average income per dependent worker become equal in absolute terms; but this as a statistical result due to the devaluation of the Italian lira in respect to the yen. From this point of view, I think it is difficult to deny the comparative advantage of Japan.

However, the dynamics of the cost of labor was more or less the same in both countries: in 1973 the cost of labor for a unit of product was about two and one-half times the cost of labor for a unit of product in 1960. More precisely, this is the result of an average yearly rate of increase of income at current prices per dependent worker of 11.7 percent in Italy and 14.1 percent in Japan, and of an average yearly rate of increase of gross national product at constant prices per dependent worker of 4.1 percent in Italy and 6.8 percent in Japan. So, whereas the dynamics of income per dependent worker shows a recovery for Japanese workers, the productivity of labor has been much higher in Japan than in Italy.

At the beginning of the period considered, the average rate of interest on bank loans was higher in Japan than in Italy; but while in Japan there was a slow decrease since then, in Italy the average rate of interest increased, so that in 1973 it was 40 percent higher than in 1960.

What have the effects of this different behavior been? The higher increase in the cost of labor compared to the increase in the cost of capital has induced Italian firms to be very careful about absorbing new labor. At the same time, the increase in the price of capital has had a disincentive effect on total investment. Therefore, from 1969 on, the rate of increase of Italian investment has diminished strongly, and the main investment has been of the labor-saving type. In technical terms, since then we have had only deepening investment and very little investment widening.

After reaching a situation of full employment in 1963 (in that year unemployment came down to 2.5 percent of the labor force), Italy has registered since then an increase not only of unemployment but also of underemployment, that is, of the people compelled to work part-time.

On the opposite side, the rate of unemployment in Japan fluctuated around 1.20–1.25 percent. The investment drive and the full employment situation in Japan are quite understandable, thanks to the lower absolute prices of capital as well as of labor in comparison with Italy.

X.

A specific factor in the stagnation of investment in Italy not taken into account by any theory is to be found in the change that has occurred in industrial relations since 1969.[5] The trade unions now have the right to interfere with the organization of work in the large as well as in the medium size firms. At present, it is not possible in Italy to displace a worker from one department of a factory to another department without the agreement of the trade union. Overtime work is severely limited. From this point of view, many Italian economists have realized that labor in Italy can no longer be considered, according to traditional theory, as an elastic factor of production. Furthermore, in recent years the Italian trade unions have claimed and obtained the right to be con-

[5] I have added this point to my paper, as it summarizes my reply to specific questions put to me from the floor by Professor Hisao Onoe, discussant of my paper, and by Professor Martin Bronfenbrenner.

sulted concerning the investment programs of firms.

As a consequence of the increasing power of the trade unions and of the decrease in the propensity to work, there has been a psychological reaction from the employer side in the form of a decrease in the propensity to invest inside the country. I conclude that the change in industrial relations must be considered a structural factor, difficult to remove.

XI.

The main effect of the different degrees of accumulation of capital in Japan and Italy can be seen during the years 1974–76.

Both countries have registered an increase in the cost of labor, in the cost of imports of raw materials, and in the cost of capital. But if the increase of the first has been more or less the same in the two countries, the cost of imports of raw materials and the cost of capital have increased much more in Italy. The average yearly rate of increase of import prices during 1972–76 has been 27.4 percent in Italy and 19 percent in Japan. But the most striking difference is in the cost of capital. While in Japan the average rate of interest on bank loans has always remained lower than 10 percent, in Italy it reached 17.74 percent in 1974 and 20.43 percent in 1976. Therefore, not only has inflation—typically, cost inflation—been higher in Italy than in Japan, but the accumulation of capital has decreased much more in Italy. The ratio between gross fixed investment and gross national product has decreased from 21.5 percent in 1960–73 to 18.4 percent in 1974–76, and the ratio between fixed gross investment in machinery and equipment and gross national product has decreased from 9 percent to 7.9 percent.

Both countries have tried to get out of the economic crisis through the export drive. But while Japan succeeded in this policy, keeping active the balance of trade as well as the balance of payments, so that we could say that Japan combines with the internal accumulation of capital an accumulation of capital from the rest of the world, Italy was compelled to go heavily into debt for the short as well as for the medium and long term.

XII.

Let me try to summarize now the main findings of this paper:

1. The radical thesis of Colin Clark that a high rate of investment goes along with a low rate of economic growth has not been verified in this specific case. On the contrary, the research has shown the existence of a direct causal relationship between capital accumulation

and economic growth.

2. The rate of economic growth is a function not only of the quantity but also and mainly of the quality of investment. That means that only investments that are not directly productive could eventually impede the acceleration of economic growth.
3. A high and well-sustained amount of fixed investment has allowed Japan to reach and maintain full employment for a long time. Stagnation of investment has been accompanied in Italy by unemployment and underemployment.
4. That does not mean that accumulation of capital is a primary factor of economic growth; it simply means that if we neglect investments, economic growth may decelerate.
5. The principle of acceleration, which makes investment depend especially on consumption, may not operate if other conditions on which investment decisions are based are not present (*e.g.*, level of profits, profit expectations, comparative cost of capital and labor, availability of financial means, psychological propensity to invest). Therefore, if it is right to pursue the goal of a welfare state based on more consumption and housing, we must remember that the modification of the pattern of the economy may bring about—as the Italian case shows—a change in the social and political conditions, which may induce a decrease in the propensity to work and to invest.[6]
6. A large amount of autonomous investment may induce a derived demand for other capital goods. Therefore, investment generates investment.
7. Given a full employment situation, an increase in investment may be obtained only through control of consumption, which means that, as classical theory claims, in such a situation consumption and investment are competitive and not complementary.
8. The accelerated inflation due to the increase of the cost of labor, capital, and imported raw materials generates an increase of unemployment and a deceleration of accumulation of capital and economic growth. The higher the rate of inflation, the more intensive the deceleration.

[6] This point has been added because the discussant of my paper, Professor Hisao Onoe, observed that domestic consumption has contributed to the growth of Italian economy. "Italy," he said, "has given some lessons to Japan, which—after the strong accumulation of capital—should modify the pattern of its economy in the direction of the welfare state."

Japanese Economic Growth and Economic Welfare

Hisao Kanamori *The Japan Economic Research Center*

Has the economic growth elevated the level of welfare in this country? In order to shed light on this question, an indicator "NNW" (Net National Welfare) has been compiled through modification of GNP. NNW is an indicator of welfare that has been obtained by excluding from GNP the damages caused by environmental pollution, the increases in commuting expenses, etc., and adding to GNP the increases in leisure hours, the domestic services of housewives, etc., as assessed in monetary value. According to this indicator, while the nation's GNP grew 8.8 percent per annum in the period from 1955 through 1975, its NNW also rose 7.4 percent annually. It may be said from this that, in postwar Japan, as GNP has expanded, welfare has increased as well. Nevertheless, the deterioration of human environment has been a major factor that has worked for lowering welfare. In the future, too, utmost care needs to be taken so that environmental disruption may be avoided, though high economic growth is desired in order to raise the level of welfare.

I. INTRODUCTION

Growth was an important economic policy objective to Japan until the first half of the 1960s. Almost no one had questions about the contribution of economic growth to heightening the nation's level of welfare. But, in the latter half of the 1960s, sharp criticism against growth rose. It was because environmental pollution, the destruction of nature, price inflation, and the devastation of public feelings had become conspicuous. Did growth contribute to the advancement of welfare, or did it lower it? It is hard to answer this question. There is no agreement of opinions about what is welfare, and to quantify the level of welfare involves numerous difficulties in terms of both principle and technique. But I think it worthwhile to try to define welfare in a way that welfare may be measured and to try to look into quantitative changes of welfare. To do so is to understand only one aspect of welfare in a wide sense of the word but, without it, discussion of welfare would get nowhere and the evaluation of growth and environmental policies would be impossible.

The need for quantitative measurement of the condition of welfare has gradually been accepted in Japan as controversy over growth has become intense. So, generally two kinds of measurements have been done. One is the establishment of Social Indicators,[1] and the other is the calculation of Net National Welfare (NNW).[2] As both have their merits and demerits, one may not be asserted better than the other. My opinion is that both are useful as the measurement of welfare. Fluctuations of NNW from 1955 through 1975 will be examined in the paper, under the definition—only for the purpose of this paper—that NNW is the synonym of economic welfare. Also looked into below will be whether or not economic growth contributed to the expansion of welfare, what contribution and what hindrance, if any, did economic growth give to welfare, and what kind of economic growth helped the expansion of welfare.

[1] The "Social Indicators" of Japan were released by the Council of the People's Livelihood of the Economic Planning Agency in September 1974. The Indicators offered a time-series comparison by interpreting into quantitative indicators the level of welfare in 1960, 1965, and 1970 in nine areas including health, education, employment, and leisure time.

[2] In May 1971, an NNW Development Committee was established within the Economic Council of the Government of Japan. This Committee announced "New Welfare Indicators: NNW" in March 1973. Professor Miyohei Shinohara of the Seikei University (then Director of Economic Institute of the Economic Planning Agency) played leading roles in the Council. The concept of this NNW is similar in nature to that of "MEW" (Measures of Economic Welfare) in "Is Growth Obsolete?"(1972) by William D. Nordhaus and James Tobin.

II. WHAT IS NNW?

NNW is a modification of GNP, announced by the Economic Planning Agency for the first time in Japan in 1973 as the measurement of economic welfare. NNW contains a number of problems both in terms of theory and the calculation procedure. Views of some people are that it is impossible to modify GNP into a significant indicator of social welfare. But my view is that it is worthwhile to create a more adequate indicator of the measurement of economic welfare than GNP, by adjusting GNP by converting into monetary value under certain assumptions such activities as leisure and domestic services rendered by housewives, which contribute to economic welfare but are not included in GNP, as they are not bought and sold in market, and by deleting from GNP such values that are included in GNP but are not contributing to the improvement of welfare as increases in commuting expenses due to the progress of urbanization, even though the assessment of such values may be difficult.[3]

Now, NNW consists of the following eight items:

(1) NNW Government Consumption,
(2) NNW Personal Consumption,
(3) Household-Related Services of Social Overhead Capital,
(4) Services of Personal Durable Consumption Goods,
(5) Leisure Time,
(6) Ex-Market Activities (Housewives' Domestic Services, etc.),
(7) Environmental Pollution, and
(8) Loss due to Urbanization.

NNW is arrived at by adding (1) through (6) and deducting (7) and (8) above.[4] What is to be included in NNW and what is to be excluded from

[3] Said Report itself accounted for the limits of NNW in detail. Also, *see* United Nations, *The Feasibility of Welfare-Oriented Measures to Complement the National Accounts and Balances* (1976). NNW includes only elements that can be somehow translated into monetary value. So, one may not be persuaded on the basis of NNW against the claim that economic growth resulted in the reduction of wild cats and snow grouse and, therefore, the reduction of welfare.

Nor does NNW deal in such important aspects of welfare as full employment, the equitable distribution of income, the stabilization of prices, and the elimination of business cycles. Therefore, NNW is but a partial indicator not only of gross welfare but also of economic welfare.

A strong voice that GNP may not be converted into an indicator of welfare is that of Edward F. Denison, "Welfare Measurement and the GNP" (1971).

[4] NNW excludes investment. That it does is based on the idea that investment is but "intermediate goods" and that what directly contributes to welfare are the consumer goods and services that investment produces. The computation in said "New Indicator of Welfare" covers 1955 through 1970. This paper follows the computation for said period, provided that, whenever statistically questionable, the computation is modified in this paper.

NNW are not only determined according to reasons, but are also predicated upon the availability of statistical data. Noise and bad odor are examples of those which are strongly detrimental to NNW, but are disregarded because of the non-availability of their statistics. This is an undeniable and important limitation of NNW. A summary description of the method of assessment follows.[5]

1. NNW Government Consumption

Government expenditures are classified by function. Expenditures for education/culture, social welfare, health and sanitation are considered the elements of welfare, and the total or current expenditures for such purposes are labelled NNW Government Expenditures. Expenditures for such purposes as judicial and police were excluded with the idea that such functions do not positively contribute to the expansion of welfare. Some may rebut that judicial, and police functions, too, contribute to welfare because their absence would mean the loss of control of public order and, therefore, less welfare. But, here, they are called "defensive expenditures" having no net addition to welfare as the plus—maintenance of public order—and the minus—loss of public order—just offset each other into zero.

2. NNW Personal Consumption

This is net balance after the deduction from personal consumption, as a component of GNP, of the cost of purchase of durable consumer goods, commuting expenses, services imputable to financial institutions, and the cost of family celebrations and other ceremonial services. Consumer durables are excluded because they are accounted for elsewhere. Ceremonial cost may be reasonably left in, but I followed the theory of Professor Miyohei Shinohara in excluding it.

Major differences between the two are as follows:

(1) That environmental preservation expense is not treated as a minus item in this paper. (Such expense is a defensive expenditure and, therefore, may be deducted from GNP. But to have it as a minus item of NNW, as it was in "A New Indicator of Welfare," is illogical.)

(2) Estimation of leisure time (which will be explained later). Thus, a substantial difference exists in the value of NNW between the two.

Also, NNW for 1975 is newly estimated. I owe much to Mr. Tatsuo Ueno of the Economic Research Division and Mr. Tetsuki Hirose of the National Income Section of the Economic Planning Agency and Mr. Kiyoshi Terui of the Planning Coordination Division, Research and Planning Bureau of the Environment Agency, who gave data support.

[5] *See* Japan Economic Research Center, "Economic Growth and Welfare: GNP and NNW," (in Japanese) 1977 for detail.

3. Services of Government's Capital Goods

Services imputable to net social overhead capital stock related to personal living such as housing, city parks, schools, and systems for the supply of water, drainage, and waste disposal. The value of such services is computed under the declining value method of depreciation at the rate of 6.5 percent applicable to net balance depreciable value over the useful life of 25 years.

4. Services of Personal Consumption Goods

The value of services imputable to net durable consumer goods stock.

5. Leisure Time

A fictitious value of leisure time is assessed by multiplying by average wage per hour the total number of hours of leisure time of the people of Japan who are 15 years of age or older, computed by whether or not at paid work, by sex, and by age group—provided that the average wage has been adjusted for differences in the value of leisure time to people of different sex and age. Whether such adjustment is justified is an open question.

6. Ex-Market Activities

The main ingredient of this item is the value, assessed at the average wage of women, of domestic services of housewives (those who are 15 years of age or older and who are exclusively engaged in domestic work)—that is, the value of domestic services rendered by female persons at paid work is excluded.

7. Environmental Pollution, Water Contamination, Atmospheric Air Pollution, and Waste Materials

The cost for the reversion of pollutants that are emitted in excess of the environmental standard back to the standard values.

8. Loss Due to Urbanization

This item represents increases in commuting time and traffic accidents. Commuting time is evaluated at average wage per hour. Traffic accidents are evaluated as the combination of average compulsory traffic accident indemnification insurance payment per accident and average property value of the killed during 1970 at an average age (of 38).

III. ECONOMIC GROWTH WELFARE

The estimated values of Japanese NNW are presented in real terms (in 1970 prices) in Table 1. In Fiscal Year (FY) 1975 (which begins on 1 April), NNW was 74 trillion yen (248 billion dollars) and GNP, 93 trillion yen (310.9 billion dollars) (conversion at the rate of 300 yen to the dollar). To make a comparison of the absolute amounts of NNW and GNP and say that NNW was lower than GNP has little meaning. GNP includes investments, whereas NNW does not. Also, the size of NNW depends largely on its definition, which determines what is to be included and what is to be excluded. What is of more interest is the rate of increase. That is, how increases in NNW (or welfare as defined by NNW) compare with increases in GNP. Or, was the antigrowth faction right in their claim that the level of welfare dropped in the past 20 years, while the economy grew rapidly? Or was the growth faction right in their claim that the faster the growth, the faster the advancement of welfare?

A comparison of the rates of increases in NNW and GNP is presented in Table 2.

As long as welfare is measured by NNW from this table, the anti-growth faction was wrong. Growth did accelerate welfare. As seen in Table 2, NNW showed steady growth during the two decades from 1955 through 1975, when it grew by average annual rate of 7.2 percent. Moreover, the average annual growth rate for five-year periods gradually rose from the 5.1 percent in the 1955–60 period to 6.8, 8.2 and 9.3 percent in subsequent five-year periods.

What caused the rise of NNW? NNW increases can be attributable to the six elements shown in Table 3. They are, in the order of importance:

1. Personal Consumption

NNW rose by average annual rate of four or five percent during said two decades only due to increases in consumption. Increases in this greatest contributor to the NNW expansion were possible as the people's income rose faster than prices, thanks to economic expansions. The claim that increase in consumption is possible without economic growth—the most naive argument of growth-opposers—is, of course, unfounded. What may be possible is an argument that, as the level of living rises, the increment of happiness that is brought about by increased consumption becomes smaller. But the moment the concept "happiness" is interjected, all quantitative arguments become impossible. When the level of living is about what it is in Japan, it is believed practical and realistic to think that increase in consumption is a basic condition to welfare expansion.

TABLE 1

REAL NET NATIONAL WELFARE

(Billions of yen, in 1970 prices)

Fiscal Year	*1955*	*1960*	*1965*	*1970*	*1975*
NNW Gov't Expend.	1,199	1,374	2,254	2,988	3,865
NNW Pers. Consumpt.	10,427	14,706	22,168	32,097	43,003
Gov't Captal Gds. Svce.	62	99	169	317	559
Pers. Durab. Consum. Gds. Svce.	91	195	755	2,342	4,187
Leisure Time	4,871	6,098	7,326	10,509	16,759
Ex-Mkt. Activ.	1,876	2,388	4,068	7,213	12,707
Environ. Pollut. (Deduction)	38	1,037	3,735	6,805	5,729
Loss Due to Urbaniz. (Deduction)	452	695	889	1,113	1,119
NNW	18,036	23,128	32,116	47,548	74,231
GNP	17,268	26,183	41,591	72,144	93,260

TABLE 2

NNW AND GNP: INCREASE RATES

(Increases in fiscal year averages in 1970 prices)

	1960/1955	*1965/1960*	*1970/1965*	*1975/1970*
NNW	5.1	6.8	8.2	9.3
GNP	8.7	9.7	11.6	5.2

TABLE 3

ELEMENTS OF NNW REAL AVERAGE ANNUAL INCREASE RATES AND DEGREE OF CONTRIBUTION

(Percentage)

	Real Average Annual Increase Rate				*Degree of Contribution*			
Fiscal Year	*1960/1955*	*65/60*	*70/65*	*75/70*	*60/55*	*65/60*	*70/65*	*75/70*
NNW Gov't Expend.	2.8	10.4	5.8	5.3	0.2	0.7	0.4	0.3
NNW Pers. Consumpt.	7.1	8.6	7.7	6.0	4.3	5.6	5.3	3.8
Gov't Capital Gds. Svce.	9.8	11.3	13.4	12.0	0.04	0.05	0.1	0.1
Pers. Durab. Cosumpt. Gds. Svce.	16.5	31.1	25.4	12.3	0.1	0.4	0.8	0.6
Leisure Time	4.6	3.7	7.5	9.8	1.2	0.9	1.7	2.2
Ex-Mkt. Activ.	4.9	11.2	12.1	12.0	0.5	1.3	1.7	1.9
Environ. Pollut.	93.7	29.2	12.6	−3.4	−1.0	−2.0	−1.6	0.4
Loss Due to Urbaniz.	9.0	5.0	4.6	0.1	−0.2	−0.2	−0.1	0.0
NNW	5.1	6.8	8.2	9.3	5.1	6.8	8.2	9.3

TABLE 4

AVERAGE MONTHLY WORK HOURS PER WORKER

Year	*Regular Work Hours*	*Overtime Hours*	*Total*
1960	180.8	21.9	202.7
1965	176.4	16.5	192.9
1970	169.9	17.8	187.7
1975	161.5	10.9	172.4

TABLE 5

LEISURE TIME

	1960	*1965*	*1970*	*1975*
Male	1,087	1,173	1,061	1,278
Female	715	711	693	805

Source:
NHK, "A Survey of Living Time of the People."
Note:
Per capita average of annual total number of hours for age group of 30–39.

TABLE 6

DOMESTIC WORK OF HOUSEWIVES

Year	*1955*	*1960*	*1965*	*1970*	*1975*
Number of housewives (Ten thousand persons)	861	1,005	1,188	1,373	1,597
Weekly average number of hours of domestic works per housewife	50.2	50.0	51.9	53.8	53.2

Source: Bureau of Statistics, Office of the Prime Minister; NHK.

TABLE 7

NET STOCK OF SOCIAL OVERHEAD CAPITAL RELATED TO HOUSEHOLD

(Trillions of yen, in 1970 prices)

Fiscal Year	*1955*	*1960*	*1965*	*1970*	*1975*
Housing	0.3	0.5	0.9	1.6	2.8
Sewer systems	0.3	0.4	0.5	1.0	2.0
Waste disposal	—	—	0.1	0.2	0.4
Water supply	0.3	0.4	0.8	1.4	2.5
City parks	0.1	0.1	0.2	0.2	0.3
Schools	1.5	1.5	1.9	2.3	3.7
Total	2.4	3.0	4.3	6.7	11.8

Source: Planning Bureau, Economic Planning Agency, Government of Japan.

2. *Leisure Time*

Growth worked to advance welfare also through increases in the length and value of leisure time. Some people say that we should take time and enjoy our lives rather than striving after economic growth. But leisure time results only from a rise in the level of income. The number of hours of work is being reduced as the level of income rises in every country.[6] Japan is no exception, as can be seen from Table 4. Leisure time is time spent for social activities, rest, reading, listening to radio or watching television programs, and other personal activities. Table 5 shows the per capita averages of annual total leisure time hours spent by Japanese men and women in the age group of 30 to 39.

3. *Ex-Market Activities*

The recent 20 years saw fair increases in the volume of housewives' do-

[6] Labor unions in Japan had traditionally demanded wage increases rather than work-hour reduction. But, since 1965, demand for shorter work hours mounted, and the five-day week system has been adopted by an increasing number of establishments. This resulted in the expansion of leisure time. But the rapid reduction in work hours, which took place since 1970 through 1975, is attributable much to the reduction of overtime hours. This was due to economic stagnation, which means that it is unreasonable to deem the increased leisure time as an element of welfare advancement. But, because of the difficulty of classifying leisure time into that which is voluntary and that which is involuntary, such involuntary leisure time has been included in the estimate NNW of this paper, for a resultant somewhat overstatement of NNW for 1975.

TABLE 8

NUMBER OF WORKERS WHOSE ONE-WAY COMMUTING TIME IS THIRTY MINUTES OR LONGER

(Thousand persons)

Year	*1955*	*1960*	*1965*	*1970*	*1975*
Cities with population of one million or more	4,151	6,005	7,654	8,508	9,076
Cities with population of 300,000 or more but under 1,000,000	705	1,190	2,040	2,580	3,617
Total	4,856	7,195	9,694	11,088	12,693

mestic work, which are labelled here as "ex-market activities" as an ingredient of NNW because, for instance, if a housewife does washing herself without sending the laundry to a cleaner's, her service contributes to NNW, even though it does not contribute to GNP. (*See* Table 6).

The reason why housewives' work increased was probably because the population has been flowing from farming households to urban households and because more wives can afford to stay home due to income increases.

4. *Services of Consumer Durables*

The numbers of washing machines, refrigerators, television sets, automobiles, and other durable consumer goods owned by people have increased. As they increased, aggregate benefits from such goods increased.

5. *Government Consumption*

Government services in the areas of education, culture, social welfare, health, and sanitation have increased.

6. *Services of Government Consumption Goods*

Social capital stock—schools, houses, water supply and sewer systems, etc.—has also increased as shown in Table 7, and benefits therefrom expanded.

NNW increased as the economy grew, and the NNW increase is attributable to the above listed elements.

TABLE 9

TRAFFIC ACCIDENTS

(Thousand persons)

Year	*1955*	*1960*	*1965*	*1970*	*1975*
Number of accidents	94	450	567	718	473
No. killed (1,000s)	6	12	12	17	11
No. injured (1,000s)	77	289	426	981	622

Source:

Traffic Bureau, National Police Agency, *Annual Statistics of Traffic Accidents.* The number of accidents for 1965 and prior years includes property damage accidents, but that for subsequent years shows the number of personal injury/death accidents only.

IV. ELEMENTS OF DECLINE OF WELFARE

The greater the economic growth rate, the greater the expansion of welfare, is a wrong assertion. As Table 2 above shows, the rates of economic growth and the rates of welfare advancement were not necessarily proportionate to each other. Welfare rose faster than economy between 1955 and 1970, whereas, between 1970 and 1975, the rate of welfare increase rose despite the lower economic growth rates. Thus, the ingredient of economic growth bears heavily upon welfare.

Two factors hindered the expansion of welfare during the period of 1960 through 1970. One was losses from the process of urbanization, and the other was environmental pollution. The effect of the former was rather insignificant,[7] while the latter was a substantial factor of decline of welfare.

1. Loss Due to Urbanization

This item represents the quantification of negative effects upon welfare of prolonged commuting time and of increased traffic accidents due to urbanization. Commuting time tends to become longer gradually as indicated by Table 8. Japanese traffic accident statistics (Table 9), however, have declined rapidly in terms of both the number of accidents and the number of killed or injured, after reaching a peak in or about 1970.

[7] Of course, the conclusion can be much different depending on what are included in loss from urbanization. On the other hand, there can be gains from urbanization that are not to be included in GNP but which are a part of welfare. Such gains are ignored in the computation here.

TABLE 10

LOSS FROM URBANIZATION

(Billions of yen, in 1970 prices)

Fiscal Year	*1955*	*1960*	*1965*	*1970*	*1975*
Due to commuting time	364	497	656	707	859
Due to traffic accidents	87	198	234	407	260
Total	452	695	889	1,113	1,119

As a result, loss from urbanization has levelled off since 1970. (*See* Table 10).

2. *Environmental Pollution*

The most detrimental factor to welfare was environmental pollution. The degree of contribution to decline in welfare was particularly high between 1960 and 1965. That criticism against rapid growth mounted in this period was no accident. Environmental pollution became a big social issue in the latter half of the 1960s.

Responsible for environmental pollution are a number of things such as water contamination, atmospheric air pollution by sulfur oxides (SO_x), soot, and/or dust, automobile exhaust gas, industrial waste, and house trash.[8] The emitted volumes of such pollutants all increased rapidly from 1955 through 1960, as shown by Table 11. The degree by which such pollutants caused decline in welfare was estimated in Table 12 under the method explained earlier.

Loss from environmental pollution reached a peak in Fiscal Year 1970 at 6.8 trillion yen, which was 9.4 percent of GNP of 72 trillion yen and 14.2 percent of NNW of 48 trillion yen. It was a serious policy mistake to have ignored the magnitude of external diseconomies caused by environmental pollution. It is believed that if investment for the expansion of production was partially diverted to the prevention of environmental pollution in this period, NNW would have been greater, even though the GNP increase rate might have been somewhat lower.

But as environmental pollution has advanced, efforts to control it have gradually become intensified. Facilities investments for the purpose of environmental pollution control as percent of total facilities investments by manufacturers and gas/power suppliers under the jurisdiction of the

[8] NO_x emitted by fixed origins is not included here, due to the unavailability of data thereof.

TABLE 11

VOLUME OF ENVIRONMENTAL POLLUTANTS

	Unit	*1955*[a]	*1960*[a]	*1965*[a]	*1970*[a]	*1970*	*1975*
BOD	ton/day	4,682	7,174	10,664	14,117	8,760	6,226
By factories		3,232	5,044	7,866	10,894	7,011	4,010
By homes		1,450	2,130	2,798	3,223	1,749	2,216
SO_2	10,000						
Sulfur	tons/year	50.7	82.9	146.6	234.5	176.7	126.5
Soot/Dust	10^3	423.6	715.8	882.4	1,136.1	1,044.4	1,064.8
Steel	tons/year	69.1	126.2	188.0	316.4	242.6	266.0
Cement		159.0	249.3	237.7	207.3	252.9	281.6
Thermal generation		185.2	310.3	305.8	275.6	205.9	235.9
Urban trash		—	—	69.1	145.4	121.2	46.9
Heavy oil boilers		10.2	30.1	81.8	191.4	221.6	234.4
Automobile Exhaustion Gas	ton/day						
NO_x			358[b]	665	1,699	1,599	2,197
Waste	10^3			192,514	302,345	308,337	143,557.4
Non-industrial	tons/year			10,135	9,785	3,166	6,115.4
Industrial				182,379	292,560	305,211	137,442.0

[a] Estimates by "New Welfare Indices" (1973).
[b] 1961 figure.

Ministry of International Trade and Industry rose rapidly since 1970 from 5.3 percent in 1970 to 19 percent in 1975, as indicated by Table 13. Also, as indicated by Table 14, expenditures of the Government of Japan and of local authorities (prefectural governors' offices and municipal mayors' offices) for the purpose of preventing environmental pollution, as well as public and private preventive facilities, rapidly increased. The degree of environmental pollution started to drop since 1970, when the emitted volume of sulfur dioxide (SO_2) and industrial waste started to reduce. This was partially due to economic stagnation, but chiefly due to the successful implementation of pollution control measures, which should have been implemented earlier.

V. EVALUATION OF THE GROWTH POLICY

We have so far seen that welfare has rapidly advanced in Japan as her economy grew, but that the way in which the economic growth was accomplished should have been different. Whether the rate of economic growth of Japan was excessively high in the past from the standpoint of welfare is a difficult question. If future welfare is to be elevated by economic growth, capital must be accumulated by constraining current consumption. But under curtailed consumption in the interest of future welfare, current expansion of welfare would be constrained. An optimum growth rate should be somewhere in between. But whether the Japanese growth rate was above such an optimum rate in the past can in no way be determined because both judgments that need to be made in order to answer this question are difficult: an "objective judgment" to assess the relative values of current consumption and future consumption and "technical judgment" to measure the effect of curtailment of consumption at a given point of time on increase in consumption at a given future time. But the fact that the increase rate of NNW has gradually risen seems to indicate that to have constrained consumption and increased investments in the 1950s and 1960s aiming at the expansion of future fruits was a wise thing. If, in those days, consumption were at a higher rate, the current NNW would have been higher then, but total welfare for the entire 20 years would have been smaller. Therefore, it appears right to assert that what was wrong was the way in which economic growth was accomplished, rather than how high the rate of growth was. Of course, the diversion of a part of investment away from production facilities to pollution control facilities, if effectuated, would have resulted in a lower growth rate. But it is unreasonable to complain about the level of growth rate itself.

Lastly, future problems will be discussed. Noteworthy in Table 2 is the fact that, while the rate of GNP increase between 1970 and 1975 was 5.2 percent or only about one-half of the rate in the preceding period, the

TABLE 12

LOSS FROM ENVIRONMENTAL POLLUTION

(Billions of yen)

Fiscal Year	*"New Welfare Indices" — NNW —* *1955*	*1960*	*1965*	*1970*	*Recent Estimates* *1970*	*1975*
Nominal:						
BOD	0	435.5	981.5	1,505.6	1,739.3	2,282.9
Factories		256.2	562.3	832.0	1,074.0	970.1
Homes		179.3	418.9	673.6	665.3	1,312.8
SO_2	0	247.7	669.6	1,797.6	1,736.6	1,798.6
Soot/Dust	24.8	76.4	158.3	400.9	370.5	570.2
Automobile exhaust gas	0	0	207.2	1,151.2	1,048.4	3,229.1
Waste			917.1	1,245.6	1,949.7	1,152.9
Non-industrial			71.2	106.7	34.6	101.2
Industrial			845.9	1,138.9	1,915.1	1,051.7
Real:						
BOD		557.7	1,142.8	1,505.6	1,733.3	1,481.6
Factories		299.3	630.7	832.0	1,068.7	639.5
Homes		258.3	512.1	673.6	664.6	842.1

SO_2		289.3	750.7	1,797.6	1,727.9	1,167.9
Soot/Dust	34.6	89.3	177.4	400.9	368.7	375.9
Automobile exhaust gas	0	0	269.8	1,151.2	1,032.9	1,906.2
Waste			1,035.4	1,245.6	1,942.0	797.5
Non-industrial			87.0	106.7	34.5	66.7
Industrial			948.4	1,138.9	1,907.5	730.8
Total:						
Nominal	24.8	759.6	2,933.6	6,100.9	6,844.5	9,033.7
Real	34.6	936.3	3,376.0	6,100.9	6,804.8	5,729.1

rate of NNW increase was higher than the preceding period at 9.3 percent. Does this mean that welfare can advance even if the rate of GNP growth is low?

When Table 3 is scrutinized for whether the rate of NNW increase was high in the period 1970–75, the answer is "yes" to some extent. An important factor that contributed to the NNW increase in this period was improvements in environment. That is, although environmental pollution had been a seriously detrimental factor to NNW prior to 1970, the abatement of pollution since 1970 became a factor for NNW increase.

On the other hand, however, four problematic aspects of NNW increase in this period should not be overlooked.

One is that the NNW increase was due to the economic growth of the past. What pushed up NNW in this period were increases in leisure time and ex-market activities, which were translated into market values at real wages per hour. Wages may not accomplish real increase if the economy remains stagnant for a long period of time. As a result, NNW advancement, which depends not only on consumption but also on leisure time and ex-market activities, would become impossible.

The second is that increase in leisure time in this period included an involuntary element of the reduction of work hours due to economic recess, as stated earlier. This means that NNW was overestimated. A prolonged low growth situation would bring about a higher level of unemployment and, therefore, a reduction in work hours. But reduction in work hours is in fact a substantial negative factor to welfare.

The third fact is that the rate of personal consumption increase was

TABLE 13

TREND OF ANTIPOLLUTION INVESTMENTS

(Investments for the Prevention of Environmental Pollution as Percent of Total Facility Investments by Manufacturers and Gas/Power Suppliers under the Jurisdiction of the Ministry of International Trade and Industry)

1965	3.1%	1972	8.3%
1966	2.9	1973	9.8
1967	3.5	1974 (Final estimate)	14.3
1968	3.7		
1969	5.0	1975 (Plan)	19.0
1970	5.3		
1971	6.5		

Source:

MITI, *Trend of Private Facility Investments for the Prevention of Environmental Pollution.*

TABLE 14

COST OF ENVIRONMENTAL PRESERVATION

(Billion yen, in 1970 prices)

	1955	*1960*	*1965*	*1970*	*1975*
Public expenditures for pollution control				34.3	80.0
National government				19.9	37.9
Local jurisdictions				14.4	42.1
Services imputable to capital stock for pollution control					
Public		32.1	94.5	219.9	532.7
Private		7.7	29.6	103.3	521.2
Total	19.9	39.7	124.1	357.4	1,133.9

TABLE 15

SUSTAINABLE NNW (IN 1970 PRICES)

(Billions of yen, %)

	Value (Billion Yen)					*Increase Rate (%)*			
Fiscal Year	*1955*	*1960*	*1965*	*1970*	*1975*	*60/55*	*65/60*	*70/65*	*75/70*
NNW_1	18,036	23,128	32,116	47,548	74,231	5.1	6.8	8.2	9.3
Net investment	1,387	4,187	6,605	16,144	18,689	24.7	9.5	19.6	3.0
NNW_2	19,423	27,315	38,721	63,692	92,920	7.1	7.2	10.5	7.8
GNP	17,268	26,184	41,592	72,144	92,972	8.7	9.7	11.6	5.2

Notes:

NNW used here were estimated by Ms Yuriko Tabase. Also noteworthy was the contribution of Doctor Kimio Uno, Assistant Professor at the Tsukuba University. The NNW estimate that I reported in my talk at the symposium titled "International Economic Symposium—in Pursuit of New Prosperity," held in October 1976, under the sponsorship of the Japan Economic Journal Press, has been up-dated for the purpose of this paper with the use of additional statistics that have since become available.

high even though the rate of GNP increase was low because the ratio of investment as a component of GNP dropped, while the ratio of consumption as a component of GNP rose. A greater part of products must be directed to investment when the rate of economic growth is on the increase and, therefore, NNW increase lags behind economic growth. Conversely, when the rate of economic growth is on the decline, the ratio of facility investments drops and, therefore, consumption need not be curtailed so much. But these are short-lived transitional phenomena. When the ratio of facility investments has descended to a certain level, consumption may no longer be increased by reduction in investment. With this in consideration, James Tobin and William D. Nordhaus estimated an index called "sustainable MEW," in addition to MEW (Measures of Economic Welfare). Following the example, net investment is added to NNW in arriving at NNW_2 (sustainable NNW) as presented in Table 15. The rate of NNW_2 increase is much lower in the period of 1970 to 1975, indicating that NNW increase in this period was at the sacrifice of investment and would not continue.

The fourth and the last point is that the abatement of environmental pollution, although partly attributable to slowdown in growth, owes to a substantial increase in expenditures for pollution control purposes as stated before. If slow growth continues, expenditures for such purposes may not be increased.

In 1975, unemployment started to increase and, in some months in the summer of 1976, real wages were lower than in the comparable month of the preceding year. From this, it can be asserted that it was only up to about 1975 that welfare could be raised under low growth. What is considered important to the future advancement of welfare in Japan is that appropriate environmental and urban measures are taken as the growth rate is being increased.

REFERENCES

Denison, Edward F. 1971. "Welfare Measurement and the GNP," *Survey of Current Business,* vol. 51, January, pp. 13–16, 39.

Nordhaus, William D. and Tobin, James. 1972. "Is Growth Obsolete?" in *Economic Growth.* National Bureau of Economic Research, General Series No. 96. New York: Columbia University Press for NBER, pp. 1–80.

United Nations. 1976. *The Feasibility of Welfare-Oriented Measures to Complement the National Accounts and Balances.* New York.

Comments

Kimio Uno (Japan)

I believe that the paper represents an earnest effort to cope with a very relevant question. I found the paper very persuasive. Based on empirical data, it clearly indicates some of the characteristics of the consequences of economic growth. This particular methodology by Tobin and Nordhaus was specifically mentioned in the plenary session by Professors Malinvaud and Modigliani. Some of the methodological problems concerning NNW estimates are pointed out in the Kanamori paper itself; especially important in my opinion are footnotes 3, 4, 6, and 7.

The paper also spells out clearly some reservations in interpreting the results; it also mentions that the welfare measured by NNW approach is not the welfare in general, and that the word "welfare" used in the paper is simply restating the NNW definition, a fair treatment.

I have nothing to quarrel with the paper per se except to point out that the decline of the economic growth rate in this country would have occurred anyway with the shift of priorities in resource allocation from the expansion of production to the improvement of environmental pollution and urban congestion.

Also, I wonder whether it is possible to judge *a priori* the goodness or badness of products or activities. For example, macroeconomic effects of pollution prevention expenditure or military expenditure or of burying a pot underground, just as Keynes has pointed out, would be different, depending on whether or not the economy is at full employment. These kinds of interrelations and dynamic consequences are not considered in the NNW framework. Does not this imply the need to prepare a statistical base, compatible with the national income account, but much wider in scope?

It is a correct decision to confine the analysis to the quantifiable aspects, as is stated at the beginning of the paper. Even within this confine, however, the scope should have been and could have been wider than the one presented here to claim NNW as a measure of economic welfare.

In the following I would like to raise some issues.

1. We must remember that macroeconomic aggregation is a fiction, an economists' invention. What exist in reality are the economic actors, including the households, the producers, and the government. Is not welfare something to do with the situation in which households find themselves? The beneficiaries and the evaluators of the society, I believe, are in the household sector.

The moment we introduce economic actors, the limits of aggregate, constant price approach become clear because the income-expenditure balance is in terms of nominal figures, and not in real terms. This means that real figures are not sufficient: nominal values are just as important as real, and sectoral figures are just as important as the aggregate.

2. It may also be necessary to compare the conflicts of interests among various social groups. Social groups could be debtors and creditors, or farmers and

employed workers, or various age groups of the population. Needless to say, under inflation, the debtors gain and the creditors lose; on the aggregate, however, the gain and the loss cancel out and is not stated explicitly. The same holds true in the case of transfers of income by agricultural price policy or by social security measures. What I want to say here is that those items do not show up in the NNW framework, which is on the aggregate, although they have direct bearings on "welfare." At the same time, they can be easily quantified. Analysis by social groups is perhaps indispensable in growth evaluation.

3. There is another drawback in analyzing on the aggregate. That is the fact that balance among the items included in NNW is not questioned. All the items included are simply added into a scalar magnitude. This procedure hides such phenomena as congestion, which is an excessive use of social capital or public goods in general.

As the income level goes up, the demand for food and clothing is more or less satisfied, and a stage is reached where the quality of life cannot be improved unless public goods, which include institutional arrangements as well as social capital, are supplied together with the private expenditure. There are numerous examples of this: housing and sewerage, and automobiles and highways are some. Perhaps we should look into the individual components of NNW for this.

4. The services rendered by social capital stock are included in NNW, and this is clearly a big step forward. But the explicit recognition of public goods would lead us to think of the ways in which those public goods are supplied in a *market economy situation*. Economic theory tells us that market mechanism fails in adequate provision of public goods. May I add that this does not automatically mean that central planning would do a better job. Anyway, the pattern of growth that this country has followed could be inherent in our economic system.

5. There exists another line of thinking, especially in post industrial economies, to guarantee some minimum level of living, regardless of cost benefit calculation through the market. This is sometimes referred to as "civil minimum" or "national minimum." Under the scheme, some socially determined minimum level of supply or access is guaranteed to the population in general, regardless of their purchasing power in the market.

At any rate, taking (4) and (5) together, the increasing importance of public goods and social security measures indicates that we have to think of the social decision-making process that takes place outside the market mechanism. Outside the market, but still within the domains of economic studies. At least we should recognize the importance of participation of the population in the decision-making process, an element of quality of life, but one that cannot be accommodated in the NNW framework. Inclusion of decision-making in our scope means that we have to deal with subjective evaluation of the participants in the process. For this purpose, the scope of NNW, which is limited to those aspects that are subject to monetary imputation, may pose a limitation.

In conclusion, I would like to suggest that NNW is helpful in judging the net products on the aggregate. NNW is basically a cost-benefit analysis of economic activities on the aggregate level, contrasting the positive and negative outputs. The implication is, therefore, if there remain some positive gains after netting out goods and bads, the growth could be not only justifiable but is desirable.

However, happy figures on the aggregate do not necessarily mean everything is all right inside. To assume so is another kind of laissez faire, expansion on the aggregate automatically fulfilling harmonious results.

Being such, NNW falls short of being an indicator of economic welfare or of the quality of life. It lacks the standpoint of the beneficiaries of the society, that is, the household sector. In NNW framework as well as in the national income framework, existence of household is of course recognized, but in a piecemeal fashion. It is represented by wage income, consumption, housing investment, etc., but not as an integral unit of an actor. The standpoint has to be turned upside down, and the socioeconomic variables pertinent to the household sector must be so arranged as to allow for sectoral analysis.

It is only then that we can include the relation between subjective evaluation and objective situation in our analysis. Subjective evaluation is an *objective* statistic when observed on the society-wide basis. It is not a random expression of one's ideas. Based on the correspondence between objective situation and subjective evaluation, we should be able to identify the needs of the society.

I may have given an impression that I am negating the production sector, but what I wanted to say is that it is clearly possible to integrate the dual standpoints in a single framework. That is, the standpoint of production as represented by GNP or NNW and the standpoint of the beneficiaries as represented by quality of life indicators or social indicators.

Finally, I repeat that NNW is a step forward and should be developed further. On the first page, Mr. Kanamori states that both are useful as a measurement of welfare, and I fully agree with the author. Clearly it is one of the two ways of reaching for a scalar magnitude that reflects the general welfare level of the population in a society. Another is the use of the subjective weights that people attach to the various aspects of social life. NNW should not be denied for lack of "theory" or lack of dimensions. Issues come first and theories can come later.

Did Technical Progress Accelerate in Japan?

Kazuo Sato *State University of New York*

Various indicators point to an acceleration of the catching-up process in the 1960s over the 1950s in the Japanese economy. Nonetheless, growth accounting exercises show no such acceleration. This contradiction is examined as regards the methodology of growth accounting—in particular, nonneutrality of technical change, measurement errors in the capital input, and non-invariance of the macro production function—all pointing to inadequacies in growth accounting. It is also observed that Japan's technology gap was largely closed by the early 1970s in the capacity of an imitator-importer, but Japan is yet far from assuming world leadership in technology as an initiator-inventor.

I. WHAT ARE THE PROBLEMS ?

Japan was technologically isolated from the rest of the world for more than a decade spanning over World War II. It faced a substantial technology gap when it re-emerged from the wartime dislocations. Although Japan made a steep ascent through the 1950s to eliminate the gap, it was in the 1960s that Japan's technological capability began to be firmly established. Its spectacular growth performance was clearly visible in its industrial exports, which expanded at a rate double that of the world trade in manufactures. Japan arrived at the stage of technological maturity in the 1960s after a take-off in the 1950s. Its innovative intensity accelerated considerably over time. This observation is corroborated by the movement of the gross fixed capital formation proportion in GNP. The ratio (in current prices) rose from 20 percent in the early fifties to 35 percent in the early seventies. With labor growth decelerating, capital deepening was intensified in the 1960s. The capital intensity could be raised at such a stupendous rate, one conjectures, only if technical progress were accelerated in the 1960s. So I reasoned in an earlier paper (Sato, 1971). Since other countries grew far less rapidly than Japan, the technology gap must have narrowed substantially in the course of the sixties. The trend acceleration, emphasized by Kazushi Ohkawa and Henry Rosovsky (1973) as a salient feature of Japan's modern economic growth, must have reached its zenith in the sixties.

In the meantime, however, sources-of-growth analysts applied their impressive tools of growth accounting to Japan's postwar growth and discovered that total factor productivity growth revealed no sign of acceleration (*see* Dale W. Jorgenson and Mitsuo Ezaki, 1973; Edward F. Denison and William K. Chung, 1976). The rate of technical change was unchanged or even down in the 1960s according to their computations. This finding led Ezaki and Jorgenson (1973) to question my thesis of accelerated technical change.[1] In another context, Tuvia Blumenthal (1976) wonders about the validity of the other thesis that Japan caught up with Western technology by the end of the sixties. My intuition and instincts tell me otherwise, but facts must be carefully examined if these alternatives are to be challenged. I shall undertake the task in what follows. I shall demonstrate that a fundamental change occurred in the direction of technical progress at about 1961 as a response to a number of new economic factors. While the rate of technical change should measure this change, growth accounting fails to do so. Narrowing of the technology gap must also be seen in the same light.

The plan of the paper is as follows. In *Section* II, we present a few indicators pointing to the acceleration of innovative intensity in contrast

[1] However, the new finding of Dale W. Jorgenson and Mieko Nishimizu in this conference volume seems to be in line with my hypothesis.

to the finding of growth accounting. In *Section* III, we take a close look at economic changes behind the growth-accounting figures. In *Section* IV, we question the measurement of the fixed capital input when technological shifts change direction as between the two decades under study. In *Section* V, we conjecture as to why such a dramatic change occurred in all sectors of the economy. The change required accelerating innovative intensity which, however, is not reflected in total factor productivity growth. In *Section* VI, we observe that Japan virtually closed the technology gap by the end of the sixties in the capacity of an imitator and importer of foreign technology. In *Section* VII, we conclude that Japan is quite behind Western nations as an initiator of new technology. In this sense, Japan's technology gap is still quite large. Japan's growth prospect, as well as its position in the international community, depends critically on how fast Japan can close this gap. Japan now finds itself at a historical watershed.

II. DID TECHNICAL PROGRESS ACCELERATE IN JAPAN?

A. Indicators of Innovative Intensity

Imagine that innovative activities can be quantified by an indicator, say z. Denote the level of total factor productivity by T. Its growth rate may be assumed to be a positive function of z. Let us examine a few global indicators of R&D activities in Japan to see how z changed. Some of them are assembled in Table 1 and graphed in Figure 1. Technology imports are generally regarded as the most critical input in Japan's technological growth in the postwar period.[2] In real terms, they grew at more than 30 percent per annum from 1951 to 1961. Since then, its growth settled down to more modest 10 percent per annum. Domestic R&D expenditure (in constant prices) followed a virtually identical path, deflecting itself at 1961 and shifting somewhat upward between 1967 and 1968. Three other indicators—the number of persons engaged in R&D, the number of patent applications, and the number of patent registrations[3]—grew at more or less comparable rates with the former two, though the inflection is not always so clear. We get a broad impression that the 1950s were a decade of take-off and the 1960s a decade of full swing as far as technological growth is concerned.

[2] Technology imports are payments for patent royalties, license fees, etc. Japanese statistics classify them into Class A (contracts running more than a year and payments in dollars) and Class B (contracts less than a year or payments in yen). The latter category is very minor.

[3] The lag from patent applications to approvals is substantial. Even after a substantial shortening due to the 1961 revision of the patent law, it is more than three years.

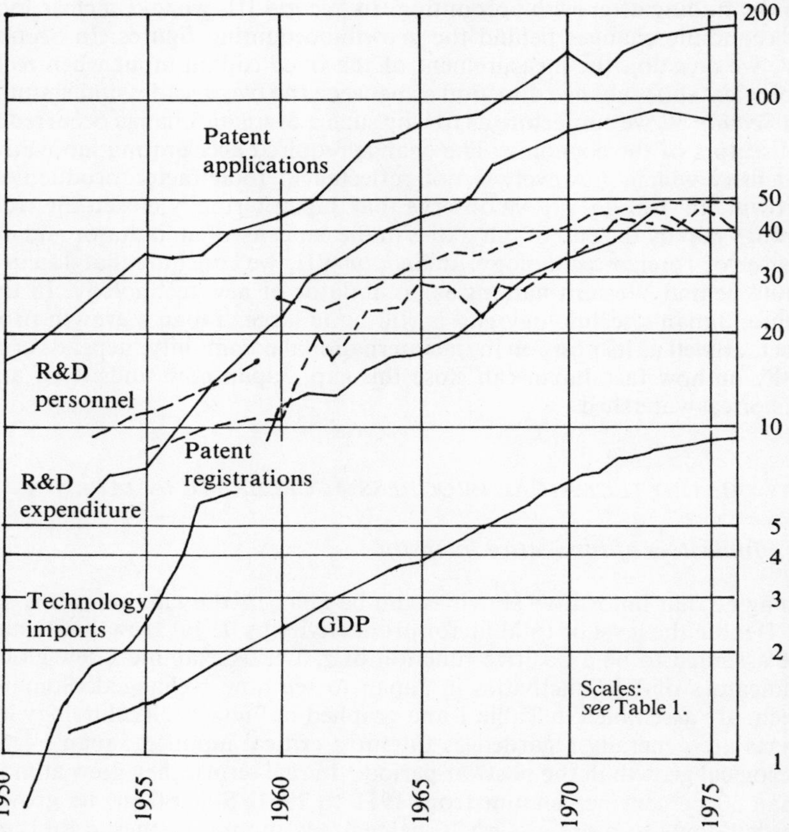

Patent applications
R&D personnel
Patent registrations
R&D expenditure
Technology imports
GDP
Scales: *see* Table 1.
200
100
50
40
30
20
10
5
4
3
2
1
1950
1955
1960
1965
1970
1975

Fig. 1. Indicators of innovative activities in Japan, 1950–76

.35
.30
.25
.20
1950
1955
1960
1965
1970
1975

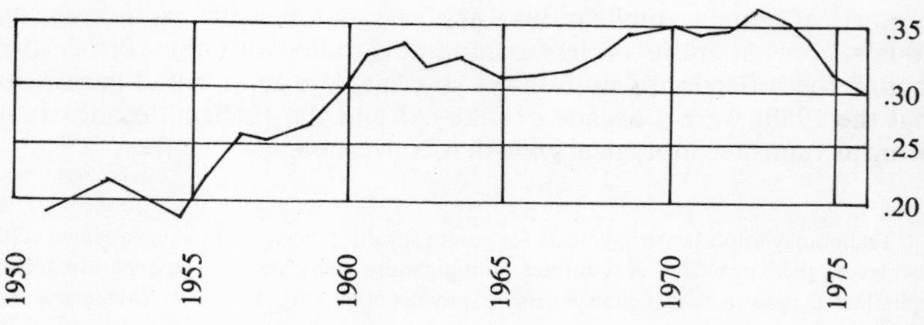

Fig. 2. The ratio of gross fixed capital formation to GNP (in current prices), 1951–76

These indicators are in absolute terms. They have to be expressed in relative terms in order to indicate our desired index z. We may compare them with real GDP, which grew at the steady rate of 10 percent. Normalized in this way, we conjecture that z rose through the 1950s, reached a peak at 1961, and followed a plateau through the rest of the years up to 1973. If our postulated relationship between total factor productivity (TFP) growth and innovative intensity is true, we should anticipate that the former followed the same pattern of change. As new techniques are mostly embodied in new investment goods, we expect that more rapid technical change induces investment to expand relative to GNP. Figure 2 shows that, in fact, the gross fixed capital formation (GFCF) proportion exhibited exactly this sort of change. The proportion was 20 percent in the early fifties, started to rise for the next few years, reached a peak at 1961, and remained at a high level close to 35 percent (though with a sag at the mid-sixties) until the latest recession. One concludes with confidence that this parallel movement between the GFCF proportion and innovative intensity is not coincidental. Indeed, one can begin with the observed behavior of fixed investment and speculate that technical change was accelerated considerably in the 1960s. The increasing share of fixed investment in GNP led to a more rapid growth of capital stock. Since labor growth decelerated in the meantime, the capital-labor ratio increased its rise in the 1960s. It should have depressed the marginal efficiency of investment and made it difficult to maintain investment at a high level, that is, unless technical change was accelerating at the same time.

B. *Results of Growth Accounting*

Growth accounting or the sources-of-growth approach became popular in Japan as elsewhere. A number of studies are by now available. We compare a few recent ones in Table 2.[4] To compare the 1950s and the 1960s, we take 1961 as the dividing year whenever possible. While the results assembled in Table 2 are generally similar, we are also struck by their diversity. Their differences are no doubt due to differences in coverage of the economy, in coverage of inputs and outputs, and in weighting formulas and weights. Clearly, the rate of technical change, as measured by TFP growth,[5] is subject to significant measurement errors. Nonetheless, with the exception of Kanamori (1972),[6] Table 2 conveys a message that the rate of technical change remained unchanged or even declined in the 1960s over the 1950s.

[4] For earlier studies, *see* the studies referred to in Table 2.

[5] TFP growth is further reduced by accounting for a number of factors, *e.g.*, economies of scale, intersector resource shifts, etc. *See, e.g.*, Denison and Chung (1976). These adjustments, however, involve arbitrary judgment.

TABLE 1

INDICATORS OF INNOVATIVE ACTIVITIES IN JAPAN, 1950–1976

Year	*(1)*	*(2)*	*(3)*	*(4)*	*(5)*	*(6)*	*(7)*
1950	1.0	....	....	16.9	....	....	....
1951	8.6	....	....	17.8	....	....	19.8
1952	13.7	....	....	20.9	....	11.1	20.5
1953	18.3	95.4	62.0	24.6	....	12.2	22.4
1954	20.3	106.7	72.6	29.4	....	13.2	21.3
1955	26.6	112.8	76.2	34.5	8.6	14.5	19.8
1956	46.9	120.7	93.3	33.2	9.4	16.1	23.5
1957	57.5	131.0	123.2	33.2	9.8	18.0	26.6
1958	62.4	144.2	149.5	38.5	10.0	18.8	25.5
1959	76.7	*161.0*	188.7	41.5	10.3	21.1	26.6
1960	114.3	243.3	226.6	43.5	11.3	24.2	30.1
1961	132.2	225.2	290.9	48.4	20.9	27.4	33.3
1962	128.4	242.6	324.0	60.1	15.7	29.8	33.7
1963	148.9	272.5	351.0	71.0	23.3	32.7	32.2
1964	161.5	289.3	399.3	75.0	23.7	36.8	32.5
1965	167.0	303.8	425.8	81.9	26.9	38.9	30.6
1966	183.7	323.0	459.7	86.0	26.3	42.8	30.8
1967	219.0	327.6	497.0	85.4	20.8	48.2	32.1
1968	269.3	356.3	629.4	96.7	28.0	53.9	33.6
1969	296.0	367.3	697.0	105.6	27.7	60.4	35.1
1970	326.6	392.2	796.4	130.8	30.9	67.6	35.0
1971	345.5	429.3	823.2	105.8	36.4	71.1	34.3
1972	379.6	426.9	875.7	130.4	41.5	77.8	34.5
1973	447.4	459.2	913.3	144.8	42.3	86.2	36.6
1974	418.2	468.1	879.5	149.3	39.6	85.1	34.2
1975	381.7	491.3	864.4	159.8	46.7	87.1	30.8
1976		487.9		161.0	39.8	92.6	30.0

Legend, Sources, Notes:

(1) Technology imports, millions of 1965 dollars.

(a) Payments for class A technology imports, deflated by (b) the index of average annual earnings of full-time equivalent employees in all U.S. industries.

SOURCES: (a) Science and Technology Agency (Japan), *White Paper on Science and Technology*, annual issues; (b) Department of Commerce (U.S.), *The National Income and Product Accounts of the United States, 1929–74* (1977) and *Survey of Current Business*, July 1977.

(2) Persons engaged in R&D (researchers, research assistants, technicians, etc.) in the private and public sectors, thousands (Japan).

SOURCE: *See* (1.a), originally from Prime Minister's Office, *Basic Survey of Research Institutions* for 1953–59 and *Survey of Science and Technology* thereafter. The two series do not overlap.

(3) Total R&D expenditure, billions of 1965 yen.

(a) Total R&D expenditure in the private and public sectors, deflated by (b) the price index, which is the simple average of (b1) the manufacturing wholesale price index and (b2) the index of cash earnings of regular workers in all industries. Weights are in rough correspondence with the cost composition in the 1960s.

SOURCES: (a) *See* (1.a). (b1) the Bank of Japan wholesale price index for manufacturing for 1957 on and Ohkawa *et al., Prices* (Tokyo: Toyo Keizai Shimpo-sha, 1966), p. 193 for 1953–56. (b2) Prime Minister's Office, *Japan Statistical Yearbook*, various issues.

(4) Patent applications, thousands.

SOURCE: *See* (1.a), originally from Patent Agency (Japan), *Annual Report*.

(5) Patent registrations, thousands.

SOURCE: Prime Minister's Office (Japan), *Japan Statistical Yearbook*, originally from Patent Agency (Japan), *Annual Report*.

(6) GDP at factor cost (excluding public administration), billions of 1970 yen.

SOURCES: 1954–73 from Kogane *et al.* (1975), extrapolated for 1952–53 by data in Denison and Chung (1976), Table 3-3, col. (1)–col. (3) and for 1974–76 by constant-price GNP.

(7) The ratio of gross fixed capital formation to GNP at market prices (current value), percent.

SOURCE: Economic Planning Agency (Japan), *Annual Report on National Income Statistics*, various issues. (1951, extrapolated by the fiscal-year data.)

This finding indicates that technical change did not accelerate in the sixties after all. If so, there must have been no relationship between TFP growth and innovative intensity. Two explanations suggest themselves readily. First, our normalization may have been wrong. These indicators should have been deflated not by real GNP but by the stock of technology. The latter may have followed the same sort of a growth path as the indicators themselves, which represent flows. This interpretation requires the stock of technology to be extremely low in the beginning of the 1950s (relative to GDP). The stock of technology *in use* may have been very low in the late forties when both production and productivity plummeted precipitously in the wake of the war devastations. But Japan regained its prewar peak productivity by 1953. This seems to reject the first explanation.

The second explanation may argue that the backlog of unutilized technology, both domestic (developed through the war years in munitions industries) and foreign, was reduced substantially in the 1950s. Therefore, innovations may have run into diminishing returns in the 1960s. Innovative activities had to be more intense to raise TFP by a given percentage.[7] However, as we shall argue later, there is no evidence that Japan closed the technology gap so much by 1961. Hence, this explanation is also not plausible.

If these explanations cannot be used to reject our hypothesized relation,[8] there remains another possibility, namely, that TFP growth measured by growth accounting is something other than our dependent variable, provided that the former is error-free—an assumption that itself has to be carefully examined. It is necessary to go beyond statistics and to see what really happened in the Japanese economy to make it step up innovative intensity.

[6] Hisao Kanamori (1972) reports an acceleration in TFP growth (4.7 percent, 1955–60; 5.9 percent, 1960–65; 8.4 percent, 1965–68). The difference is due to higher output growth in later years in his estimates.

[7] If ϱ is the marginal product of the stock of technology, the contribution of z (deflated by GDP) to TFP growth is ϱz (*see* Zvi Griliches, 1973, p. 64). The present explanation then is equivalent to saying that a rise in z was offset by a fall in ϱ.

[8] A third explanation is possible, though. TFP growth may consist of the contribution of R&D and an autonomous element. In the 1960s, the former rose but the latter may have fallen. Harry Oshima, the discussant of the present paper, observed that Japan's rapid postwar growth owed a great deal to the democratization of its institutions and that its effect may have waned by the 1960s. This comment is along the lines of the third explanation. However, while institutional reform is a precondition for the take-off, it does not directly work on TFP growth. It can raise TFP growth only by improving the economy's capacity to absorb new techniques.

III. WHAT LIES BEHIND TOTAL FACTOR PRODUCTIVITY GROWTH?

A. Clearing Some Conceptual Problems

As a leading practitioner of growth accounting, Denison consistently prefers net national income as the output measure and measures factor shares net of capital consumption allowances.[9] Others take GNP and factor shares gross of depreciation.[10] These conceptual differences can make a great deal of measurement differences.[11]

There are two reasons why the gross concept should be chosen. The first is a theoretical one. Fixed capital goods remain in production services so long as they can earn positive quasi-rents, *i.e.*, gross value added less wages. Capital goods are mostly owned by users. Capital cost is already sunk and owners find it advisable to employ their capital goods until their quasi-rents vanish. In this, GDP is the decision variable. The second is a practical one. Capital consumption allowances, as recorded in national accounts, are usually derived from business accounting records (or imputed by national income statisticians) and do not necessarily have any relation with actual wear and tear of capital goods. As such, these numbers are arbitrary. It is better to minimize arbitrariness, since it is a major source of measurement error. In Japan, during the two decades under study, the depreciation rate rose quite significantly (from 7 percent of GNP in 1952 to 15 percent in 1972) because accelerated depreciation was enforced. When we rely on the gross concept, the fixed capital input receives more weight and all others less weight. Measurement of the capital input becomes more important quantitatively.

B. A Look at Sectoral Developments

It is necessary to make sure that the results of Table 2 are not influenced by the fallacy of composition. For this purpose, we compare sectoral growth of output (GDP at factor cost),[12] of employment (persons engaged in production), and of capital (gross nonresidential fixed capital

[9] *See* Denison (1967).

[10] *E.g.*, Jorgenson and Ezaki (1973).

[11] This can explain why the contribution of inventory growth is so preposterously large in Denison-Chung estimates of Table 2. Is plant and equipment not the transmitter of technical change?

[12] Real value added should be derived as a Divisia index or by the double-deflation procedure as its approximation (Sato, 1976). Japan's national income statistics do not provide constant-price GNP by sector. Even the study referred to depends on gross production indexes or single-deflated value added.

TABLE 2

GROWTH ACCOUNTING: COMPARISON

(percent)

Growth rates	Jorgenson-Ezaki 1953-61 (w)[d]	Jorgenson-Ezaki 1961-31 (w)	Kosobud 1953-61	Kosobud 1961-68	Kanamori 1955-60	Kanamori 1960-68 (w)	Denison-Chung 1953-61 (w)	Denison-Chung 1961-71 (w)
Economy	*private + gov't enterprise*		*All business*		*All economy*		*nonresidential business economy*	
Output[a]	9.9	10.6	11.7	11.7	8.9	10.9	9.86	10.18
TFI[b]	6.0 (100.0)	6.6 (100.0)	4.3	4.9	4.2	4.0 (100)	3.88 (100.0)	4.43 (100.0)
Labor	2.8 (65.7)	2.4 (60.5)	...	...	2.1	.9 (69)	2.30 (76.3)	1.75 (73.6)
Capital & Land	3.0 (34.3)	4.4 (39.5)	...	...	2.1	3.1 (32)	1.58 (23.7)	2.08 (26.4)
TFP[c]	3.9	4.0	7.4	6.8	4.7	6.9	5.98	5.75
S & E Stock[e]	9.5	13.4			7.0	11.3	7.32	11.23

Sources: Jorgenson and Ezaki (1973), Kosobud (1974), Kanamori (1972), Denison and Chung (1976).

[a] Output is measured by GNE in Jorgenson and Ezaki, Business GDP in Kosobud, national income in Kanamori, and nonresidential business national income in Denison and Chung (1976).

[b] Total factor input.

[c] Total factor productivity.

[d] Weights.

[e] Nonresidential structure and equipment stock. Jorgenson and Ezaki adjust capital growth for "quality change," which is the difference in growth between the Divisia and fixed-weight index. The figures above are for quantity (fixed-weight). For capital and land together, the growth rates are as follows: 1953–61—quantity 4.8 percent, quality 3.9 percent; 1961–68—quantity 8.8 percent, quality 2.4 percent.

[f] The contributions of capital are subdivided as follows:

	1953–61	*(w)*	*1961–71*	*(w)*
Inventories	.75	(6.6)	1.00	(7.6)
Structures & equipment (nonresidential)	.83	(13.0)	1.68	(14.0)
Land	0	(4.1)	0	(4.8)

See Denison and Chung (1976), Table 4–6, p. 38.

stock). We divide the private economy into three major sectors—agriculture (including forestry and fisheries), manufacturing, and the rest (excluding public administration). Table 3 gives average growth rates for the two decades respectively.[13]

It is clear that all sectors behaved quite similarly. Capital deepening is evident in all sectors, the rate of increase more or less doubling in the 1960s over the 1950s. The capital-output ratio declined in the 1950s (except in agriculture) and increased in the 1960s. Apparently, all sectors must have been subject to the same set of influences. It is therefore safe to work with the aggregate economy.

C. Shifts of the Macro Production Function

For the time being, let us pretend that there was a well-behaved neoclassical production function for the private economy as a whole and see what the data tell us about its shifts over time. Table 4 summarizes the data for the two decades. The first significant feature that we can detect is the substantial decrease in labor's share of GDP through the 1950s and its subsequent stability. Another feature is that, while average labor productivity grew at roughly the same rate in the two decades, the capital-output ratio fell in the first decade and rose in the second decade. Thus, if we start from the same point A in Figure 3, the economy shifted to a new position B_1 in the 1950s and to another B_2 in the 1960s. In the former, both capital and labor improved their productivities. The unit isoquant passing through B_1 is a uniform move toward the origin, suggesting that technical change was more of the Hicks-neutral type in the 1950s.[14] In the 1960s, however, while labor improved productivity, capital reduced productivity. This change, together with the particular change in labor's share, seems to indicate the labor-augmenting type of technical change. TFP growth is supposed to compare shifts of the production function in the two situations. Graphically speaking, we draw the ray OA and verify its intersection points with the two unit isoquants at A_1 and A_2. OA_i / OA is equal to $1 - TFP$ growth. Table 4 suggests that A_1 and A_2 are virtually coincident. Then, it visually follows from the diagram that the production function must have shifted much faster in the 1950s than in the 1960s above the ray OA. Our data hardly seem to support my thesis of accelerated technical progress! But before jumping to a hasty conclusion, let us examine if the capital input is being measured correctly.

[13] The first sub-period begins with 1954 since the GDP series are available only from that year in the source cited.

[14] As the capital-labor ratio rose and labor's share fell, the elasticity of substitution must have been above unity.

IV. HOW SHOULD THE FIXED CAPITAL INPUT BE MEASURED?

The thorniest problem in growth accounting is the measurement of the capital input. Theoretically, the capital input is represented by flows of capital services. But they are not directly measured. Furthermore, even if we can measure machine-hours like man-hours, there is the practical problem of how to weight services of diverse equipment and structure. Therefore, we usually take the capital stock estimates and assume that capital services are in proportion to stocks with adjustment made for capacity utilization. Needless to say, this expediency does not solve our problem because it implicitly assumes that capital goods are homogeneous. The only concession that is made for heterogeneity of capital is the allowance for depreciation by physical wear and tear that lowers physical productivity of machines. Usually, the depreciation goes further to cover economic depreciation, *i.e.*, gradual obsolescences (Denison, 1957). By taking the net or depreciated capital stock, we hope that the capital input is more adequately approximated. However, the hope may be a mere wishful thinking completely alien from reality. We shall consider this critical point in more detail later.

If one wishes to avoid this sort of arbitrary imputation, one must stick with the gross capital stock, which is the undepreciated replacement value of all capital goods in existence. Practical difficulties aside, this stock can be estimated by census-taking. However, the gross capital stock (*GKS*) does not measure capital services if capital goods really depreciate. In a steady-state world, this ignorance makes no difference since both the gross and net stock grow at the same rate. This, however, is a cold comfort since the world is never in a steady state.

In Japan, the Economic Planning Agency (EPA) embarked on an ambitious task of census-taking of national wealth supposedly at the five-year interval since 1955.[15] Its estimates, however, were limited to net capital stocks (*NKS*) except for the latest 1970 census which published both *GKS* and *NKS*. Net stocks were estimated by applying the table of remaining values based on predetermined depreciation rates for individual types of assets at replacement cost. However, for constructing its time-series of nonresidential business fixed capital stocks, the Economic Planning Agency reconverted the 1955 Census-of-Wealth benchmarks to the gross basis. It extends the *GKS* estimates from these benchmarks according to the following formula:

$$\Delta GKS_t = I_t - (R_t - S_t)$$

where I = GFCF, R = removals of assets, and S = purchases of used assets. $R - S$ is therefore net removals. The EPA estimates R and S from

[15] There are censuses of 1955, 1960, 1965 (very partial), and 1970, whose results have been published.

TABLE 3

AVERAGE ANNUAL GROWTH RATES

(percent)

Sector	*period*	*Y*	*N*	*GKS*	*Y/N*	*Y/GKS*	*GKS/N*
Private	1954–61	10.99	1.80	8.02	9.19	2.97	6.22
economy	1961–71	9.99	1.29	11.97	8.70	−1.98	10.68
Agricul-	1954–61	4.56	−2.14	6.60	6.70	−2.04	8.74
ture	1961–71	1.27	−4.90	9.90	6.17	−8.63	14.80
Manufac-	1954–61	16.10	4.46	11.47	11.64	4.63	7.01
turing	1961–71	12.05	3.17	13.95	8.88	−1.90	10.78
Others	1954–61	10.88	3.81	6.08	7.07	4.80	2.27
	1961–71	10.40	3.19	10.78	7.21	−0.38	7.59

Sources:
Y, N: Kogane *et al.* (1975); *GKS*: EPA (1977).

Symbols:
Y = GDP at factor cost in 1970 yen
N = Persons engaged in production
GKS = Gross nonresidential fixed business stock in 1970 yen (including provisional construction accounts).

TABLE 4

FACTOR SHARES AND GROWTH RATES, PRIVATE ECONOMY

(percent)

	1954–61	*1961–71*	*1952*	*1961*	*1971*
Factor Shares					
Labor	67.24	63.24	72.73	62.79	62.41
Capital					
Inventories	6.11	6.53	4.96	7.34	7.10
Fixed equipment and structures	22.51	26.12	17.33	26.17	26.57
Land	4.14	4.11	4.98	3.70	3.92

TABLE 4 *(Continued)*

Growth Rates		
GDP (*Y*)	10.99	9.99
Quasi-rents (*R*)[a]	23.95	25.70
Man-hours (*L*)	2.37	.90
Gross fixed business capital stock (*GKS*)	8.02	11.97
Inventories	11.82	12.72
Land	0	0
Capital and land (*TK*)[a1]	7.72	10.76
Y/L	8.62	9.09
Y/GKS	2.97	−1.98
R/GKS	15.93	13.73
Implicit deflator—*GFCF/GNP*[b]	.50	−3.10
Total factor productivity (*TFP*)[c]	5.32	5.49

Sources:

Denison and Chung (1976), Kogane *et al.* (1975), EPA (1974a).

Notes:

Factor shares are derived from Denison and Chung (1976) by adding back depreciation to Net National Income and returns to fixed business capital stock. Man-hours are the product of number of persons engaged (excluding public administration) given in Kogane *et al.* (1975) and hours of work given in Denison and Chung. GDP at factor cost is from Kogane *et al.* (1975), excluding public administration. Gross capital stock is from EPA (1975a). Inventories and land are from Denison and Chung (1976).

[a] Real quasirents are derived by the formula

$$g(R) = \frac{1}{1-a}\, g(Y) - \frac{a}{1-a}\, g(L)$$

where $g(x)$ is the growth rate of x and a is labor's share. For the explanation, *see* Sato (1977b). This entry is shown since it is free from the measurement problem of capital inputs.

[a1] Combined index of *GKS*, inventories, and land.

[b] On the basis of these deflators, the nominal *GKS/GDP* ratio was: 1.91 (1954), 1.63 (1961), 1.42 (1971).

[c] Computed by the following formula:

$$g(TFP) = (1-a)\,(g(R) - g(TK)).$$

quarterly corporate income statistics.[16]

If capital goods become less and less productive as they age, they are to be retired and scrapped as they reach certain age or vintage, say, T years. If the correspondence is exact, net removals at year t should be equal to GFCF at $t - T$. For the period under study, we find the correspondence for quinquennial averages shown in Table 5.

This table indicates that T remained roughly constant at 11 years, say, judging from the correspondence of net removals and GFCF. Net removals maintained a more or less stable share of *GKS*. The corresponding double-declining-balance depreciation ratio is 18 percent ($=2/T$). Apparently, the bookkeeping depreciation ratio was accelerated to catch up with the more realistic value.[17] Both the *NKS* and depreciation figures based on the latter are not appropriate for the purpose of growth accounting, though they are meaningful economic aggregates for other purposes, *e.g.*, business investment decisions. This means that national income figures are off. By lucky coincidence, however, bookkeeping depreciation and national income expanded at similar rates. Hence, the output growth rate is not affected much by the choice of the output concept. However, this does not apply to the *NKS*. By the same happy coincidence, the *NKS* and the *GKS* grew also at about the same rates (1954–61, 7.42 *vs.* 7.31 percent; 1961–71, 11.23 *vs* 11.77 percent).[18] But, then, it implies that capital services grew at the same rate as the *GKS*. Could this be true?

Let us see if the stability of T implies that technical change was also stable. The answer is not necessarily. It depends critically on which way technical change is being biased. To make the point clear, let us take the putty-clay model. I_v is GFCF in year v. For simplicity, assume no change in factor proportions over time (*i.e.*, its age). Its quasi-rents per unit of I_v at year t is given by

$$QR_v(t) = \frac{Y_v}{I_v}\left(1 - \left(\frac{w}{p}\right)_t \frac{L_v}{Y_v}\right)$$

where w/p is the real wage rate. Notation should be self-explanatory. The latest vintage has

$$QR_t(t) = \frac{Y_t}{I_t}\left(1 - \left(\frac{w}{p}\right)_t \frac{L_t}{Y_t}\right)$$

[16] *See* T. Noda *et al.* (1966) for the description of the method in detail.

[17] If the depreciation rate were as reported in the table, the economic life of fixed assets would have shortened from 28 years to 12 years. This much shortening can occur if and only if old vintages are becoming more and more uneconomical, that is, if technical change is accelerating. Hence, the *NKS* estimates are inconsistent with the finding of no acceleration of technical change.

[18] Hence, our use of the *GKS* makes little difference from that of the *NKS*. The growth rates are from Denison and Chung (1976).

TABLE 5

NET REMOVALS OF NONRESIDENTIAL FIXED BUSINESS ASSETS AND GFCF IN ALL PRIVATE INDUSTRIES

	Net removals GKS (mid-year) %	*Depreciation NKS (mid-year)* %	*Net removals (annual)*	*GFCF (annual)*
			billions of 1965 ¥	
1952–56	2.63	7.17	477	...
1957–61	2.23	9.35	537	553[a](1946–50)
1962–66	2.55	13.20	1001	992 (1951–55)
1967–71	2.95	16.78	1880	1940 (1956–60)

Source: Denison and Chung (1976), Table I-2.

[a] Crude estimate based on the discontinued national income series in EPA, *White Paper on National Income*, 1963 (1965).

The marginal vintage, *i.e.*, the oldest vintage that is to be retired, is determined by

$$\frac{Y_{t-T}}{L_{t-T}} = \left(\frac{w}{p}\right)_t,$$

provided that vintage labor productivity is monotonically increasing. T is constant if the current real wage rate and the average labor productivity (APL) of the marginal vintage grow at the same rate. Imagine that the APL of the current vintage is growing at a faster rate. For all the vintages in operation as a whole, the real wage rate is rising somewhat slower than the overall average of the vintage APLs. Labor's share in gross value added must be falling. This must be what happened in the 1950s. Since the capital-output ratio was falling, the apparent rate of return on capital (fixed) continued to rise. In the 1960s, on the other hand, both the real wage rate and the vintage APLs increased at the same rate and labor's share was kept stable. The capital-output ratio was rising. Hence, the apparent rate of return fell, but investment goods were becoming cheaper relative to GNP so that the rate of return in monetary terms did not decline.

If capital services are to be weighted in a way to reflect their economic obsolescences, then the weights should be given by

$$R_v(t) = QR_v(t)/QR_t(t),$$

which assigns the weight of unity to the newest vintage and the weight of zero to the oldest vintage. However, the weights are not necessarily monotonic since they depend not only on vintage labor productivities but also on vintage capital productivities. The latter changed the direction of

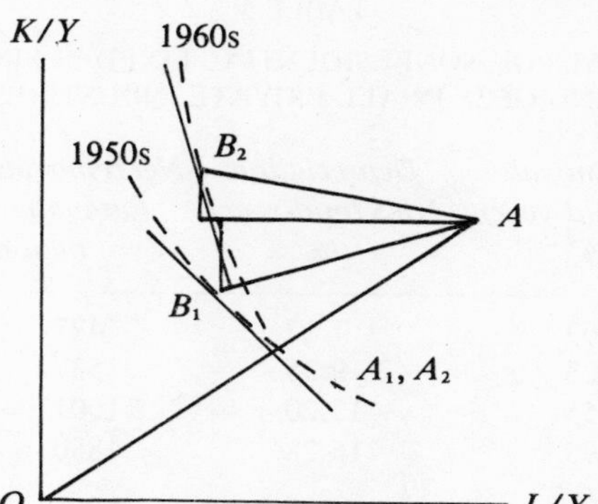

Figure 3

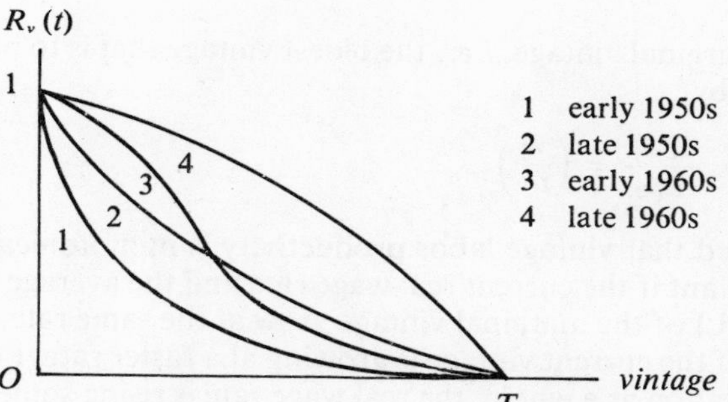

Figure 4

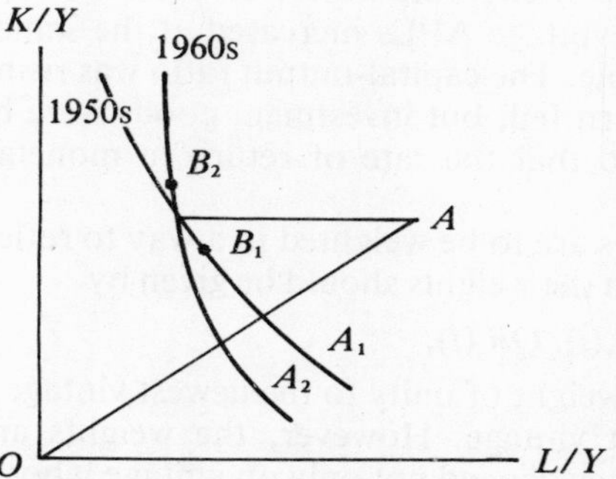

Figure 5

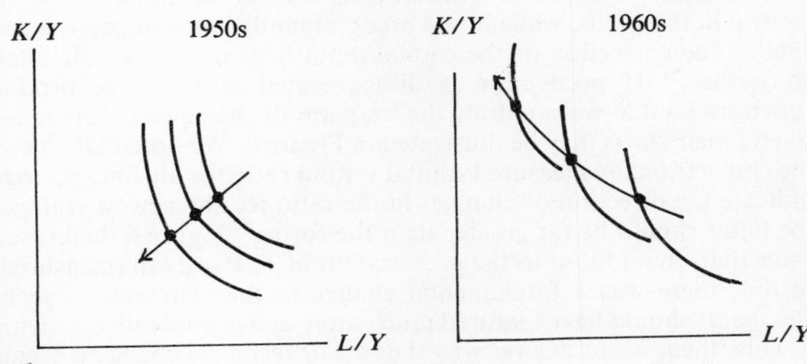

Figure 6

change at 1961. Considering the distribution of vintage productivities at each instant of time, we speculate that the weights altered their shape as illustrated in Figure 4. The shape for the early fifties is reinforced by the well-known existence of excess capacity, which made the contribution of idle capital stocks, presumably of old vintages, to be zero.[19]

The double-declining balance depreciation method presupposes a curve monotonically declining to the right and convex to the origin. As the depreciation is accelerated over time, the curve tends to be pushed toward the origin. In other words, a complete opposite from Figure 4 is assumed.

An important point is that, in transition like in the early 1960s, the weight system changes in such a way that capital services are no longer in proportion to the *GKS* or the *NKS*. It is not hard to see that the gradual elimination of excess capacity through the 1950s pushed the weight curve outward, making the capital input growth greater than measured by the standard method. On the other hand, in the transition of the 1960s, the capital input growth is overstated by the standard method and should be reduced. A numerical exercise indicates that the reduction can easily be more than 10 percent. It is probable that the capital-output ratio growth if measured by our way did not materially differ between the two decades. In Figure 3, B_1 is raised and B_2 lowered vertically toward the horizontal line through A. Figure 3 is thus revised to Figure 5. Now, OA_2 is shorter than OA_1 so that TFP growth was greater in the 1960s than in the

[19] *See* the MITI index of manufacturing capacity utilization (1953, 80.5; 1961, 90.3; 1971, 94.5). This index seems to be a good representation of the entire economy, judging from its correspondence with the Japan Economic Research Center (JERC) estimates of actual/potential GNP for the 1960s (periodically reported in JERC, *Quarterly Economic Forecast*).

1950s.[20]

We already conjectured that technical change was more or less Hicks-neutral in the 1950s, while it was predominantly labor-augmenting in the 1960s. The correction of the capital input does not materially alter this conjecture.[21] If production is disaggregated into vintage production functions so that we can draw the isoquant of the newest vintage successively, their shifts may be illustrated in Figure 6. We note that changes in the conventionally measured capital-output ratio for all vintages together indicate the direction of changes in the ratio for the newest vintage, but the latter should be far greater than the former. Figure 6 therefore indicates that, in addition to the acceleration in TFP growth (measured correctly), there was a fundamental change in the character of technical change. It should have required much more active innovative activities.

This, then, is our answer why the rate of technical change is seemingly insensitive or even falling from the 1950s to the 1960s, while there was every indication of intensified innovative activities. The latter was essential in altering the bias of technical change so as to make capital deepening easier.[22] This is how the GFCF/GNP ratio could rise so much without adversely affecting the rate of return to capital. If there was no change in the direction of technical progress, the economy could not have maintained its high rate of growth into the 1960s when labor growth substantially decelerated.

V. *WHY DID THE CHARACTER OF TECHNOLOGY CHANGE IN THE 1960s?*

We have argued that there was a fundamental change in the character of technological shifts at 1961. Why and how was this change induced to

[20] It is well known that the rate of technical change depends on the direction of measurement unless technical change is Hicks-neutral (Sato, 1976). In the present instance, the rate is less in the 1960s than in the 1950s when measured in the northwest direction. But the segments above *B*'s are unobserved, and this sort of measurement makes no sense.

[21] Of importance here is that the shift still remains more labor-augmenting in the 1960s than in the 1950s after this correction.

[22] Misgivings are expressed on the concept of factor substitution (*see* John R. Hicks, 1975; Joan Robinson, 1975). We can assume that all changes in factor proportions are technological changes rather than factor substitution. This view, however, does not necessarily eschew the concept of the *ex ante* production function.

Also, even with *ex post* fixed proportions for vintages, the sum total of vintages can generate a macro production function with which growth accounting can be applied. An important problem arises when the macro production function itself changes its shape since growth accounting implicitly assumes its invariance. Quasi-rent adjustment as discussed in this section can be regarded as a way to deal with this situation. These points are discussed in Sato (1975, 1977).

occur? We consider this point below.

Japan's rapid economic growth altered its industrial composition very rapidly. Labor moved out of agriculture *en masse*. The share of the primary sector in the economy's labor force declined from 40 percent in 1953 to 30 percent in 1960, to 20 percent in 1968, and 13 percent in 1974. A steady massive outflow of labor depopulated rural areas. However, in relative terms, the contraction was modest in the 1950s (−2 percent per annum from 1951 to 1961). The contraction accelerated since 1961 (−5 percent per annum), halving agricultural population in less than a decade and a half. In the 1950s, migrating workers were younger and unattached. In the 1960s, older farmers heading families started to leave their farm work. But through the period, acreage under cultivation remained stable. Land productivity would have fallen substantially if the loss in labor was not offset by increases in nonlabor inputs. In the 1950s, technological innovations in agriculture depended more on the chemical and biological types than on the mechanical and engineering types.[23] Coupled with the relatively modest contraction of the labor input, the pressure for capital deepening was still modest. The capital-output ratio slightly increased.

The situation changed dramatically from 1961 on. In the succeeding 10 years, the capital-output ratio doubled and the capital-labor ratio quadrupled. Mechanization spread through the successive introduction of new agricultural instruments and machines ranging from inexpensive paddy cultivators and weeders (1970 unit value: $12) to power cultivators and tractors ($270), farm trucks and three-wheeled trucks ($1500), and finally to combines ($1500). One can observe the familiar process of diffusion in the adoption of these machines, exhibiting the well-known logistic curve in every instance. Paddy cultivators and weeders were already near their saturation point by 1961. Other small instruments like power dusters and sprayers were still in the active diffusion process in the early 1960s. More expensive farm trucks approached the saturation point in the late 1960s. The latest craze has been combines.

Thus, in agriculture, technical change turned toward capital deepening via mechanization in the 1960s to compensate for the continuing depletion of young qualified labor as well as the diminishing importance of chemical and biological technology as a source of innovations. Following orthodox economic theory, one can emphasize changes in the factor price ratio that prompted capital deepening. The substitution of capital for labor, however, requires a variety of new capital goods and should be regarded as a result of shifts of the production function rather than movements along a given production function. These shifts required different sorts of innovative activities in the 1960s than in the 1950s.

[23] *See* Yujiro Hayami (1973) and Ohkawa and Rosovsky (1973, ch. 4).

Let us now turn to the nonprimary sector, particularly manufacturing. The mass exodus of labor from the farming sector enabled the former to maintain labor growth at 3 percent per annum in the 1960s (*see* Table 3). However, labor migrating out of agriculture became less and less of prime quality since older farmers had to be recruited in the 1960s. The decline in the labor quality had to be compensated for by more capital deepening. The strong demand for qualified labor led to excess demand for young workers entering the labor markets after graduation from high school. It tended to raise the entering wages faster than the average rate in the 1960s. Since both large and small firms compete for young trainable labor, this wage pressure permeated through the nonprimary economy. It was thus highly desirable to develop more capital-using techniques.

In contrast, no fundamental change was required in the nature of technology in the 1950s. Technical change that improved both capital and labor productivity was more amenable to existing productive facilities. Initially, there was a process of regaining productive efficiencies lost in the aftermath of the war. Technological developments that had been accumulated during the war years provided the backlog of unutilized indigenous techniques ("improvement engineering" as Ohkawa and Rosovsky [1973] call them). Technology imports grew by leaps and bounds, but imported techniques were of old types (prewar vintages). Old capital goods could be modified to meet new needs. This is clear in the significant position of transfers of second-hand equipment from large firms to small firms through the 1950s. While large firms paying higher wages found it economical to replace old equipment with new as real wages continued to rise, small firms could find the former's discards still profitable since their wage rates were as little as 1/3 of the former. Purchases of used assets were a significant proportion of small firms' investment expenditure.[24] But, here again, the situation changed dramatically around 1960. The ratio of purchases of used assets to *new* assets is estimated by the Economic Planning Agency as shown in Table 6. Since used assets were probably marked down considerably below what they were really worth, the ratio underestimates the real importance of transfers of used assets in the 1950s. Of importance to us is the fact that even small firms stopped buying old capital goods from large firms. While the process was not abrupt, we can take 1960 as the dividing point.[25] This devel-

[24] The ratio of used to new asset purchases was 2/3 in manufacturing establishments with less than 20 employees and 4 percent in those employing more than 1000 workers in 1955 (MITI, *Census of Manufactures*).

[25] The ratio reported in Table 6 registered a sharp reduction at 1960 in all industries except for manufacturing and construction. In manufacturing, the ratio gradually fell through the 1950s in smaller establishments, finally reaching the stationary level by 1961 (for those with 10–19 workers, the level is about 12 percent).

TABLE 6

THE RATIO OF PURCHASES OF USED ASSETS TO NEW FIXED INVESTMENT, ALL CORPORATE FIRMS

	All industry %	*Manufacturing* %
1955–59	10.5	5.2
1960–64	2.8	2.4
1965–69	3.4	2.6
1970–72	3.0	1.9

Source: EPA (1974).

opment could have been predicted. Small firms were suffering from difficulty of recruiting young workers. They had to staff their employees with older men who were not too readily adaptable. They had to pay wages competitive with large firms at least for younger employees. Hence, the need for modernization was particularly acute for smaller firms. They could no longer rely on techniques of older vintages. They could survive only by increasing their innovative activities. Firms, both large and small, had to become more aggressive in the search for new techniques to cope with the new developments in the economy. Technology imports were maintained at a high level as a ready-made source of new techniques and new products though their *growth* could not be maintained at the level of the 1950s.[26] Indigenous activities in R&D also kept pace mainly at the initiative of private firms,[27] though the interchange of indigenous techniques seems to have been on a limited scale.[28]

All these discussions point to a fundamental change in innovative intensity from the 1950s to the 1960s. The change is something that cannot be measured by total factor productivity growth.

[26] A little over half of technology imports were from the U.S., one-third from five European countries (France, West Germany, the Netherlands, Switzerland, and the U.K.), and 10 percent from the rest of the world.

[27] In Japan, private firms share about 60 percent of R&D expenditure (including government and universities), while the ratio was 30 to 40 percent in the West.

[28] *See* Science and Technology Agency, *White Paper on Science and Technology*, various issues.

VI. DID JAPAN CLOSE THE TECHNOLOGY GAP?

I now examine my second hypothesis, *i.e.*, that Japan caught up with the West in technology by the end of the 1960s and virtually closed the technology gap.[29] By its nature, the comparison is readily meaningful only in commodity production, particularly manufacturing.[30] For the latter, statistical evidence indicates that the hypothesis is correct.

Kenzo Yukizawa[31] compares gross physical labor productivities between the United States and Japan for 60 manufactured commodities on the basis of both countries' Censuses of Manufactures. While productivity differentials were widely scattered, the average stood at Japan/U.S. = 0.32 in 1958, 0.34 in 1963, 0.50 in 1967, and 0.67 in 1972. As is well known, the productivity differential is greater between large and small firms in Japan than in the United States. Therefore, when the comparison is limited to large firms, the Japan/U.S. productivity differential must have been almost unity by 1972.[32]

A corroborating piece of evidence is available in a 1974 questionnaire survey conducted by the Science and Technology Agency on R&D activities of 488 private firms (with equities exceeding one billion yen) asking for firms' self-assessment of their levels of basic technology *vis-à-vis* Western counterparts. The proportions of replies in the total sample broke down as follows: more advanced, 12 percent; comparable, 59 percent; less advanced, 23 percent; very much behind, 6 percent.[33]

The same point is revealed by other statistics. In the composition of technology imports, as reported by the Science and Technology Agency, the percentage of completely new techniques (*i.e.*, with no similar techniques imported previously) decreased from 70 in 1961 to 45 in 1965, and to 26 in 1970 of all Class A technology imports (by number of cases).[34] Also, the percentage of imported techniques, which are competitive with indigenous techniques, rose from 45 in 1965 to 69 in 1970 of total Class A technology imports.[35]

[29] The same view is expressed in Ohkawa and Rosovsky (1973), p. 237.

[30] Recall that technology imports are virtually all in manufacturing.

[31] Kenzo Yukizawa, "Relative Productivity of Labor in American and Japanese Industry and Market Size, 1958–1972," in this conference volume.

[32] Comparing values per person engaged in manufacturing establishments with 1,000 or more workers with those in all establishments, we find the following for 1972: Japan—gross value, 1.70; value added, 1.61; U.S.—gross value, 1.19; value added, 1.22. The correction on this basis sets the Japan/U.S. differential at .96.

[33] Science and Technology Agency, *White Paper on Science and Technology*, 1975, p. 217.

[34] Economic Planning Agency, *Economic Survey*, 1966–67, p. 67 and Science and Technology Agency, *Indicators of Science and Technology*, 1972, pp. 126–27.

[35] Science and Technology Agency, *Indicators of Science and Technology*, 1972, pp. 126–27.

These statistics all support the view that Japan closed the technology gap by the beginning of the 1970s.[36] There seems to be no doubt about it.[37]

VII. QUO VADIS, JAPAN?

It is generally agreed that Japan owed a great deal to technology imports in pulling it out of the rubble of the war devastation and remaking it into an economic superpower in the short span of a quarter century. Technology imports were cheap when compared with their payoff. Japan was enabled to economize on the public support of R&D. Japan remained technologically imitative and adaptive and did not lack in the absorptive capacity with its highly trained skilled workers and engineers. But, by the end of the sixties, Japan had absorbed as much as it could readily absorb.

However, closing the gap as an imitator and adaptor is far from closing the gap as a creator and initiator.[38] Japan's weakness in the latter capacity implies that it has a great deal of distance to cover before it can attain the position of the world's leading supplier of technology.

[36] For the economy as a whole, GDP per capita of the labor force in 1970 is estimated at 50 for Japan with the U.S. at 100 (after correcting for differences in purchasing power parity) according to Denison and Chung (1976, p. 9). However, value-added productivity can be misleading in international comparisons. In manufacturing, the ratio of value added to gross value of production is lower in Japan (.35) than in the U.S. (.47). If Yukizawa's estimate of the Japan/U.S. differential in 1972 is corrected for this difference, it reduces from .67 to .50, a number more in line with the economy-wide differential. A reason for the low value added/gross value ratio in Japan is the fact that Japan pays much less in wages than the U.S. as gross productivity differentials go. The latter is more appropriate for our comparison.

With the production-function approach, Jorgenson and Nishimizu conclude in their paper in this conference volume that Japan and the U.S. were on the same isoquant in 1973, implying that the technology gap was completely closed in that sense. This finding, while it reinforces our position, does not mean that Japan can approach the U.S. merely by increasing capital intensity. This process itself requires technical change.

[37] Blumenthal (1976) challenges this thesis: If the gap were narrowing, how could technology imports expand so much? The 1972 value is more than the total for 1949–59. The gap must have been widening. Thus, he asserts that "the fallacy of the catching-up argument is in the implicit assumption that the technological level of the source countries is constant" (p. 251). As the latter has been expanding, Japan's technology imports could increase without narrowing the gap. His argument, however, can be valid if and only if the source is growing faster than Japan's absorption. This is unlikely. In fact, the deflection in the growth of Japan's technology imports at 1961 suggests that the source must have become smaller relative to Japan's needs.

[38] Blumenthal (1976) actually has this in mind when he says that Japan has not closed the technology gap.

TABLE 7

APPLICATIONS FOR PATENTS, 1964 AND 1974

(percent)

Nationality of applicants	(1) *Applications to own country*[a]		(2) *Applications to foreign countries*[a]		(3) *= (2)/(1)*	
	1964	*1974*	*1964*	*1974*	*1964*	*1974*
Japan	234	190	9	33	5.2	27.5
U.S.	100	100	100	100	130.2	158.7
West Germany	117	48	50	66	55.8	219.0
U.K.	36	32	34	28	125.1	140.6
France	25	20	19	22	100.2	179.0
Switzerland	8	8	19	21	326.1	400.7
Netherlands	3	3	11	10	464.0	485.7
Canada	3	3	5	5	231.1	282.1

Source:

Science and Technology Agency, *White Paper on Science and Technology*, 1966, p. 259, and *Indicators of Science and Technology*, 1977, p. 19.

[a] U.S. = 100.

This point is nowhere more evident in the balance of technology trade. In 1971–73, the ratio of technology exports to imports was the U.S. (9.30), the U.K. (1.04), France (1.02), West Germany (.38), and Japan (.13),[39] even though Japan's modest position is a vast improvement over its past.[40]

The same point is observed also in patent statistics. Two-way matrices of patent applications by applicants' nationality and countries applied yield the information in Table 7. If the number means anything, there is apparently no dearth of talent in Japan. There are more than twice as many Japanese applicants as Americans—but with a significant difference. The Japanese are extremely inward-looking. Even though they

[39] Science and Technology Agency, *Indicators of Science and Technology*, 1977, pp. 20–21.

[40] The ratio was: 1956–59, 1.0 percent; 1960–64, 5.2 percent; 1965–69, 10.9 percent; 1970–74, 13.4 percent; 1975, 22.6 percent. Another indicator is the contribution to technical assistance to LDCs. As percent of GNP, it was 0.014 percent in Japan, and 0.072 percent in DAC countries (average) in 1973. *See* Science and Technology Agency, *Indicators of Science and Technology*, 1977, p. 19.

have been increasing their applications with foreign patent offices in recent years, the proportion is in no way comparable with the Americans and the Europeans.[41] There is not a shortage of researchers in Japan.[42] Nonetheless, Japan is behind other advanced nations in the ability to create new technology. The Science and Technology Agency continues to lament this feature for the past several years in its annual *White Papers.* Though Merton J. Peck and Shuji Tamura (1976) speculate that "Japan might conveniently be first in being second—that is, the most successful follower" (p. 583) by continuing to rely on imported technology in at least the next decade, it should be difficult for Japan to maintain its "followership" and yet to keep up economic growth at a reasonable level. Even limiting ourselves to Japan's near future, it is difficult to predict its growth performance.[43] It critically depends on Japan's technological resourcefulness. As is clear from the transition from the 1950s to the 1960s, the character of innovations determines not only total factor productivity growth but also the investment requirement. While economic growth per se may be no longer the objective, the problem is that a free-market economy either grows or stagnates.

In the longer run, we note that Japan has yet far to go in international exchange of technology. This, however, touches upon one of the most fundamental cultural traits of the Japanese, namely, its insularity. The Japanese have been an importer, imitator, and adaptor for so many centuries that they have maintained highly sensitive antennas that receive new scientific developments abroad to satisfy their voracious appetites for foreign cultures. But their reciprocation has been of a limited scope so far. The time now has come to change all this, as the technology gap has finally been closed. Japan has to move out of the familiar path of importing foreign techniques from advanced countries and of exporting goods manufactured from them to LDCs. Whether Japan can succeed in transforming itself into a technological leader or not will determine Japan's future.[44] Thus, the coming decades will be the watershed in

[41] One reason, the Science and Technology Agency notes (*White Paper*, 1965, p. 179), is that Japanese inventions (more heavily on utility models) were internationally unattractive, and Japanese firms must have felt that it was not worth bothering with going through formalities of foreign applications. It also suggests inferior quality of Japanese patents.

[42] Researchers per 10,000 of population were 36 (U.S.S.R., 1974), 25 (U.S., 1974), 23 (Japan, 1975), 19 (West Germany, 1974), 11 (France, 1971), and 8 (U.K., 1969). *See* Science and Technology Agency, *Indicators of Science and Technology*, 1977, p. 3.

[43] Innovative indicators in Figure 1 seem to have hit the ceiling in 1973. The inducement to invest has thereby been weakened and the economy faces the difficult task of balancing saving and investment.

[44] At the Congress, Harry T. Oshima noted that Japan still has a great deal of work cleaning up its own backyard, *i.e.*, its agriculture and small-scale enterprises, so that it need not go outside. However, can a high-income country like Japan afford to remain so insular?

Japan's modern history.

REFERENCES

Blumenthal, Tuvia. 1976. "Japan's Technological Strategy," *Journal of Development Economics*, vol. 3, September, pp. 245–55.

Denison, Edward F. 1957. "Theoretical Aspects of Quality Change, Capital Consumption, and Net Capital Formation," in *Problems of Capital Formation*. New York: National Bureau of Economic Research, pp. 215–61.

———. 1967. *Why Growth Rates Differ*. Washington, D.C.: Brookings Institution.

——— and Chung, William K. 1976. *How Japan's Economy Grew So Fast, The Sources of Postwar Expansion*. Washington, D.C.: Brookings Institution.

Ezaki, Mitsuo and Jorgenson, Dale W. 1973. "Macro Seisansei Henka no Sokutei, 1951–1968," in *Nihon Keizai no Choki Bunseki—Seicho, Kozo, Hado*. Edited by Kazushi Ohkawa and Yujiro Hayami. Tokyo: Nihon Keizai Shimbun-Sha, pp. 87–127.

Griliches, Zvi. 1973. "Research Expenditures and Growth Accounting," in *Science and Technology in Economic Growth*. Edited by B.R. Williams. London: Macmillan, pp. 59–83.

Hayami, Yujiro. 1975. *A Century of Agricultural Growth in Japan*. Tokyo: University of Tokyo Press.

Hicks, John R. 1975. "Revival of Political Economy: The Old and the New," *Economic Record,* vol. 51, September, pp. 365–67.

Japan, Economic Planning Agency, Economic Research Institute. 1974. *Fixed Nonresidential Business Capital in Japan* (June 1954–March 1974). Mimeo., December.

———. 1977. *Private Business Fixed Capital Stock, 1955–1975*. March.

Jorgenson, Dale W. and Ezaki, Mitsuo. 1973. "The Measurement of Macroeconomic Performance in Japan, 1951–1968," in *Economic Growth: The Japanese Experience since the Meiji Era*. Volume I. Edited by Kazushi Ohkawa and Yujiro Hayami. Tokyo: Japan Economic Research Center, pp. 286–361.

Kanamori, Hisao. 1972. "What Accounts for Japan's High Rate of Growth?" *Review of Income and Wealth*, vol. 18, June, pp. 155–72.

Kogane, Y., *et al.* 1975. "Seisan Butsuryo o kiso to shita Sangyo betsu Jisshitsu Kokunai Soseisan no Suikei ni kansuru Kenkyu," *Keizai Bunseki*, no. 57, August.

Kosobud, Richard. 1974. "Measured Productivity Growth in Japan, 1952–1968," *Japanese Economic Studies*, vol. 2, Spring, pp. 30–118.

Noda, T., *et al.* 1966. "Soshihon Sutokku no Suikei, Minkan Kigyo, 1952–1964," *Keizai Bunseki*, no. 17, March.

Ohkawa, Kazushi and Rosovsky, Henry. 1973. *Japanese Economic Growth*. Stanford, Calif.: Stanford University Press.

Patrick, Hugh and Rosovsky, Henry, eds. 1976. *Asia's New Giant, How the Japanese Economy Works*. Washington, D.C.: Brookings Institution.

Peck, Merton J. with the collaboration of Tamura, Shuji. 1976. "Technology," in Patrick and Rosovsky (1976), chap. 8, pp. 525–85.

Robinson, Joan. 1975. "The Unimportance of Reswitching," *Quarterly Journal of Economics*, vol. 89, February, pp. 32–39.

Sato, Kazuo. 1971. "Growth and Technical Change in Japan's Nonprimary Economy, 1930–1967," *The Economic Studies Quarterly*, vol. 22, April, pp. 38–54 (in Japanese); *Japanese Economic Studies*, vol. 1, Summer, 1973, pp. 63–103 (in English).

———. 1975. *Production Functions and Aggregation*. Amsterdam: North-Holland.

———. 1976. "The Meaning and Measurement of the Real Value Added Index," *Review of Economics and Statistics*, vol. 58, November, pp. 434–42.

———. 1977. "Theoretical Issues in Production Accounting," to appear in Kazushi Ohkawa, ed., the conference proceedings volume of the Second Asian Regional Conference on Income and Wealth.

Comments

Harry T. Oshima (U.S.A.)

I find it very puzzling that Professor Sato (and in the paper by Dale Jorgenson and Mieko Nishimizu of this session) identify the residual (or total factor productivity) with technical progress. It may be that I have been working too long in Southeast Asia where I regard institutional factors as the major constraint in the introduction, use and spread of new technologies. But I thought that the detailed growth accounting procedures of Dennison have shown that many factors are involved in total factor productivity. By taking only the private sector, the government sector is left out, but the government sector is the main sector acting as the change agent in institutions so that the omission of institutional changes is a serious matter. In this note, I would like to speculate on the importance of institutional changes in Japanese growth, which in their interrelations with technical changes, labor, and capital may account for a large part of the residual. Incidentally, several other papers in this session do dwell heavily on matters such as differential structure (Ohkawa), industrial relations (Bronfenbrenner), industrial organizations (Imai and Uekusa), welfare institutions (Kanamori) in explaining Japanese growth. I will focus on the growth of the small-unit sector of Japan where institutional changes have large impact on the residuals (defined to be in Professor Sato's paper as total factor productivity).

The small-units or small enterprises are especially important in the primary and tertiary industries which in 1950 engaged 60% of the total labor force. Professor Sato's Figure 1 depicting the growth of patent registrations, technological imports, R&D expenditures, etc. seems to me to pertain only to the larger enterprises (defined to be those employing 30 workers or more). For all of nonagricultural industries even in 1963, firms with less than 30 workers employed 53% of the labor force in nonagriculture and this situation was virtually unchanged in 1972 (52%).

I speculate that institutional changes (which make possible the efficient use of labor, capital and technologies) slowed down in the 1960s compared to the 1950s. The late 1940s and the 1950s were the formative years for postwar Japan from an institutional point of view. Old institutions were modified and new ones formed, all of which went into the institutional framework within which not only economic activities must take place but also social, political, cultural and other activities. Vast transformations took place which I feel had extensive impact on output and especially on total factor productivity as these are not included in the growth of the *quantity* of labor and capital. These influences were largely felt in the late 1950s and the early 1960s, so that by the late 1960s the *growth* in the residuals as a result of institutional changes was minimal and this may account for the fall in the residuals.

To get a notion of the impact of institutional changes on economic efficiency, let me give a few illustrations, since these cannot be measured even roughly. In agriculture, starting with the land reform in the 1940s, rural institutions were democratized so that the cooperatives and other farmers' associations began to function more and more for the large numbers of *small* peasants, and not mainly for the large farmers and landlords. Agricultural and home extensions, education, health, irrigation and electricity became more widely available to the smaller farmers. The extensive system of credit, purchasing and marketing cooperatives brought economies of scale to very much larger groups of farmers. All this meant that the *opportunities* to increase agricultural incomes became widely available to the rural population, compared to the prewar decades. The democratizing movement spread to the farm housewives whose capabilities improved. It is difficult to see how much of the costs of these changes could be included in the growth rate of labor and capital, especially since the government sector is excluded. The success of rural development in Japan during the 1950s was impressive, especially if one were to compare with rural development programs anywhere in Asia.

Or take the matter of management and industrial relations. The Japanese model which evolved in the 1950s based on past experience proved to be most appropriate for nations in the earlier stages of growth where skilled manpower for modern industries is always scarce and is the major bottleneck to rapid growth. I am not speaking here about formally educated manpower, which is easy to develop in a short time through schools and colleges. I refer to the training of semiskilled, skilled, technical, and professional manpower. The system of permanent commitment, seniority, bonus and retirement payments, participatory and consensus decision-making and so on are intended to develop trained manpower quickly and efficiently through extensive in-service training in the early stages of development where vocational schools are inadequate and people are too poor to pay for good schools. The Japanese system is intended not only to retain the trained manpower but to win the loyalty, devotion, and diligence of those trained by the firm and to develop their initiative, innovativeness, and willingness to work together. I believe that most of the impact of this unique model of management was already part of the residual during the first part of the 1960s.

Finally, take the vast network of mass communication which largely developed during the 1950s, in particular the system of TV. I am told by mass communication specialists that Japan patterned its educational TV on the BBC of England around early 1950; but by the middle of the 1960s, NHK had far outstripped the BBC by nearly every criteria for evaluating educational TV. If one looks at Japanese TV programs solely from the point of output and the residual, mass communication experts consider it as one of the world's best system of continuing education (*i.e.*, academic, vocational, social, cultural, and political). But by the late 1960s, perhaps because of affluence, the entertainment programs seem to be growing more rapidly than the educational.

Let me therefore conclude that the residual contains not only the benefits from technological changes but many others. I have mentioned only a few which I assume may be important in explaining the failure of the residual to grow faster in the 1960s over the 1950s: the fuller utilization of labor per unit of time and of capital capacity as full employment was reached in the late 1950s and, more important, the vast transformation of institutions to meet the new circumstances of the first decade of postwar Japan. In the latter, I have mentioned only three which seem to have direct impact on the residual, rural institutions, management practices, and mass communications. But others better informed of Japan may be able to list others, such as the improved functioning of the bureaucracy, the efficiency in household activities, and so on.

I agree with Professor Sato that there are many limitations in growth accounting, but I am less concerned than he because I do not interpret it as a measure of technological change. Since I think of the residual as a much better measure of efficiency than output per worker, I believe it is a valuable technique. But like all techniques, the limitations are many and they must be thought of as a first step in a series of steps in the analysis of various problems.

Thus with regard to the last point of Professor Sato, I find it difficult to agree, even though he may be right, that "Japan has closed the technology gap."[1] It does not follow from this that Japan's major task must now be to become the creator of new technology, to become the "technological leader." As he points out, the small unit sector of Japan is still very large, perhaps employing about one-half or more of the labor force. As Professor Sato points out, it is only in the large-enterprise sectors that Japan's productivity can match the best in the world; but what of the *technology* of the small-unit sector?

Let me take a few minutes to argue the view that the transformation of the small sector into bigger units is even more important for Japan than trying to become a technological leader. Let's take agriculture, first. Small farms comprise almost the entirety of Japanese agriculture, and as Keizo Tsuchiya points out in his recent book, *Productivity and Technological Progress in Japanese Agriculture* (Tokyo, 1976), output per worker is low and cannot be increased much more with *small-machine technology*. To convert Japanese agriculture to large-scale technology of the Western type, there must be consolidations of small farms into large farms plus heavy expenditures by the government for irrigation and land improvements; otherwise output per worker cannot rise to Western levels. And the major constraint here is not so much new technology or capital, but institutional: how to get farmers to agree to land consolidation.

The *S* (service) sector in Japan measured by employment is as large as that of the U.K., Australia and Canada, about 40% in 1970. The subsector of *S*, commerce, is the largest in the world, equaling the 22% of the U.S. and its growth rate of employment between 1960 and 1970 was the highest, 4% compared to the 2% for the U.S. (Data from ILO, *Yearbook of Labour Statistics.* 1972). The S sector in Japan is dominated by small units with meager sales and operated by family labor, and by Western standards, its efficiency per worker is very low, as in agriculture. The major obstacles to conversion to Western levels are also institutional, not technological.

The third small-unit sector using labor-intensive technology is the secondary sector, especially in construction and manufacturing; and though the proportion of the labor force engaged is not as large as in the other two sectors, it is still substantial, perhaps one-fifth to one-fourth.

It seems to me that this dualism or gap between the large-scale and small-scale sectors must be lessened substantially soon, if Japan's economy is to continue to grow in a sustained fashion. Instead of widening the gap by improving the technology of the already highly productive large-scale sectors, the differences in

[1] In what sense is this gap defined? Is it the gap of certain industries or all industries of the economy?

output per worker between the two sectors should be narrowed, if the distribution of income and expenditures is not to worsen, if the gap in production costs between the two sectors is not to widen, and if labor is to be released in large numbers for the expansion of the large-scale sector.

But the transformation of the small-scale and labor-intensive sector is not going to be an easy task, as it is a product of monsoon Asian economy, whose structural and institutional characteristics are so different from the Western economy. This transformation to large-scale, Western type of economy is necessary if per capita incomes are to rise to the highest levels comparable to the West.

Elsewhere[2] I have attempted to show that the great densities of monsoon Asia (the so-called "teeming millions") originate in the historical evolution of rice-paddy agriculture, which requires enormous amounts of labor during planting—because of the need to transplant in order to get high yields—and during harvesting because of the use of small knives and sickles, rather than the scythe, as in wheat agriculture. The monsoon winds concentrate rain only in certain portions of the year, so that during these slack periods the peasants must seek employment in nearby industries, which because of the seasonality must be small and labor-intensive. And these small units of agricultural and industrial production entail small units of distribution channels, so that the tiny farms in monsoon Asia (most of them no larger than 2 hectares in most parts of monsoon Asia) are the basic source of small industry and commerce. Over the centuries, this type of economy has developed its own set of structures, social values, and institutions, some of which are worth retaining even in large-scale economies, but many of which must change if large-scale technologies are to be introduced.

I believe that Japan's great contribution in the course of the past century and in particular during the postwar quarter century was not only the impressive efficiency of several of its large-scale industries, but perhaps more important, to open the possibilities and to point to the directions on which Asian economies, caught in the trap of centuries of low incomes by the great densities, can now move out of poverty to higher levels of per capita incomes. This took decades of the most innovative technological progress in agriculture, starting with the short-stalk, dwarf variety of rice in the 1920s and going along with this progress the innovations in rural institutions, all of this accelerating in the first decade and half of the postwar period. The upshot was that rice yields rose to six tons per hectare while at the same time releasing large numbers of the best manpower to urban industries. (Taiwan and South Korea appear to be moving in the same directions Japan did in the 1950s, and now Southeast Asia may be.)

But the 1960s witnessed the difficulties of converting the small-scale sectors into large sectors, so that severe labor shortages occurred, as Professor Sato notes (p. 21). The obstacles were largely institutional—the refusal of farmers to consolidate their holdings or to form larger operating units, the strength of traditional organizations, structures, and arrangements in the retail and personal services sectors, the indigenous tastes, consuming patterns, and ways of living, and so on. Perhaps one can speculate that the failure of the residual to accelerate may have been due to the slowing down of institutional changes in the 1960s compared to the 1950s, and that if the residual is to grow more rapidly again, innova-

[2] *See* my "Unemployment and Underemployment in Monsoon Asia," *Philippine Economic Journal* (Spring, 1971).

tions in institutional arrangements, particularly pertaining to the primary and tertiary sectors, may have to be speeded up. A new wave of institutional transformation is called for, comparable to the Meiji period and the first postwar decade and perhaps this has already started with the new welfare policies established in the 1970s and the emerging changes in foreign relations. I do not mean to imply by all this that technological innovations for the large-scale enterprises should be avoided. Certainly, they will be required for those industries which must meet challenges from international competitive forces; but the focus need not be to become the "technological leader" of the world. In sum, the catching up process has been relatively simple, dealing as it did with the secondary sector. But Japan has far to go in catching up with the West as to the primary and tertiary sectors whose relative "backwardness" (*i.e.,* small-scale mechanization and organization) will require major institutional innovations in the setting of monsoon Asia.[3]

[3] Time does not permit comments on the data, except to note that the Japanese statistical systems grew rapidly during the 1950s, and by the 1960s they were virtually completed. In the early 1950s this system was in its infancy and the data on employment and capital were extremely poor for growth accounting purposes, especially with respect to the small-unit sectors. Elsewhere I have attempted to show that a good statistical system is a vital part of an institutional framework for economic growth, affecting not only planning and policy-making by the Government but also the efficient operation of private enterprises, households, educational institutions such as schools and mass media, and so on. Here one can note the great importance of interactions and interrelations.

Another type of interactions is between institutional changes and the intensity and quality of work per hour. With the democratization and improvement of rural institutions and industrial relations, the intensity and the quality of work per hour probably rose in the 1950s, reaching a peak by the early 1960s, so that by the latter 1960s a saturation point may have been reached since the intensity and quality of worker per hour cannot increase substantially forever.

Summary of Discussions

On the sessions dealing with the topics of "Comparative Analysis of Economic Growth, Industrial Organization and Growth Accounting"

I. GENERAL SUMMARY (Chairman: Mark Perlman, U.S.A.)

Focus appeared to divide itself between those measuring relative differences between Japanese and other nations' economic growth rates and then trying to locate the proximate purely economic factors primarily responsible *and* those concerned with socio-political considerations which provided the climate for these changes or which were the obvious immediate results of those changes. This difference in focus went beyond the well-known division in Academia of National Accounts-Econometric versus Marxian, since many of the former group have gone on to speculate as to historical institutional causes like the sources of R&D and the impact of unions, etc.

The historical impact of inflation on growth is becoming a more visible assembly point for very recent Japanese analytical work. It is still less important as a topic than the direction and size of technological flows, however. In regard to the latter the eradication of the "technology gap" leads to speculation whether Japan, once a technology imitator, will become a technology exporter or whether it will turn increasing portions of its energy to improving its domestic standards of living. If the latter, interesting work has been done (and creatively criticized at this Congress) on measurements of net national welfare. The topic of inflation, of course, ties in with research interest in domestic Japanese living standards as well as sources of saving for investment.

Interest in the impact of governmental policy regarding concentration in ownership remains great but is undergoing some changes in emphasis. While evaluation of success of these policies is still of leading interest, thinking has turned to the efforts of R&D, to the role of subcontracting to allied but non-wholly owned subsidiaries, to market behavior, and even to the invulnerability of multinational corporations.

II. SUMMARY OF DISCUSSIONS (Rapporteur: A. Hernadi, Hungary)

Ohkawa's paper on "Past Economic Growth of Japan in Comparison with the Western Case" was discussed by *Kazuo Shibagaki* (Japan), who, referring to the differences in views of Marxist and non-Marxist economists in Japan, suggested that there should be scientific cooperation between the two groups and raised the point that *Ohkawa*'s usage of the terms "backward nations" and "latecomers" had better be clarified more by specifying the criteria for such a classification. *Shibagaki* went on further to suggest the need for differentiation even within each of these groups, referring to the theory of a Japanese Marxist Kozo Uno, who regarded Germany as a "latecomer" and not Japan. By placing emphasis on the fact that Japan's industrial structure showed discontinuity, *Shibagaki* found it important and necessary to examine what made Japan's concentration in heavy and chemical industries possible. *Ohkawa* replied to this briefly by saying that he did not intend to give Japan as an example of development for the present-day developing countries and that rather he saw Japan's case as a special one. His emphasis was mainly on the narrowing down of the technological gap and the trend acceleration in the case of Japan.

Martin Bronfenbrenner's paper on "A 'Marginal-Efficiency' Theory of Japanese Growth" was first commented on by *Tuvia Blumenthal* (Israel), as reproduced in the text. He was followed by *Hugh Patrick* (U.S.A) who called attention to the fact that inflationary process had not been uniform and that at different periods in history there were different reasons for it. Then, *Morris Beck* (U.S.A.) pointed out that inflation could decrease and increase consumption in different sectors of the economy at the same time. Other comments came from *Robert Wood* (Japan), who emphasized that Japan's domestic capital investment was both higher and more effective because her advanced financial system was able to make funds consistently available to the projects that had a payoff greater than the interest rate; *Shigeo Minabe* (Japan), who queried what kind of role voluntary saving would have on the future rate of growth of the Japanese economy; *Rémy Prud'homme* (France), who made a point that France's case had been similar to Japan's in all respects except for the high rate of growth; and *Christian Sautter* (France), who added a word to the effect that French economic growth had been, among OECD countries, second only to Japan.

Bronfenbrenner answered firstly to *Blumenthal* by saying that unstable path of economic development actually came off as a compromise as Domar stated and that voluntary savings were difficult to judge by statistics and it might be said as a matter of guess that people could not consume all they wanted to. He went on to reply to other questions by saying that he agreed that inflation had not been uniform possibly because the Japanese had a special way of handling depressions as epitomized by the phrase: "Throw money at the problem, then throw even more money!"; that actual government spending was more than the anticipated taxes; and that French trade unions were not only political but they did really have a psychological motivation similar to that of the Japanese National Railroad Union.

The main discussant on *Jorgenson-Nishimizu*'s paper on "U.S. and Japanese Economic Growth, 1952–1973: An International Comparison" was *Hugh Patrick* (U.S.A.), who prefaced his remarks by saying that because of the shortage of time he would give only some points of criticism, and not compliments or evaluation. They were that (1) the paper did not explain cyclical performance, (2) it did not take into account synergistic interactions among factor inputs and technological change, (3) it did not incorporate changes in the relative degree of monopoly or oligopoly in the economy, and (4) it did not seek explicitly to explain the differences in rates of growth of factor inputs or in total factor productivity. Then, he addressed two questions to the authors, namely: (1) How would their results be altered by inclusion of the general government sector? and (2) How to interpret the purchasing power parity estimates, especially for output?

Apart from *Jorgenson*'s replies with reference to various tables in the paper, he made a point that the unit cost of capital input was much higher in Japan than in the U.S., while the reverse was the case for the unit cost of labor input. In summarizing, he said that Japan had really caught up with the U.S. in respect to the level of technology, the difference being in capital intensity, which was still double in the U.S. compared with Japan. On the first of the two questions raised by *Patrick*, *Jorgenson* replied that it was the productivity in the private sector that really counted.

Kenzo Yukizawa's paper on "Relative Productivity of Labor in American and Japanese Industry and its Change, 1958–1972" was commented on by *Masaru Saito* (Japan), as reproduced in the text; and then there was an intervention by *Donald J. Daly* (Canada), who queried if it did not make a difference whether one used gross output calculation or net output calculation and also indicated his surprise on the result that the service and trade sector in Japan was closer to the U.S. than manufacturing. *Yukizawa* replied to *Saito* by saying that as far as the dual structure was concerned, export trade of Asian countries had expelled the small businesses of Japan from trade; and to *Daly*, he explained that he used input-output tables instead of net output calculations.

"Industrial Organization and Economic Growth in Japan" by *Ken'ichi Imai* and *Masu Uekusa* was commented on by *Soichiro Giga* (Japan), as reproduced in the text, followed by an intervention by *Mark Perlman* (U.S.A.) who asked what the interrelation was between R&D and the concentration or firm-size. The authors of the paper replied, first indicating their general agreement with *Giga*, but then pointing out certain problem areas where they differed on the matter of emphasis, such as (1) in the postwar Japan market forces went beyond planning in a greater degree than *Giga* implied; (2) the Anti-Monopoly Law had never been a component of the so-called 'Japan Incorporated' and its role should not be overestimated; (3) today's groupings of business firms were significantly different from the *zaibatsu* of the past; and (4) the internationalization of the antitrust policy would be a necessary step, though not a satisfactory one. To *Perlman's* question, the authors replied by saying that the firm-size might contribute to the increased research inputs in the economy's more innovative sectors, but the concentration by itself did not.

Roberto Zaneletti's paper on "Capital Accumulation and Economic Growth: A Comparison of Italy and Japan" was first commented on by *Hisao Onoe* (Japan), who, after raising a general point to the effect that it required some explanation why investment was said to have contributed so much to the economic growth of Japan, went on to assert that Japan's adaptability had been proved more in structural changes and that therefore it was not enough to speak about the volume or value of investment expenditures. The problem for Japan was, according to *Onoe*, the transformation of the country into a welfare state in the immediate future to come, and Italy would serve as a good lesson in the sense of warning what type of things to be avoided. He further queried if one reason for the capital flight from Italy might not have been the fact that the rate of investment there was smaller than that of saving. *Onoe* was followed by *Martin Bronfenbrenner* (U.S.A.), who suggested that the paper could have been improved by estimating the capital coefficients and also that it might have been useful to discuss the role of the Italian trade union movement in connection with the problem under review.

The author responded to the above comments first by emphasizing the point that Italy also was highly export-oriented in her economic activities and then went on to say that Japan's 1964–67 export drive was more successful than that of Italy because of the different patterns of change which their export prices took. He agreed that a shift for welfare economy would be a good idea except that care should be taken on the question of the budgetary balance of the state. It was pointed out that Italy's deficit financing policy led to a 13 percent indebtedness to the GNP by 1976. On the question raised by *Bronfenbrenner*, the author called attention to the fact that trade unions in Italy were very powerful to the extent of demanding consultation on investment programs and that labor was no longer an elastic factor, but a fixed factor, in the process of production.

Hisao Kanamori's paper on "Japanese Economic Growth and Economic Welfare" was first commented on by *Kimio Uno* (Japan), as reproduced in the text, and followed by *Rémy Prud'homme* (France), who raised a specific question as regards the way the environmental-pollution costs were calculated. The author's answers consisted of (1) a statement to the effect that Net National Welfare (NNW) was meant to measure only a part of economic welfare and that even then it was better than nothing; (2) that the sectoral approach, as suggested by *Shigeto Tsuru*, had difficulties of its own and therefore aggregate functions had to be accepted; and (3) a somewhat inconclusive response to *Prud'homme* that what was deducted from NNW was "the money spent on anti-pollution."

On the last paper of the day by *Kazuo Sato* on "Did Technical Progress Accelerate in Japan?" it was *Harry Oshima* (U.S.A.) who made a comment which is reproduced in the text. The author replied by saying that (1) there was no other measurement than the residual, so that was what he had to use; (2) it was difficult to separate the contributions made by technology imports and the domestic R&D; (3) he agreed on the need to develop the technological level further at home before taking it abroad, since small scale industries were backward in this sense; and (4) he disagreed on the point that Japan should remain inward-looking.

PART TWO

ENVIRONMENTAL AND RESOURCES CONSTRAINTS

Appraisal of Environmental Policies in Japan

Rémy Prud'homme *Université de Paris*

An environmental policy has been developed in Japan since 1967. This policy is appraised in terms of (i) results, (ii) costs, (iii) approaches. Results are found to be quite impressive in terms of pollution abatement, but not in terms of amenities. Costs are shown to be relatively high, but their macroeconomic consequences appear to be negligible. Approaches followed by Japanese policy-makers are shown to be completely different from approaches prescribed by conventional environmental economics: objectives were decided without much consideration of costs, and direct controls were preferred to pollution charges.

I. INTRODUCTION

Japan, like most countries, has a long history of environmental pollution. The well-known case of a copper mine located at Ashio, which was seriously damaging the environment in the eighteen-eighties,[1] is but one example of environmental disruption associated with industrialization. What is interesting to note is that Japan, unlike many other countries, quickly developed antipollution policies. As early as 1877, Osaka Prefecture issued an ordinance entitled "Regulations on Control of Manufacturing Plants," which even today could be regarded as quite progressive. Attention was paid to the selection of industrial sites that would minimize environmental disruption; in 1895, for example, the Besshi copper refining plant was transferred, for environmental reasons, to Shisaka Island in the Inland Sea. Antipollution measures were taken, such as production cutbacks in case of emergency, or erection of high smokestacks (a 156-meter-high smokestack was constructed at Hitachi mine in 1914). Compensation of pollution-related damages also took place. Japan, therefore, has traditionally been active in the field of environmental policies.

This tradition, however, was somewhat forgotten in the prewar and postwar periods, just at a time when the pressure of economic growth on environmental quality increased greatly.

This pressure has been, and still is, greater in Japan than in any other country. Its main determinants are well known: high levels of economic activity, high rates of growth, high degree of spatial concentration, etc. Some of them are illustrated by Table 1.

In the postwar period not much was done to resist that pressure. Japan gave priority to industrial development. Ambitious targets were set (income-doubling 10-year plan), and no effort was spared to reach them. As a matter of fact, Japan tried so hard that actual growth rates were higher than planned growth rates, but this meant that alternative or competing goals—such as environmental protection—were to a large extent sacrificed. In the fifties and in the early sixties, this strong commitment to industrial growth, and its corollary, a relative neglect of environmental quality, were shared by most segments of the Japanese society. This attitude, which was to change drastically at the end of the sixties, explains why there were practically no implicit or explicit environmental policies in the two postwar decades.

It is, therefore, not surprising that, by the late sixties, Japan had become one of the most polluted countries in the world. Ambient concentrations for a number of pollutants (particularly sulfur dioxide (SO_2), carbon monoxide (CO), nitrogen oxides (NO_x), mercury, cadmium), as

[1] *See* Norie Huddle and Michael Reich, *Island of Dreams* (1975), pp. 25–33, for an account of the unsuccessful fight of Ashio area residents to halt copper pollution.

TABLE 1

SELECTED ECONOMIC INDICATORS PER KM² OF INHABITABLE AREA[c] (JAPAN AND SELECTED OECD COUNTRIES, 1974 OR 1975)

	GNP[a] *($10^6$$ U.S.)* *(1975)*	*Industrial*[a] *Output* *($10^6$$U.S.)* *(1974)*	*Energy*[a] *Consumption* *(10^3 TOE)* *(1974)*	*Number of*[b] *Automobiles* *(1974)*
Japan	6.05	2.04	4.12	331
U.S.A.	0.32	0.09	0.36	27
U.K.	1.04	0.26	1.00	80
France	0.87	0.25	0.47	47
Italy	0.81	0.24	0.66	74
Sweden	1.67	0.44	1.09	69
Netherlands	3.10	0.83	2.38	146

Sources and Note:

[a] OECD.

[b] International Road Federation, *World Road Statistics*, 1975.

[c] Inhabitable areas are defined here as utilized agricultural areas + urban areas + non-utilized agricultural areas.

well as pollution-induced damages, were high. Several lethal cases of environmental pollution occurred.

Changes in environmental quality led to changes in societal attitudes, reflected in changes in political pressures, which brought about changes in governmental policies. Similar changes took place in most developed countries, but were probably more dramatic in Japan than elsewhere.

The policy response was quick and strong. In the late sixties and in the early seventies, a number of important decisions were taken by the Diet, the Administration, and the Courts. The legislative foundations of this policy were set in 1967 when a Basic Law for Environmental Pollution Control was enacted and in 1970 when a special session of the Diet was devoted to environmental issues, at which no less than 14 pollution-related laws were revised or enacted.

The executive created a new institution in 1971 to carry out this policy: the Environment Agency. This agency, however, is not the only agency or ministry interested in environmental policy. In particular the powerful Ministry of International Trade and Industry (MITI), which has created a specialized Directorate (called Industrial Location and Environmental Protection Bureau), plays an important role in policy formulation and implementation.

The judiciary also played an important role in the development of en-

TABLE 2

SO_2 POLLUTION, 1965–74

	Ground Concentration Levels[a] *(parts per million)*	*Monitoring Stations Meeting Quality Standards (%)*[b]
1965	0.057	n.a.
1966	0.057	n.a.
1967	0.059	48
1968	0.055	59
1969	0.050	67
1970	0.043	72
1971	0.037	87
1972	0.031	34
1973	0.030	46
1974	0.024	69

Source: OECD.

[a] Averages over 15 stations in operation since 1965.

[b] Quality standards for 1967–71 are different from quality standards for 1972–74.

TABLE 3

CO POLLUTION IN TOKYO, 1965–74

	First Source (parts per million)	*Second Source (parts per million)*
1965	3.2	
1966	2.9	
1967	3.1	
1968	3.6	6.2
1969	4.4	5.9
1970	5.7	3.4
1971	4.7	3.1
1972	4.3	2.3
1973	3.7	2.7
1974	3.5	n.a.

Sources:

First Source: Environment Agency, *Quality of the Environment*, 1975, p. 74; Second Source: Tokyo Metropolitan Government, *Tokyo Statistical Yearbook,* 1973, p. 461; average over four general monitoring stations.

vironmental policies. In 1967–69, victims of Minamata disease, of Itai-itai disease, and of pollution-related asthma sued the companies that they held responsible for the diseases. In these four lawsuits,[2] which were widely covered by the media, and became known as the "four major lawsuits," the courts decided in the early seventies in favor of the plaintiffs and found the polluters responsible. The impact of these decisions on attitudes, laws, and practices was great.

It is this policy that we shall try to appraise in terms of its achievements (*Section* II), of its costs (*Section* III), and of its approaches (*Section* IV).

II. ACHIEVEMENTS

It is of course difficult to find out whether an environmental policy has been successful or not. There are no overall indicators of environmental quality; and even if there were, their evolution could not be entirely attributed to the efficiency of environmental policies. It is nevertheless necessary to try to see what has been achieved by the effort developed in Japan to protect the environment.

A clear and important point is that the situation concerning pollution by sulfur dioxide has improved significantly in the past few years, as shown by Table 2. Yearly average levels of SO_2 concentrations decreased since 1967 and were, on average, lower by 50 percent in 1974. As a result, the percentage of monitoring stations at which ambient air quality standards were met increased rapidly.

The situation concerning pollution by carbon monoxide has also improved significantly in the past few years. In 1975 ambient air quality standards were met at all monitoring sites. Available data for Tokyo comes from two different sources, by which CO pollution is measured differently; both sources show a significant improvement in recent years (*see* Table 3).

On the other hand, the available data for nitrogen oxides, which is not very reliable, show an increase in yearly average concentrations for a set of monitoring stations. Increases in maximum hourly concentration averages in cities like Tokyo and Osaka are also reported.

Trends in water quality are less clear. Pollution by organic matters, as measured by biochemical oxygen demand (BOD) in rivers over recent years, does not seem to have significantly changed, as shown in Table 4. Available data for a number of rivers do not suggest a very clear pattern of change either; pollution levels varied from year to year; for some rivers, things improved; for others they did not. By and large, pollution by organic matter did not improve much.

As regards water pollution by harmful substances such as cadmium or

[2] There were two suits filed by victims of Minamata Disease in different areas.

TABLE 4

WATER POLLUTION,[a] 1971–74

	BOD in Rivers	*COD in Sea Water*
1971	77	82
1972	76	84
1973	76	84
1974	78	84

Source: Environment Agency.
[a] Percentages of samples meeting quality standards.

TABLE 5

POLLUTION BY HARMFUL SUBSTANCES,[a] 1970–74

	1970	*1974*
Cadmium	2.80	0.37
Cyanides	1.50	0.06
Organic phosphorous	0.20	0.00
Lead	2.70	0.37
Chromium (sexivalent)	0.80	0.03
Arsenics	1.00	0.27
Total mercury	1.00	0.01[b]
Alkyl mercury	0.00	0.00
Total	1.40	0.20

Source: OECD.
[a] Percentages of samples exceeding quality standards; data is based on more than 16,000 samples for 1974.
[b] 1973.

mercury, a very clear improvement took place, which is shown in Table 5. The percentage of samples exceeding quality standards decreased dramatically from 1.4 percent in 1970 to 0.20 percent in 1974.

The evolution of environmental quality does not reflect correctly the efficiency of environmental policies. The efficiency of a policy should be assessed by comparing what would have happened without the policy with what actually happened. Had no environmental policies been developed in Japan, discharges associated with production and consumption activities would have increased at the rate at which production and consumption increased. So would have ground concentration levels.

TABLE 6

FOSSIL FUEL CONSUMPTION AND SO_2 GROUND CONCENTRATION, 1968–74

	(1) Fossil Fuel Consumption[a] (Million TOE)	(2) SO_2 Concentration[b] (parts per million)	(3) $\frac{(2)}{(1)} \times 10^6$
1968	211	0.055	260
1974	439	0.024	55
Change	+103%	−56%	−79%

[a] OECD.
[b] Table 2 above.

This point can best be illustrated in the case of SO_2. As shown in Table 6, fossil fuel consumption, which is the major source of SO_2 pollution, increased by about 100 percent between 1968 and 1974. But SO_2 pollution, that is SO_2 ground concentration levels, was halved instead of doubling. Pollution per unit of polluting activity was therefore decreased by as much as 80 percent.

There are relatively few cases in which the relationship between a particular pollutant and a particular polluting activity is straightforward. But broad economic magnitudes can be taken as indicators of polluting activities. Their increase is a measure of an increase in environmental pressure. This increase has slowed down in the seventies: the industrial production index even decreased in 1974 and 1975. Over the period 1970–75, however, most economic indicators increased in Japan, and generally increased more than in most other countries. The stock of automobiles, for instance, increased by more than 70 percent over this period. Trends in environmental quality, which range from dramatic improvements to slight deterioration, but are on the whole rather favorable, and, inasmuch as one can judge, as favorable or more favorable than in other countries, must be appreciated against this background. Pollution abatement policies in Japan therefore appear to have been fairly efficient.

Pollution levels, however, are only one aspect of environmental quality. Other aspects, often called "amenities," refer to quietness, beauty, privacy, social relations, and similar elements of the "quality of life," which are difficult to define and to measure, but which are essential. In the field of amenities, not much was attempted, and even less was achieved.

It is likely that, in the seventies, just as in the sixties, "levels of amenities" will continue to decrease. Bigger and bigger plants have been constructed, and it may well be that social diseconomies of scale (to be measured in terms of amenities) are associated with the familiar private economies of scale. Natural landscapes, as well as man-made gardens, equally cherished in Japan, were threatened or destroyed by urban, industrial, and transportation developments. Efforts were made to introduce various forms of land-use planning: their very number suggests that they were not very efficient.

This would explain why successful pollution abatement policies did not eliminate environmental discontent, which is about as strong now as it was in the late sixties.

III. COSTS

A second question can be raised about Japanese environmental—or rather, as we have seen, pollution abatement—policies: how much do they cost, and what are their economic consequences?

Surprising as it might be in a country where statistics are usually good and plentiful, costs of pollution control in Japan are poorly known and have to be estimated.[3]

Table 7 gives estimates of antipollution investments as a percentage of GNP.

Table 7 shows that antipollution investments are high in Japan—significantly higher than in any other country—and that they have increased rapidly over recent years. Antipollution investments by private enterprises amounted in 1974 to about 4 percent of total investments by private enterprises. This figure is an average for all enterprises (including nonmanufacturing enterprises), and the value of this ratio is much greater for some sectors, such as thermal power plants, oil, pulp and paper, chemicals, and iron and steel.

There is no published information on operating costs, nor on the stock of accumulated antipollution equipment. We nevertheless attempted to figure out the economic cost (depreciation of the stock of antipollution equipment, plus opportunity cost of capital, plus operating costs) of pollution abatement in Japan in 1975 and arrived at a figure that amounts to 1.7 percent of GNP.[4]

Pollution control costs, particularly economic costs, are a measure of the drain on resources exerted by pollution abatement policies in a country; capital and labor resources that are being utilized to abate

[3] *See* R. Prud'homme, *Environmental Policies in Japan* (1977), Appendix I, for a description of various methods utilized to arrive at estimates.

[4] *See* Prud'homme (1977), Appendix II.

TABLE 7

ANTIPOLLUTION INVESTMENT COSTS AS A PERCENTAGE OF GNP, 1970–75

	Antipollution[a] Investment by Enterprises	*Antipollution[b] Investment by Government*	*Total Antipollution Investment*
1970	0.4	0.6	1.0
1971	0.5	0.8	1.3
1972	0.5	1.0	1.5
1973	0.6	1.0	1.6
1974	0.7	1.0	1.7
1975	1.0	1.0	2.0

[a] Prud'homme (1977).

[b] Ministry of Home Affairs.

Note:

Figures for 1974 and 1975 for local government were not available and have been estimated on the basis of 1970–73 data.

pollution are not being utilized to manufacture consumption goods, nor productive equipment. Pollution control expenditures have an opportunity cost. They are a price to be paid for environmental quality.

This view is correct, but in real life things are somewhat more complex. What is expenditure for one enterprise will be income for another enterprise. Depending upon the state of the economy, upon businessmen's behavior, upon fiscal policy, and more generally upon the whole economic machinery, pollution abatement expenditures will modify a number of economic flows, which will modify other flows, etc. In dynamic terms, environmental expenditures should not only be seen as losses of resources, but as modifications of economic flows. The net results of these modifications on GNP, on prices, on employment, on trade, etc., are difficult to predict, but the important point is that they are not necessarily negative. In some cases economic consequences may even be quite positive. The main justification for pollution abatement policies is of course that they decrease pollution and pollution costs, thereby increasing welfare; but it is important to note that, in practice, the macroeconomic consequences of these policies are not as bad as is often said. It is widely recognized that GNP is not a good indicator of welfare, and environmental policies should be carried out even if they were to reduce GNP. It is nevertheless interesting to know whether they do reduce GNP or not.

The study of the past behavior of the Japanese economy cannot be utilized to assess macroeconomic consequences of environmental policies. It would be too easy to say that since 1970 the Japanese economy has done as well as, or better than, most other economies and to conclude that environmental policies have had no adverse impact. The reasoning would not be correct. Environmental expenditures have only been one of the many factors that influenced the behavior of the Japanese economic machinery and its results. No cause-effect relationships between pollution control costs and macroeconomic magnitudes can be established on the basis of past experience.

The only way to establish such relationships is to utilize macroeconomic models to simulate the development of the economy with and without pollution abatement policies. All other factors being thus kept constant, differences in outputs can be related to differences in inputs. Several such models have been utilized in Japan. All of them embody questionable assumptions and utilize unreliable data; they are therefore open to criticism, and their results should be handled with care; but they provide the only tentative answers to the very important question: what are the economic consequences of environmental policies?

One model, developed by Professor S. Shishido and A. Oshizaka,[5] combines a Leontief-type input-output model and a Keynesian-type macroeconomic model. It studies, over a period of six years, the effects of two types of environmental policies: a "harder" policy and a "softer" policy, which are defined in terms of pollution abatement. First, antipollution investments required to achieve these pollution abatement targets are estimated; the amount calculated to meet the "softer" target,[6] 1.5 percent of GNP, is quite comparable to the amount actually invested in antipollution equipment. Then, the impact of those additional investments on GNP, prices, etc., is estimated by the model.

The most striking result of this simulation is that GNP and employment are practically *un*affected by environmental policy. More precisely, GNP is "reduced" by the deflationary price effects associated with nonproduction investment and, at the same time, "increased" by the expansionary income effect resulting from these antipollution investments. In the first years, the positive effect is even greater than the negative one; later on they balance each other: for the sixth year, GNP is raised by 0.1 percent (over what it would be in the absence of antipollution policy) by the "softer" policy.

A second interesting result of the simulation relates to the changes in the structure of output and employment that are brought about by environmental policies. The share of output and employment increases in primary and fabricated metals and in general and electric machinery, whereas it declines in the food industry, textiles, and electricity. This is a

[5] *See* Shuntaro Shishido and Akira Oshizaka, "Econometric Analysis of the Impacts of Pollution Control in Japan" (1976).

[6] Public works for sewages are excluded.

rather surprising result: antipollution policies benefit polluting industries! This is because polluting industries such as the steel industry are the main beneficiaries of antipollution investments, although they incur high pollution abatement costs themselves; for them the income effect is greater than the price effect.

The model also yields information about prices. Antipollution policies do raise the price level; but this increase is modest. It is estimated to be, over six years, 1.9 percent in the case of the "softer" policy, hardly more than 0.3 percent per year. Of course, much higher increases appear in certain sectors, such as automobiles (5.9 percent), electricity (6.2 percent), primary iron (7.6 percent), or pulp and paper (7.7 percent).

A second model has been utilized by Professor Y. Murakami and J. Tsukui.[7] Unlike the Shishido and Oshizaka model, which is demand-oriented, this model is supply-oriented. It describes the economic "path" necessary to maximize consumption over time under a certain number of constraints. These constraints relate to capital, labor, production functions, foreign trade, etc. Pollution abatement is introduced as an additional constraint. Here again two types of pollution abatement policies are introduced: a "harder" policy, which eliminates completely (100 percent) five major pollutants (SO_2, NO_x, BOD, industrial waste, household waste), and a "softer" policy, which eliminates partially (75 percent) the same pollutants. The difference between the paths described by the model with and without pollution abatement policies indicates the consequences of these policies.

The main results of the model are given in Table 8, for the "soft" type of environmental policy.

They again suggest that the overall impact of environmental policies is not great. Output and consumption—which are what matters—are not very significantly affected by pollution abatement. Over a period of 10 years, consumption is only reduced by −0.2 percent, *i.e.*, is practically unchanged. The impact of the "harder" environmental policy (complete elimination of pollution) is hardly more worrying: output is reduced by 3.2 percent (instead of 0.5 percent) after 10 years, and by 6.5 percent (instead of 3.0 percent) after 15 years.

A third model, utilized by the Economic Planning Agency, also suggests that the macroeconomic impact of "stricter" environmental policies would not be very great.

None of these three models is entirely satisfactory; the Shishido-Oshizaka model, which is purely Keynesian, ignores constraints on resources; the Murakami-Tsukui model incorporates many doubtful hypotheses; the Economic Planning Agency model seems rather crude; all suffer from the lack of reliable data on the structure and magnitude of

[7] *See* Yasuke Murakami and Jinkichi Tsukui, "Economic Costs Prevention of Pollution: A Dynamic Analysis of Industrial Structure" (1976).

TABLE 8

CHANGES[a] IN MACROECONOMIC VARIABLES GENERATED OVER TIME BY A SOFT ENVIRONMENTAL POLICY,[b] AS ESTIMATED BY MURAKAMI AND TSUKUI

	After 5 Years (%)	*After 10 Years (%)*	*After 15 Years (%)*
Output	−2.8	−0.5	−3.0
Consumption	−1.4	−0.2	−3.1
Gross investment	+26.9	+1.8	−12.9
Productive investment	+23.1	−3.2	−16.1
Housing construction	+11.1	−2.5	−16.1
Resources import	−16.8	−2.0	+3.7
Exports	+1.8	+8.2	+10.7

Source: Adapted from Murakami and Tsukui (1976), Table 5–1, p. 19.

[a] Magnitude with environmental policy, minus magnitudes without environmental policy, divided by magnitude without environmental policy.

[b] 75 percent elimination of pollution generated without environmental policy in terms of SO_2, NO_x, BOD, industrial waste, and household wastes.

pollution abatement costs. Their results must therefore be interpreted with great care. It should nevertheless be pointed out that they strike an optimistic note. Although the models do not take into account the positive economic consequences of pollution abatement, which are the main rationale for environmental policies, they suggest that the adverse economic consequences of pollution abatement expenditures are not very significant.

IV. APPROACHES

The most interesting questions about the Japanese experience in pollution abatement relate to the approaches utilized. Professional economists know, or think they know, how to deal with pollution problems. They know where to go (to the pollution level at which the marginal cost of pollution equals the marginal cost of depollution) and how to get there (by taxing pollution emissions). The Japanese policy-makers have completely ignored those principles.

They rely basically upon the setting of ambient standards by the national government, coupled with the setting of emission standards by local governments. For most pollutants, rather strict ambient standards are established at the national level, by the administration under very general guidance given by law. In order to achieve these ambient standards, sets of emission standards are basic requirements, which are often made more stringent by local governments. In many cases, local governments engage in plant-by-plant negotiations and agreements to define emission allowances for each major polluting facility in the area.

In setting ambient as well as emission standards, Japanese policy-makers do not embarrass themselves with economic computations. The familiar proposition that a balance must be struck between environmental costs and economic benefits has even been officially rejected. The 1967 Basic Law stated that "efforts shall be made to balance pollution control against the needs for economic development." Although this provision referred only to environmental conservation and not to health protection, it was violently attacked, and was eventually deleted in 1970.

This lack of economic rationale also appears in the "antibusiness" attitude that prevails, and which is particularly surprising on the part of legislatures and governments which are not, on other accounts, inimical to the business community. Industry does not appear as a source (of environmental degradation) that has to be stopped, but as a villain that has to be castigated. Hence, the success of the polluter-pays-principle (P.P.P.) in Japan.

The polluter-pays-principle states that the costs of pollution abatement should be borne by polluters, who will in most cases pass them on in the prices of their goods, rather than by governments: it is a "consumer-pays-principle," as opposed to a "taxpayers-pay principle"; it was introduced in 1972 by the OECD as an economic principle, which would (i) prevent distortions in international trade and (ii) improve the allocation of resources. The P.P.P. immediately became very popular in Japan, and it is often referred to. But the economic objectives of the principle, and the mechanisms by which they are achieved, are not always well understood. For many, the principle just means that polluters are guilty and must be punished. In short, it is understood as a "Punish Polluters Principle."

In Japan, pollution abatement is a must, a moral obligation, backed by very strong political pressures. Stringent measures are taken with only limited knowledge of dose-effect relationships and with even more limited knowledge of economic costs and consequences involved.

It is interesting to note that this approach works rather well. Many apparently "irrational" decisions proved to be particularly wise decisions. Automobile emission standards are a case in point. The standards for NO_x exhausts set in Japan were so stringent that they were generally considered "unmeetable"; yet, in a matter of years, several automobile manufacturers developed the necessary techniques. The standards had actually been set without a detailed examination of whether they could be met; had such an examination been conducted, the standards would probably have been less strict, and low-pollution cars would never have been developed. Similarly, the bans that were imposed on some toxic chemicals (such as PCBs) or processes (such as mercury-using processes for manufacturing caustic soda), and which were thought at first to create insuperable difficulties for the industries concerned, actually led to technological innovations.

What happened is that science and technology provided answers to most of the problems raised. The possibilities of science and technology are such that they extend the frontiers of rationality. A rational decision is a decision that takes into consideration costs and benefits; the trouble is that it is not only very difficult to determine benefits (*i.e.,* damages avoided), it is often impossible to estimate costs. Costs of processes that have not yet been invented cannot be estimated. They are said to be very high or even infinite ("it can't be done"), but may well turn out to be reasonable. And it is a decision based upon such overestimated costs that is irrational. The Japanese experience in the field of pollution abatement lends support to the idea that to a large extent it is not technology that should constrain policy choices, but policy choices that should constrain technology.

The Japanese practice runs also counter to conventional economic principles in its preference for direct administrative controls over indirect economic incentives. There are only two instances of pollution charges: a tax based on SO_2 emissions in the framework of the so-called Compensation Law Scheme and a special landing fee that varies according to the noise emitted by the aircraft. They do not play a very important role.

Direct and often detailed interference in the economic operation of enterprises is the main tool of environmental policies in Japan. Businessmen receive "orders" from central or local government bureaucracies as to the amount of pollution they are allowed to discharge and the types of processes they can choose. In other words, Japan relies to a large extent upon planning rather than market mechanisms to reduce pollution. The market is a system where information is transmitted by prices, not by directives, and where action is motivated by profit, not by compliance. It is widely recognized that pollution is a case of "market failure," *i.e.*, that uncorrected market mechanisms breed pollution. This leaves policy-makers with the following choice: (i) keep the market mechanism and correct it by modifying the prices (which means in practice taxing pollution) or (ii) reject the market mechanism and replace it by planning. Japan basically follows the second course. It may increase costs. It certainly increases the power of the administration. But it works.

Economic theoreticians, who are usually very fond of compensations, will find some solace in the 1973 Compensation Law Scheme. This unique system provides that all the persons who have lived or worked in designated areas and who suffer from designated pollution-related diseases are deemed to be pollution-victims and entitled to receive monthly payments. The money is paid by polluters, in the form of a tax based on the volume of sulfur oxides released (for 80 percent) and by road-users or taxpayers in the form of a transfer from the automobile tonnage tax (for the remaining 20 percent).

V. CONCLUSIONS

Many lessons can be drawn from the Japanese experience in environmental policy.

A first lesson is that it is possible to reduce pollution levels. Japan did it, at least for the pollutants for which it developed and implemented crash programs: SO_2, CO, cadmium, mercury, and PCB. There is no inherent contradiction between economic growth and environmental quality, and it is possible to increase production while decreasing pollution—provided appropriate mechanisms are developed.

A second point is that such mechanisms, although costly in terms of economic resources, can easily be borne by the economy of countries such as Japan and do not have a great impact on their growth.

A third point is that the protection and development of "amenities" are as important as the abatement of pollution. One should not concentrate all efforts on pollution abatement—as did Japan—and neglect other aspects of environmental quality—which are more difficult to define and to tackle, but which are not less important.

One other point is that direct administrative guidance works. It may be less cost-efficient than indirect economic incentives (although this is only proven in theory), and it may require, or bring about, an omnipotent bureaucracy, but it is at least an alternative tool worth some consideration.

A final point is that traditional cost-benefit analysis is of little help for decision-makers faced with "technological uncertainties." They have to bet that new processes or new products will appear at reasonable costs. They will only do it under strong political pressures. The Japanese experience suggests that technology follows and that boldness pays. It lends support to the motto that appeared on the walls of the Sorbonne, in May 1968: *"Soyez réalistes, demandez l'impossible."*

REFERENCES

Huddle, Norie and Reich, Michael. 1975. *Island of Dreams: Environmental Crisis in Japan.* Tokyo: Autumn Press.

Murakami, Yasuke and Tsukui, Jinkichi. 1976. "Economic Costs of Prevention of Pollution: A Dynamic Analysis of Industrial Structure," paper presented at the International Conference for Environmental Protection, organized by *Nihon Keizai Shimbun,* Tokyo, May 26–28.

Prud'homme, Rémy. 1977. *Environmental Policies in Japan.* Paris: OECD.

Shishido, Shuntaro and Oshizaka, Akira. 1976. "Econometric Analysis of the Impacts of Pollution Control in Japan," paper presented at the International Conference for Environmental Protection, organized by *Nihon Keizai Shimbun*, Tokyo, May 26–28.

The Environmental Protection Policy in Japan

Ken'ichi Miyamoto *Osaka City University*

In 1970, the Japanese law system was more fully developed than that of other countries. The development of Japanese environmental policy was epoch-making after 1970. This change was a result of the pressure of the antipollution movement and the public opinion. However, the damage of environmental pollution has not decreased. Although the law system and administrative organization were arranged more than 10 years ago, and its budget increased rapidly, the damage from pollution has not decreased. It is the fault of environmental policy itself and based on economic structure in Japan. In this paper, I consider these problems.

I. BRIEF HISTORY OF ENVIRONMENTAL PROTECTION

In 1964, Professor H. Shoji and I published *The Terrible Pollution Problems [Osorubeki Kogai].* At this time, the word *"Kogai"* was not in the Japanese dictionary, and my friends asked me, "What is *Kogai*?" I had to explain the ABC's of *Kogai*. But after only 10 years, *Kogai* had become a fashionable word, and people began to call all social disasters *"Kogai"*; by the beginning of the 70s, the newspapers reported daily on *Kogai*.

In the 1960s the Japanese economy had not yet developed from the Meiji Era. *Kogai* could be established as accompanying the rapid growth of the economy, along with Minamata Disease, Itai-Itai Disease, Yokkaichi asthma, PCB toxin, photochemical smog, traffic noise, ocean oil pollution, and so on. Year by year, new *Kogai* had occurred repeatedly without the solution to old *Kogai*; the damage had extended from metropolitan areas to local areas and from the human damage to natural and cultural damage. Japan had become the test site of *Kogai* generated by the rapid growth of capitalism. The Japanese were like guinea pigs.

In 1960, a new type of the pollution problem occurred in Yokkaichi City, which was the site of the first petrochemical complex, an oil refinery, and a power station on the coast of the Pacific Ocean near Nagoya City. The central government, local governments, and industries had hidden the fact of *Kogai*, because the petrochemical complex introduced by regional development policy in Yokkaichi was the leading component of rapid economic growth.

When I researched this petrochemical complex in 1962, the representative of the oil refinery factory said to me, "The cause of oil-polluted fish is not factory drainage, but oil flowing out of tankers sunken during the Second World War." The big company charged the cause of pollution to the ghost of war. The answer was the same in regard to Minamata Disease. Furthermore, the vice-manager of another petrochemical factory answered, "The smell of this factory is not unpleasant because our factory produces aromatic materials." What a cooked-up story! If the volume of "aromatic materials" leaking out from the pipe is over the allowed limit, the health of those who breathe the odor will suffer.

At the beginning of the 60s, the big companies camouflaged the clear pollution as mentioned in the cases above because the government and the big companies had emphasized that the new petrochemical complex would not cause as much pollution as the old coal complex. Many researchers working for government and big companies insisted upon this opinion. The researchers of city planning at Tokyo University made the Yokkaichi city plan in 1960, writing that Yokkaichi was a new industrial city, sunlit and green, where there was no *Kogai*. They translated this proud plan into English and delivered the research results to the rest of the world at the same time that the disease was occurring.

In the fall of 1960, the researchers of Nagoya University and Mie University investigated the pollution problems in Yokkaichi at the request of the local government. As a result, they presented a paper in mid-1961 that said that the cause of Yokkaichi's asthma was the influence of SO_2 smoke exhausted by the petrochemical complex and that polluted fish were caused by factory drainage. But the local governments did not publish these research papers, but put them in a locked safe instead.

In the fall of 1961, the labor union of the Yokkaichi city government published the secret facts of *Kogai* in the National Conference of Researching Local Autonomy *[Jichiken]*. Some journalists and researchers learned of the Yokkaichi *Kogai* from the appeal of the Yokkaichi city government labor union, and little by little they began to investigate the new type of *Kogai*. *Kogai* opinion and movement were initiated by a courageously acting labor union. However, such cases are rare; labor unions in Japan are very loyal to their own companies or the government and try to hide pollution.

However, the Yokkaichi *Kogai* made a strong impression on public opinion; people no longer believed the Tokyo University researchers who said that the building of a new factory complex would not mean *Kogai*, only rapid economic growth.

The turning point of postwar pollution problems was made when the antipollution movements opposed the introduction of a petrochemical complex in the Mishima-Shimizu and Numazu areas. This Mishima-Numazu Movement, in which the progressive mayor of Mishima also contributed personally, insisted upon "No More Yokkaichi" and gathered citizens through holding several hundred study meetings on *Kogai*. The researchers and teachers in this area participated in the movement, researching environmental evaluations, at the cost of only about $300. They published the evaluations in which they insisted that *Kogai* was possible along with this development. The Ministry of International Trade and Industry sent a research group to this area which was made up of the famous researchers of pollution problems in Tokyo. They investigated for three months and made an environmental evaluation at the cost of $56,000—for the first predevelopment evaluation; this expenditure was charged to the national budget.

The local research group had little money but was on the spot. Accordingly they were able to research the possibility of *Kogai* in an original manner with citizen cooperation. For example, they used *Koinobori* for monitoring air streams. *Koinobori* are carp-shaped kites put up with long poles in every house having a boy for one month prior to May 5th for Boys' Day Festival. The local research group assisted by high school students then investigated *Koinobori* at check points at given times. By recording the direction of the tail of the *Koinobori* on a map, they made a precise air stream map.

By contrast, the central research group, having more money, used other government organizations; they investigated the weather by military aircraft! After the investigation, the central research group published an EIS in which they stated the opinion, "There is no chance of *Kogai* occurring according to the data."

The two EIS were opposed. The local research group proposed a discussion meeting on the environmental evaluation, and one was held in Tokyo between the two groups. At this meeting, which was not political but purely scientific, the central research group could not answer some questions from the local research group. The central group's basic fault was pointed out by the local research group. A member of the central group said, "I didn't insist that *Kogai* could not occur." The EIS of the central group was not actually written by the researchers, but by an administrator of the government.

As the result of this meeting, citizen movements have grown larger. In September of 1963, the mayors and the city legislature of the two cities appealed for the halting of the proposed petrochemical complex. Then the company group, the Ministry of Trade and Industry, and Shizuoka Prefecture gave up their plan because of the local government opposition. The citizen movements won completely against a big company and the central government—the first time in postwar Japan.

There are three main features of this successful movement:

The first feature: They performed research on the spot themselves and published the results of their evaluation. They held study meetings using their evaluation; specifically a high school teacher taught about the terrible pollution problems using 2,000 slides. They held a lecture series related to environmental problems and regional development. The largest lecture had gathered 3,000 citizens. On rare occasions, the small study meeting had only five people in attendance, but the local research group held meetings untiringly 500 times during nine months. So, the Mishima-Numazu type of citizen movement was named "environmental education."[1]

The second feature: In postwar Japan, the constitution and local government law guaranteed local autonomy in decisions concerning local development. Therefore if the local governments, which depend on the citizens' opinion, opposed the development plans of big companies and the central government, development could not take place. The Mishima-Numazu Movement was different from the famous Ashio Kodoku Movement in prewar Japan. The Ashio Movement presented its need before central government in Tokyo, but the local movement was weak, stacked against the central government and, the Ashio Movement was finally defeated. The Mishima-Numazu Movement was successful

[1] Akio Nishioka, "Citizen Movements and Environmental Education—A Report From Numazu City" (1976).

because it was staged on home ground.

The third feature: This movement gathered the citizens of many different branches and different classes. Teachers, laborers, medical doctors, fishermen, managers of local companies, housewives, students, and so on participated. I would like to mention particularly that the *Chonaikai (Jichikai)*, Farm Cooperative, Fishermen's Union, Medical Association, etc., roots of conservatism in Japan, participated in the antipollution movements. *Chonaikai* is a formally organized community association of individuals, who are really forced to participate. In 1965, I wrote the English paper titled, "Grassroots of Conservatism" in which I pointed out that the local, political, commercial, and cultural organizations composing *Chonaikai* are typically the root of conservatism. Participation of these old local community organizations in antipollution movements started the decline of the grassroots of conservatism. At the same time, the Mishima-Numazu Movement began to become the grassroots of democracy.

After the victory of the Mishima-Numazu Movement, antipollution movements extended all over Japan. These movements win against big corporations. For example, Miyazu in Kyoto stopped the building of a power station. Sakai City in Osaka stopped the expansion of an industrial complex. Amagasaki in Hyogo stopped the operations of two old power stations. Usuki in Oita stopped the introduction of a cement factory. These movements used new, original methods, such as diaries of health conditions, simple monitoring instruments, epidemiological investigation of the citizenry, and so on. At the present time, the body of antipollution movements is about one thousand, grouped in Japan. There are about 80 bodies in the Osaka Prefecture, where citizen movements are the strongest in Japan.

At the same time the citizen movements were opposing the pollution problem, they were beginning to change local governments. By the end of the 1960s, almost all governors of prefectures and mayors of big cities in metropolitan areas changed from conservative to progressive, supported by the citizen movements. As a result, the environmental protection policy was strengthened.

Furthermore, public opinion against pollution aroused by the movements opened closed doors in cities governed by big factories where citizen movements could not act. In 1968 the victims of the second Minamata Disease in Niigata took the polluter to court. Soon after that, Yokkaichi victims and Itai-Itai victims took the polluters to court, and the victims of the first Minamata Disease took the polluter, Chisso Corporation, to court too.

As result of a tearful movement in the beginning of the 70s, these victims of the four pollution diseases won victories in court, and the companies compensated victims with amounts ranging from about \$27,000 to \$67,000 per victim.

These court decisions had a great influence not only on the polluters, but also on the environmental policy of the central government. For example, the Yokkaichi court decision traces the fault to the regional development policy, which gives many benefits to big companies, causing great damage to residents. The court decision stated that the fault was in industrial location and that this was caused by the failure of environmental evaluations before the development took place. Because of this decision, the government had to reconsider the National Development Plan *[Shinzenso]*.

I will now go back five years.

In 1967, the central government established the "Basic Law of Environmental Pollution Control" (Basic Law). It was called the constitution of environmental policy, but the purpose of it was actually to protect the polluter (industry) from antipollution movements. Evidence of this is seen in Section 2 of this law, concerning the purpose; namely, that it aims at harmony between economic development and the protection of the life environment. In other words, the environmental policy was held within the limits of profitability of private companies.

In 1969, the Tokyo Metropolitan Government established the epoch-making Environmental Protection Act. This Act advocated the supreme priority of environmental protection and rejected the harmony theory of the central government. It acknowledged the environmental right as a basic right of citizens in Tokyo, and it levied a heavier burden on the polluter than the national law. For example, if a corporation caused citizens to suffer because of pollution, the Metropolitan Government would stop the water supply to the polluter.

However, in Japan, national law has priority over the local act. Under the traditional thought of law, the local act cannot decide more rigid environmental standards than national law. Therefore, in the Tokyo case, the central government opposed and put pressure on the Tokyo Metropolitan Government to change its Environmental Protection Act. But Tokyo did not change this act, and it was supported by its citizens' movement and the assistance of other progressive local government.

By 1970, many drastic pollution problems had occurred throughout Japan. Cadmium pollution, PCB toxin, photochemical smog, new mercury poisoning, lead poisoning, and so on were widespread. The newspapers published environmental pollution news daily. Due to the seriousness of the situation, everyone realized that the Basic Law of the central government, which stood on the polluters' side, could not protect citizens from public hazards caused by pollution. Public opinion forced the national law to be changed.

By the end of 1970, the central government had revised the Basic Law. The new law insists upon the supreme priority of environmental protection, like the Tokyo Act, and approves authority of local governments on environmental policy. At the same time, fourteen environmental laws

TABLE 1

LOCAL GOVERNMENT CHANGES IN ENVIRONMENTAL POLICY

(million $)

	1961		*1974*	
	Prefecture	*City, Town, Village*	*Prefecture*	*City, Town, Village*
Environmental section	14	16	47	765 (1,145)*
Administrator		300	5,852	6,465
Budget		39	1,167	2,012
Reduction in sewage disposal cost		0.6		1,279
Local environment act	6	1	47	346

* Not an environmental agency, but a special administrator for environmental problems.

established and strengthened the environmental quality standards. After the decision of the Yokkaichi court in 1973, the Pollution-Related Health Damage Compensation Law (Compensation Law for short) was established.

Let us look at the statistics that indicate changing environmental administration. Table 1 presents the changes in administration and public finance of environmental protection by local governments. In 1961, only 14 prefectures of the 47 prefectures in Japan and 16 cities of 3,500 local entities had an environment section in their local government. There were only 300 special administrators for environment in Japan, and its environmental budget totaled only about $39 million.

In 1963, I had researched the environmental policy of Nagoya City, which is the third largest city in Japan and a center of heavy industry. At this time, the southern part of this city was heavily polluted. But the environmental section was then only a small branch of the section on legal epidemic protection (such as amoebic dysentery, cholera, etc.), and there were only two administrators. One of them was a graduate from the literature faculty of Nagoya University, whose speciality as a student was aesthetics. The other person was a young girl just out of high school. These were the administrators to protect 2 million citizens against many big polluters! However, Nagoya's bad situation was typical of all Japanese local governments until the end of the 60s. *Kogai* extended inevitably.

After thirteen years, in 1974, as Table 1 shows, all prefectures and 1,145 local governments had special environmental agencies, and these agencies represented one of the largest bureaus in government. The total number of administrators for environmental policy is 12,317 persons, and the total budget $3.18 billion. Compared with the 1961 figures, the administrators increased forty times and the budget increased seventy times. Such drastic change is not found in any other administration section in postwar Japan.

The Environment Agency established in 1972 united the separate branches that had dealt with environmental problems in several ministries, and the budget of the National Environmental Policy increased from $37 million in 1963 to $1.3 billion in 1975 (outside the national loan to companies of $2.6 billion).

Furthermore, as a supplement to the national law and local environmental acts, the Environmental Protection Agreement between polluters (or potential polluters) and local governments or residents was concluded. This Agreement does not carry legal weight, but does carry social and moral weight. On occasion, it is more effective than law and act; by the end of 1974 there were 7,096 such agreements made.

Those persons approved by government to oversee pollution of private industries exceed 14,000, and specialists selected through a national examination for environmental protection exceed 120,000.

Fifteen years ago, there were few researchers investigating environmental problems, and fewer economists—perhaps four or five including me. But today, many researchers are working on environmental problems. Almost all national universities include lectures on environmental problems, and some arranged special branches for environmental problems. The National Laboratory on *Kogai* was established a few years ago, and local governments have 83 laboratories for special environmental research.

Private investment for environmental protection facilities by corporations increased after 1970. In the mid-1960s private investment in environmental safeguards was small and its share in total equipment expense cannot be calculated. In 1972, the private investment for environmental protection equipment exceeded $3 billion, representing 8.3 percent of all private equipment investment. In recent years Japanese investment for environmental protection has exceeded that of the United States; but in the serious depression, investment decreased rapidly. The Japanese law system is probably more fully developed than that of other countries. Thus, it can be said that the development of Japanese environmental policy is epoch-making after 1970. And remember this change was a result of the pressure of the antipollution movement and public opinion.

However, the damage from *Kogai* has not decreased. The suffering attributable to pollution increases year by year, and designated pollution areas are increasing. In 1971, victims designated by the government numbered 3,233. In 1976, that figure increased to 33,466 as shown in Table 2. In 1973, the fishing damage cost increased to $14 million, caused by 313 pollution cases. This cost is three times that of 1969.

Of course, as the result of strengthened control, the value of SO_x under the environmental standard and the BOD (biochemical oxygen demand) on spot monitoring of some rivers decreased in many areas. But the value of NO_x increased everywhere, and photochemical smog is occurring frequently in metropolitan areas and at factory sites. Ocean pollution has grown larger; the red tide is occurring during almost all seasons on the coast of the Inland Sea.

Although the law system and administration organization was arranged more than 10 years ago, and its budget has increased rapidly, the damages of *Kogai* have not decreased. What is the cause? It is the fault of environmental policy itself and based on economic structure in Japan.

Next, I will consider environmental policy itself.

II. RECONSIDERATION OF ENVIRONMENTAL POLICY IN JAPAN

According to the Japanese experience, the policy for the environmental protection has four steps:[2]

(1) Finding of actual damages and investigation of causes,
(2) Relief of victims (compensation and restoration),
(3) Regulation to cut down pollution (containing, monitoring),
(4) Measures for prevention.

These four aspects are interrelated and must be carried out concurrently; but in the industrialized nations such as Japan where serious environmental pollution is occurring, step (1), finding of actual damage, should be taken first. For this step, it is necessary to perform an epidemiological investigation by social-medical research and find the causes of damages if any.

(1) In Japan, epidemiological investigation is lagging when compared with its necessity. For instance, we don't know yet how many patients of Minamata Disease there are or when the mercury poisoning will end in Japan. The Central Government officially recognizes 796 persons as patients of the first Minamata Disease, but over 2,000 persons have requested investigation for designation as patients, and they have waited for a long time. The research group of Kumamoto University has discovered victims taken ill as recently as 1971. Needless to say, we don't know the number of patients suffering from air pollution. The central government published a figure of about 30,000 as patients suffering air-

[2] Ken'ichi Miyamoto, "Policy for Preservation of the Environment" (1976a).

pollution-caused maladies. But if epidemiological research for damages from air pollution is conducted on a nationl scale, I think the number of air pollution victims will exceed 500,000 persons.

Two years ago, I was sent to 20 countries by the Environmental Pollution Research Committee (chaired by Shigeto Tsuru, Professor Emeritus of Hitotsubashi University) to investigate pollution problems in those countries. I found in the countries I visited that either epidemiological research was not being done or if done, its study was within limits of given conditions. When we investigated the health condition of schoolboys near four oil refineries at Torrance in the Los Angeles metropolitan area, they complained of respiratory symptoms or of an unpleasant odor from factories in certain wind directions. We asked the officers of the Los Angeles A.P.C.D. about epidemiological investigation of air pollution from oil refineries. They answered, "Epidemiological study is mainly the work of the state. We haven't made studies, but we think that air pollution is caused almost entirely by motor vehicles and not factories." After a few days, one specialist of our group visited the state health bureau and researched the available epidemiological studies on air pollution. He discovered only a paper referring to Yokkaichi asthma in Japan.

In 1975 we investigated twice two Indian reservations, White Dog and Grassy Narrows in Canada. The Canadian Government had prohibited commercial fishing in 1970 and had told tourists not to eat the fish they caught. Two tourist camps closed voluntarily because they had found that fish were polluted with high levels of mercury. Almost all the Indians lost work because of this mercury pollution. However, the Reed Company, the polluter, did not compensate anyone. The government made loans for only part of the lost income from fishing to Indian fishermen for three years. The most important problems are the testing of Indians' health condition, the stopping of eating fish, and investigation of the causes. The government had tested the mercury levels in the blood of the Indians, but neither the national nor the Ontario governments had tested the health condition of all the Indians at the time we were there. Indians without work had to depend on polluted fish for food. Our group included a medical doctor who was a specialist on Minamata Disease. We had tested health conditions of 98 persons and discovered seven patients who had a high possibility of Minamata Disease. It seemed to me that the pollution problems in this area were the same as those experienced by the Japanese until fifteen years ago.

Once the damage is found, to prevent damage, the cause of pollution should be found and the responsibility of polluter proven: However, it requires many years to prove the cause of any specific pollution and to make the polluter take the responsibility. It took seventeen years to find the cause of Minamata Disease and fully compensate a portion of the victims by Chisso. The victims of the Yokkaichi pollution had to wait for twelve years to receive compensation.

This difficulty is caused by a barrier of military, commercial, and governmental security, which has to be overcome. In particular, commercial security prevented the finding of the cause of damage. Two years ago I sent out questionnaires to all companies in an industrial complex in order to study industrial structures containing environmental problems. This investigation was assisted by the Osaka Prefecture. In this case we received answers to our questions from small industries only. All the big businesses (Shin-Nittetsu, Hitachi, Osaka Sekiyu, Kansai Denryoku, Mitsui Koatsu) we sent questionnaires to did not reply. This undemocratic attitude is shown toward all private scientific research, including newspaper research.

We also researched the economic and environmental policy of Takaishi City where an industrial complex is located. The city rejected our research because it benefitted the antipollution movement. This city, wanting to protect the industrial complex, had hidden the report on health tests of schoolboys carried out by Kansai Medical College showing a high rate of respiratory disease among schoolboys. The situation of Yokkaichi 15 years earlier was repeated. There are many more such cases. Under the market economic system, commercial and technological security has been the absolute right of business. But if this situation continues, it will not protect us from *Kogai*, and environmental science cannot really develop.

(2) The Japanese had much experience with compensation for damages suffered by pollution from the Meiji Era. If compensation is not made by the polluter, protection policy will be useless, and victims cannot survive in a community where pollution occurs. In many countries, if the polluter is found responsible for damage in court, a monetary compensation is usually offered to the sufferers. In Japan there is a system under which the government gives compensation for living and medical expenses to victims whom the government has acknowledged as being sick from *Kogai* and who are living in the polluted area if the area is approved by the pollution law. This system was established just after the Yokkaichi *Kogai* court decision.

The amount of compensation is based on the conditions of illness and the highest benefit is 80 percent of the average labor wage. Let us see Table 3. The number of beneficiaries is 33,466 as indicated in Table 2. The living and medical cost of this law is paid by the polluter. For example, the compensation for the Kumamoto Minamata patients is paid by the Chisso Company, Niigata; compensation to victims of Minamata Disease is paid by Showa Denko. The payers of water pollution are specified; but the payers of air pollution damages are not specified. An air polluter pays a rate according to its exhaust volume of SO_x. This pollution debt is gathered by a special semipublic association within the financial circle, and payment to the patients is made through the mediation of the government. The name and cost of the polluter is not published.

TABLE 2

POLLUTION-CAUSED DISEASES IN JAPAN

(February 1976)

	Victims	
	Alive	*Dead*
1st Minamata Disease	796	158
2nd Minamata Disease	590	32
Itai-Itai Disease	63	65
Arsenic poisoning	56	2
Air pollution related disease	31,961	625
Total	33,466	

Note:

These are Environment Agency certified victims.

Many victims of air pollution are living in the following areas:

	Alive	*Dead*
Osaka	8,792	141
Tokyo	5,576	2
Amagasaki	4,996	131
Kawasaki	2,723	109
Nagoya	2,407	23
Sakai	1,397	25
Yokkaichi	1,141	90
Kitakyushu	1,078	21

TABLE 3

ANNUITY BY COMPENSATION LAW

— A CASE —

(1975, $/month/person)

Age	*Man*	*Woman*
–18	$153	$158
50–54	$484	$206

Other costs of the law, such as rehabilitation of patients, are paid partly by the government.

This system is said to be an effective system because the victims will be freed from the financial burden and long-time involvement in a lawsuit. At the same time, however, the system is questioned because it obscures the responsibility of the polluter, and under this law the compensation is paid by government as social security. These benefits are therefore not true compensation as in a civil case. Fundamentally, this law was made with the purpose of checking the establishment of the law court of *Kogai*. The lawsuits for *Kogai* have step by step created a more serious burden upon industry and have not protected the polluter. For example, just after the Yokkaichi decision, lawsuits for compensation sprang up all over the polluted areas. The big companies wanted to stop dealing with the court system on *Kogai*. Then the government became the representative of the polluter by establishing the Compensation Law. This is the true purpose of this law, which the Japanese government showed with pride throughout the world. The effectiveness of this law is clear. Although the Compensation Law does not prohibit lawsuits by victims, the victims prefer the simple system more than the complex law court which costs money and a great length of time. After the Compensation Law was established, the lawsuits for compensation decreased.

I recommend that this law should be revised and that victims, including beneficiaries of this law, take the polluter to court as before. Although the compensation system in Japan has many faults, I have something to say to the people of other countries, such as the U.S., Canada, France,[3] and so on, who have no compensation system yet: I hope some type of compensation system will be established in these countries soon.

(3) How should the polluter be regulated? There are two major methods of regulating the polluter, which have to be carried out concurrently. One is to regulate the sources of pollution and another is regional planning with social capital formation and land utilization.[4] If the regulation for the source of pollution is very severe, the effectiveness of regional planning will be very limited. The Sumitomo Shisakajima copper factory is the best example of sulfur pollution beginning over half a century ago. In 1904, the Sumitomo Company removed its factory to an unpopulated inland site lacking water and 20 km distant from the Shikoku mainland because of the pressure of a strong residents' movement similar to that of Ashio. But the smoke reached the mainland across the sea 20 km away and caused damage to the agriculture. The resident movement continued for half a century against Sumitomo. Sumitomo had to pay over $4.24 million (1930 price) as compensation

[3] In France, a compensation system concerning damage from traffic pollution problems was established recently, but no general compensation law similar to Japan's was enacted.

[4] Owing to limited space, I could not write on the regional development in Japan. *Cf.* Ken'ichi Miyamoto, "Regional Development, Public Works and Environment" (1976b).

TABLE 4

ENVIRONMENTAL QUALITY STANDARDS COVERING AIR POLLUTION

Toxic Substance	*Time period/ average*	*Standard*	*Target period*
SO_x	daily average	0.04 ppm	5 years
	1-hour average	0.1 ppm	(after 1973)
CO	daily average	10. ppm	as soon
	8-hour average	20. ppm	as possible
NO_2	daily average	0.02 ppm	5–8 years (after 1973)
Suspended particulate matter	daily average	100μg/m³	as soon as
	1-hour average	200μg/m³	possible
Photochemical oxidants	1-hour average	0.06 ppm	
HC	3-hour average		

Note: From Environmental Laws in Japan.

from 1910 to 1939, and the agreement reached between the residents and Sumitomo stated that if the pollution could not be decreased by technical improvements, the plant's production could not be increased. Because of this agreement, Sumitomo began to improve the pollution protection facilities, and as a result, in the 1930s, SO_x of existing smoke was within the environmental standard established in postwar Japan. When the pollutant level of the factory decreased because of technological improvements, the long distance from the mainland was effective.

In regulation of the sources of pollution, there are problems of standards and methods. The main items of the Japanese environmental standards are shown in Table 4. Probably these standards are the most rigid in the world.

These standards were reached according to trial and error. In the beginning of the 60s the standards were vague. The government considered it only an administrative goal, and therefore it came to be called the "irresponsible target." At the present time, the standards are very severe as a result of environmental science advances, but many questions remain. For example, the rigid standards have resulted in postponement of establishment of acceptable levels for NO_x and noise because of economic and technical considerations. Mixed pollution such as $NO_x + SO_x$ + dust of chemical matter is not regulated.

When we talk of standards, we really mean those effective for protection. In many cases, the limiting or curtailment of an operation through legal or administrative order is effective for regulating pollution. Although this method is legally recognized in all countries, it has seldom been utilized. In Japan this method of injunction was written into national law and the local acts. The Environmental Protection Agreement between the company and local government or residents in Japan mentioned above approves injunction if a company breaks the agreement.

However, there are very few recent well-known injunction cases. Three years ago, Oita Prefecture stopped the operation of Sumitomo Chemical Company after explosion accidents had occurred twice. Two years ago, Okayama Prefecture stopped the operation of the Mitsubishi Oil Refinery, which spilled 40,000 kiloliters of heavy oil into the Inland Sea. These cases are rare and mainly involve accidents.

Probably the most important injunction from the standpoint of prevention is the court decision on the *Kogai* issue of Osaka Airport. The Osaka Airport site is near residential housing, similar to the Los Angeles airport. The residents suffered from noise, air pollution, smell, and fear of air accidents. They took the Ministry of Transportation as owner and controller of this airport to court. In the fall of 1975, the High Court of Osaka recognized the residents' need and ruled on these points. The decision stated that the government had to pay approximately $1600 per person for compensation for past damages. There were several thousand plaintiffs, but the number of residents in the same situation exceeded one million. Air traffic was to be stopped from 9:00 p.m. until 7:00 the next morning. The government was ordered to pay $30 per month per person until the noise standard agreed upon by the Ministry and the plaintiffs was accomplished. This decision is epoch-making—particularly important is the stopping of air traffic during midnight hours. The court, in approving an injunction against a public works, had recognized that environmental rights have priority over the prescribed public interest, which in this case actually benefited the airline companies.

(4) Damages caused by environmental pollution have irreversible absolute losses, such as damage to human health. When the absolute loss occurs, no monetary compensation will be sufficient for restoration. Therefore, in an environmental policy, the prevention of pollution should be emphasized.

For a preventive policy, the economic structure and economic policy have to change fundamentally. These changes will be discussed later.

In order to prevent pollution, assessments are required before area development takes place. In Japan, the first assessment for a development plan was in 1964 in Mishima-Numazu area as mentioned before, and just before the Yokkaichi decision, the government decided that assessments were necessary before building of public facilities would be approved. At the present time, development in the Setouchi area and reclaiming works need assessments by law, but in Japan, assessment law is not established like the National Environmental Policy Act of 1969 in

TABLE 5

COMPARING SUBSIDY TO PPP FOR ENVIRONMENTAL PROTECTION

(1974 fiscal year, million $)

SUBSIDIZING POLICY			PPP	
(1) Tax exemption for pollution protection facilities of private companies		263	(1) Burden of Compensation Law	13
(a) from national tax	163			
(b) from local tax	100			
(2) Public expenditure for industrial pollution protection work		60	(2) Private expenditure for industrial pollution protection work	73
Total		323	Total	86
(3) Public loan to private companies for their industrial pollution protection policy		735		

the United States. Environmental assessment in Japan has been proposed to the Diet twice. However, the Ministry of International Trade and Industry and the main financial circles (Keidanren, Nikkeiren, Steel Company Union, Power Company Union) were strongly opposed to the Assessment Law. Their main reason for opposition was citizen participation in the drafting of the law. This draft includes required publishing of EIS, the holding of public hearings as occasion demands, and answering citizens' questions. The way this draft is written, it is questionable as to whether citizens will have complete participation in the contents of the assessment. The draft calls for the preservation of the environment, but actually it will protect the developers from antipollution movements because the developer will be able to insist upon not being held responsible for the consequences of a development project once an assessment plan is passed. Although in principle I am for environmental assessment, this type of environment assessment draft leaves much to be desired.

Although environmental policy has progressed further than in the 1960s, its basic feature as protection of private companies from antipollution movements has not changed.

Finally, there is another point I want to present. From 1970 PPP (Polluter-Pays-Principle) was introduced to environmental policy. But, subsidies to polluters exceed the burden of PPP.

TABLE 6

COMPARISON BETWEEN THE NEW (2ND) NATIONAL DEVELOPMENT PLAN AND THE 3RD NATIONAL DEVELOPMENT PLAN (ESTIMATE)

(Targets of 1985)

	2nd Plan	*3rd Plan (estimate)*
Steel	180 million t/year	178 million t/year
Oil refinery	491 million kl/year	387 million kl/year
Petrochemicals	11 million ton/year	7.8 million ton/year
Oil imports	506 million kl	440 million kl
Factory sites	300 thousand ha.	230 thousand ha.
Industrial water	107 million m³	75 million m³

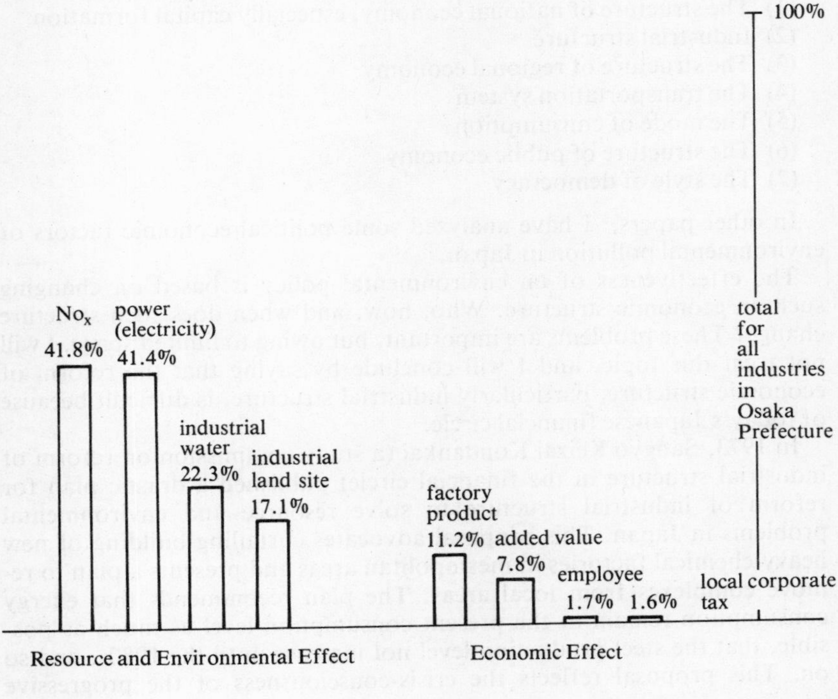

Fig. 1: Sakai-Senboku Industrial Site—The Share of Heavy-Chemical Industries in relation to All Industries in Osaka Prefecture (1974)

In the Yokkaichi case, as a result of a long tearful movement spanning 12 years, eight victims won lawsuits and received $320,000 as compensation from six big companies. But, in the same year, three of the six big companies who were defendants in the lawsuit were exempted from about $330,000 of real estate tax as a subsidy to invest in environmental protection facilities.

I can present evidence of similar situations all over Japan. Let us look at Table 5. The polluters paid about $86 million, but they received about $323 million from government as subsidies. These subsidies represent 3.8 times the PPP cost in the public sector. Furthermore, the public loan fund lent $735 million to the polluters.

III. THE ECONOMIC POLICY FOR ENVIRONMENTAL PROTECTION

Kogai in Japan is caused basically by economic structure. I have isolated seven economic and political factors that cause pollution.

(1) The structure of national economy, especially capital formation
(2) Industrial structure
(3) The structure of regional economy
(4) The transportation system
(5) The mode of consumption
(6) The structure of public economy
(7) The style of democracy

In other papers,[5] I have analyzed some political-economic factors of environmental pollution in Japan.

The effectiveness of an environmental policy is based on changing such an economic structure. Who, how, and when does such structure change? These problems are important, but owing to limited space, I will not treat this topic, and I will conclude by saying that the reform of economic structure, particularly industrial structure, is difficult because of today's Japanese financial circle.

In 1973, Sangyo Keizai Kondankai (a study commission on reform of industrial structure in the financial circle) published a drastic plan for reform of industrial structure to solve resource and environmental problems in Japan. This proposal advocates curtailing building of new heavy-chemical factories in metropolitan areas and presents a plan to remove complexes from local areas. The plan recommends that energy consumption remain at the present consumption level as much as possible, that the steel production level not increase until the 1980s, and so on. This proposal reflects the crisis-consciousness of the progressive

[5] Ken'ichi Miyamoto, "Japan's Postwar Economy and Pollution Problems" (1975). Also, "Japanese Capitalism at a Turning Point" (1974).

group of the financial circle.

After the oil crisis, the ideas in this proposal fell to the side. In 1975, the Ministry of International Trade and Industry published *The Long-Run Perspective of the Industrial Structure.* In this book, industrial structure is not changed fundamentally in the long run.

Because of the antipollution movement, the Central Government began to reconsider the 1973 *Shinzenso* [the New National Development Plan]. In the process, the National Development Agency presented a proposal for reform of the industrial and regional structure, proposing that the target of development not be economic growth but welfare. In the stagflation after the oil crisis, this drastic proposal disappeared.

As for evidence, let us look at Table 6. The economic target of the Third National Development draft is the same as the New (Second) National Development Plan. If this plan passes through the Diet, environmental pollution will continue all over Japan, particularly in local areas.

Finally I will point out an important economic conclusion.

The statistics in Figure 1 point out the total economic effect of heavy-chemical industrialization in Osaka Prefecture, the largest industrial area in Japan. This prefecture introduced a steel-chemical complex in Osaka Bay after World War II, and as a result, became the most polluted area in Japan. Figure 1 shows that heavy-chemical industrialization has not effectively increased the level of regional economy compared with the level of environmental damage. Such industrial structure uses many resources and pollutes the environment, but income, employment, and tax do not increase at the same rate. This conclusion is not only found in Osaka Prefecture, but all over Japan and all over the world.

We stand now at a turning point of industrialization.

REFERENCES

Miyamoto, Ken'ichi. 1966. "Grass-Roots Conservatism," *Journal of Social and Political Ideas in Japan*, vol. 4, August, pp. 100–106.

———. 1974. "Japanese Capitalism at a Turning Point," *Monthly Review*, vol. 26, December, pp. 15–29.

———. 1975. "Japan's Postwar Economy and Pollution Problems," *Osaka University Business Review*, vol. 26, July.

———. 1976a. "Policy for Preservation of the Environment" in *Science for Better Environment*. Tokyo: Asahi Evening News.

———. 1976b. "Regional Development, Public Works and Environment" in *Science for Better Environment*. Tokyo: Asahi Evening News.

Nishioka, Akio. 1976. "Citizen Movements and Environmental Education: Report from Numazu City" in *Science for Better Environment.* Tokyo: Asahi Evening News.

Shoji, Hikaru and Miyamoto, Ken'ichi. 1964. *Osorubeki Kogai [The Terrible Pollution Problems].* Tokyo: Iwanami Shoten.

Resources in Japan's Development*

Yasukichi Yasuba *Kyoto University*

In the last hundred years, Japan moved from a moderately resource-rich country to a resource-poor country. In the transition period before World War II, the resource problem at least looked vital and was utilized as an excuse for imperialistic expansion. A freer trade and lower ocean transportation costs freed Japan from the resource problem in the 1960s. In recent years, however, there have been some unfavorable changes that may threaten the present patterns of production and trade, involving the super–long-distance transportation of bulky commodities.

* A slightly different version of the paper was presented to a Conference on Resources in Economic History held at Bellagio, in April 1977. A shortened version will appear among the Edinburgh papers for the 1978 Conference of the International Economic History Association.

I. THE INITIAL CONDITION

Tokugawa Japan has often been described as a Malthusian economy in which the labor-to-land ratio was so high that the increase of population was suppressed by "vice and misery." It is quite true that population was kept stable through much of the latter half of the Tokugawa period, but recent studies have considerably revised this dismal picture.[1] The revisionist interpretation emphasizes the commercialization and the rise in the standard of living in the countryside in the face of the brutal force of population pressure on land.

Indeed, land was scarce; so scarce as to cause what Akira Hayami called an "industrious revolution,"[2] which represented, among other things, the substitution of human labor for animal work, the use of purchased fertilizer, more frequent weeding, the spread of double-cropping, and the intensified use of labor for cash crops and nonagricultural activities. Perhaps the most important of these nonagricultural activities toward the end of the Tokugawa period was that associated with silk. In the seventeenth century, silk was the major import of Japan, but the later transformation set the stage for the post-Perry Japan, which emerged as the exporter of raw silk.

The importation of silk and other luxuries during the Tokugawa period was made possible by the abundant supply of mineral resources. The supply of widely acclaimed gold was rather limited in reality, but silver and copper continued to be the major exports in Japan's restricted trade conducted through Nagasaki, Ryukyu, and Tsushima.[3] Iron was not in abundant supply, but production from iron sand was sufficient to satisfy the limited demand of the basically agrarian society.

The preceding short account of the initial condition on the eve of modern economic growth should not be interpreted as suggesting that Japan was already a moderately prosperous country at the end of the Tokugawa period; the fact remains that average productivity and the standard of living were still quite low.[4] Land, technology, and access to foreign trade continued to be major factors limiting further growth.

[1] An excellent summary in English of these studies can be found in Kozo Yamamura and Susan B. Hanley (1971).

[2] Report to the 1976 Conference of the Socioeconomic History Association, Akira Hayami (1977). *See* also Hayami (1973), pp. 91–106.

[3] Recently, new light has been shed on the neglected trade through Ryukyu and Tsushima by the research of Kazui Tashiro (1977). An earlier paper on the Tsushima trade was published in English (Tashiro, 1976).

[4] A recent estimate by Kazushi Ohkawa tentatively put the average per capita GNP in 1965 dollars as $140 for 1874–78 (Ohkawa, 1976, p. 3). While this is much higher than the previous figure of $74 in 1874–79 (Kuznets, 1971, p. 24), it is still considerably lower than that for other presently-industrialized countries at the beginning of modern economic growth.

II. EXPORT-LED GROWTH, 1859–99

The Treaty of Amity and Commerce of 1858, which put an end to the severe restriction of foreign trade, changed the entire picture drastically. Both exports and imports increased rapidly, with exports exceeding $10 million by 1863 and imports passing that figure two years later.[5] Silk accounted for 50 to 80 percent of exports, followed by tea and silkworm eggs. Cotton, seafood, vegetable oil, and wax were other, and rather minor, export items. Cotton fabrics and woolen fabrics were by far the most important imports, each accounting for 20 to 50 percent of total imports. Sugar, firearms, and vessels were also major imports.[6]

Some of these major exports and imports were subsequently replaced by others. Thus, the exports of cotton and silkworm eggs did not last long. Cotton soon became a major import along with cotton yarn, whose imports expanded most rapidly in the late 1860s and 1870s. Iron products and kerosene also became major imports by 1880. Copper, whose exports were obstructed by the Shogunate, and coal, which had not been widely used before, became major export items in the early Meiji period.

In 1880, about 20 years after the beginning of unrestricted trade, exports amounted to ¥29 million ($26 million) and imports to ¥49 million ($45 million). The composition of major traded goods in this year, shown in Table 1, may surprise an observer of today's Japan, but judging from the level of per capita income and the level of technology at that time, the general picture is a reasonable one. If we add some items that are missing because of the selection of a particular year, namely rice on the export side and ships on the import side, the list will show fairly well the overall characteristics of the Japanese economy then.

It is clear that at this stage natural resources were not the major limiting factor of economic activity. The Japanese economy took advantage of the newly-given opportunities and benefited by importing products embodying technology and capital, scarce factors then, in exchange for primary exports. Japan's strength in this period was in labor and in some mineral resources. None of the major exports in 1880 was highly land-intensive. Raw silk, tea, and waste silk used land for tea and mulberry bushes, but they were at the same time labor-intensive semi-manufactured goods. Tangles (seaweed), sardines, and pottery were also labor-intensive products. Copper and coal represented the mineral riches of Japan. Major exports and imports other than coal were high-price (or, at any rate, medium-price) commodities that were traded mainly with distant industrialized countries of Europe and America.

At this time, the high cost of ocean transportation was a major obstacle to the expansion of Japan's trade of bulky commodities such as

[5] Takashi Ishii (1944), pp. 52–53.

[6] Ishii (1944), pp. 83–188.

TABLE 1

MAJOR EXPORTS AND IMPORTS OF JAPAN, 1880

(1) Exports			(2) Imports		
(a) Items	(b) Quantity (L.T.)	(c) Value (1,000 yen)	(a) Items	(b) Quantity (L.T.)	(c) Value (1,000 yen)
1. Raw silk	1,863	8,607	1. Cotton yarn	16,886	7,700
2. Tea	17,909	7,498	2. Woolen cloths	—	5,792
3. Waste silk	831	1,291	3. Cotton cloths	—	5,523
4. Tangles	15,676	697	4. Kerosene	14,896[b]	1,400
5. Sardine	2,314	648	5. Wrought iron[c]	23,334	1,079
6. Potteries	—	475	6. Rice	11,710	434
7. Copper	1,545	474	7. Cotton	863	171
8. Coal	131,963	460	8. Rails	3,453	163
	(286,252)[a]	(1,086)[a]			
Total exports		28,396	Total imports		36,626

Source:
Nihon Boeki Seiran (1935), passim.

Notes:
[a] Includes sales to ships; [b] In 1,000 gallons; [c] T's, angles, etc.

coal. Fortunately, the productivity of ocean transportation of Japanese ships rose at a fast rate of about 3 percent per year in this period. Intensified competition, an improvement in port facilities and the relative decline in related factor prices also helped. As a result, freight rates on Japan's trade fell relative to other rates and also relative to the export prices. The freight rates of exported coal from Nagasaki to Shanghai fell dramatically from 84 percent of the export price in 1880 to 35 percent in 1900 and then to 20 percent in 1910.[7]

Before the Sino-Japanese War of 1894–95, the export of coal, inclusive of sales to ships, exceeded in value the import of petroleum products, while the export of copper offset the import of iron so that the net deficit in metals was small. Thus, there was little reason to worry about the scarcity of resources at that stage. There was optimism even about iron ore. A committee appointed by the government in 1891 to study the

[7] Yasuba (1977a). The annual growth rates of total productivity between 1879 and 1912, taking as inputs ships' capital service, repairs and insurance, fuels, and labor, turned out to be between 2.5 and 3.7 percent, depending on sources and measurements.

TABLE 2

TERMS OF TRADE AND ASSOCIATED INDICATORS, 1857–1900

	(1) Terms of Trade[a]	(2) Exports (million yen)	(3) Ratio of Exports to GNP (%)
1857	100.0	—	—
1865	278.3	—	—
1875	493.1	19.0	—
1880	441.4	29.0	—
1885	520.4	37.9	4.7
1890	556.4	57.8	5.5
1895	683.3	138.9	9.0
1900	592.9	217.1	9.0

Sources:
(1) Miyamoto, Sakudo, and Yasuba (1965), p. 553, and Yamazawa (1975b), p. 539; (2) Ohkawa, Takamatsu, and Yamamoto (1974), p.176; (3) The value of exports from Ohkawa *et al.* (1974), p.194.

Note:
[a] The internal terms of trade in 1857 were compared with the external terms of trade in later years. The terms of trade in Miyamoto, Sakudo, and Yasuba were linked at 1877 with Yamazawa's new estimate of the terms of trade.

feasibility of establishing an iron industry found "an abundant reserve of ore exceeding 15 million tons" in four mines alone and concluded that "as the demand for iron ore arises with the establishment of an iron factory, discoveries of huge reserves will be made in places hitherto unknown to people."[8]

The optimism did not mean that Japan was endowed with the cheap supply of iron and steel. The average price of iron in 1886–90 was £9 12s per long ton, 80 percent higher than in Britain.[9] Consequently, every effort was made to economize on the use of iron. An effort to economize on the use of capital was also made. Thus, a host of new tools and machinery created by grafting such imported technology as gears, belts, flying shuttles, and water power to indigenous tools appeared and were widely used in this period. Such was the case with *zakuri* silk reeling tools, more advanced reeling machines made of wood, *garabo* (cotton

[8] *Yawata Seitetsusho Enkakushi,* quoted in Seiichi Kojima (1945), pp. 217 and 275.

[9] The price of "iron" for Japan taken from *Nihon Keizai Tokei Sokan* (1930), p. 1109 and the price of common bars for Britain taken from Brian R. Mitchell and Phyllis Deane (1962), p. 493.

TABLE 3

OUTPUT OF VARIOUS INDUSTRIES AS A MULTIPLE OF THE "STANDARD OUTPUT," 1914 AND 1935

	1914	*1935*
Factory production as a whole	1.26	1.46
Food, beverages and tobacco	1.22	1.26
Textiles and clothing	1.32	1.92
Pulp and paper	2.31	.85
Chemicals	1.31	1.21
Coal and petroleum products	.52	1.07
Metals	.87	.87
Machinery	4.85	4.29
Transport equipment	3.17	2.27
Mining and manufacturing	1.08	1.17
Agriculture	1.05	.89
Services	.91	.95
Per capita income (in 1951 U.S. $)	113	209

Source:

Chenery, Shishido, and Watanabe (1962), pp. 102, 120.

Note:

The "standard output" (value added) is the hypothetical output of each industry corresponding to the actual per capita income and population in 1914 and 1935. It is based on Chenery's cross-section analysis for postwar years. (Chenery, 1960).

spinning) tools and various new types of looms. When Western-style cotton spinning was introduced in the 1880s and the 1890s, an effort was made to economize on capital by adopting a two-shift operation and by moving the machinery at a faster speed than in England. Again, when power looms were adopted at the beginning of the twentieth century, it was the domestically made wooden or wood-and-metal machines that spread rapidly.[10]

Since the opening of the trade, the terms of trade tended to move in Japan's favor, as Table 2 indicates. They improved 178 percent between 1857 and 1865 and 77 percent more during the following decade. In fact, the terms of trade continued to improve until almost the end of the nineteenth century. The impact on income in early years was limited because of the small size of trade in relation to GNP. Later, the trade dependency ratio approached 10 percent, and the impact on income may have been

[10] A pioneering analysis of the process of capital-saving can be found in Gustav Ranis (1957).

very substantial. However, because of the great socioeconomic disruption that took place in the intervening years, we shall refrain from making a simple evaluation of the gains. Nevertheless, it is important to realize that the expansion of exports went hand in hand with improving terms of trade. This contrasts sharply with the pattern in later years when the growth of trade and production tended to be accompanied by deteriorating terms of trade. Thus, it should be fair to describe Japan's growth in this period as export-led.

III. STRUCTURAL CHANGES, PRESSURE ON RESOURCES, AND IMPERIALISM, 1899–1945

The increase in population and in per capita income, the change in commercial policy, the influx of foreign capital, and the shift to militarism brought rapid structural changes, which tended to increase demand for natural resources from abroad.

The rate of natural increase of population gradually rose until it exceeded 1.0 percent toward the end of the nineteenth century. Part of the increased population flowed out to Hokkaido, Karafuto, Taiwan, and Korea. However, emigration to the United States was frustrated by immigration policy there, and the outflow to other countries did not reach significant proportions until the 1930s. Even the net outflow to Hokkaido was in the order of only 30 to 50 thousand per year, at most 5 to 7 percent of the annual increase of population in the first decade of the twentieth century.

It is not clear whether there was an acceleration in the growth of real output per capita. According to the most up-to-date estimate by Ohkawa and associates, the annual rates of growth of per capita output were 2.3 percent and 2.2 percent, respectively, during the upswing periods of 1887–97 and 1904–19, and 0.7 percent during each of the downswing periods of 1897–1904 and 1919–30.[11] All the same, a higher level of per capita output surely meant a greater demand for resources.

The most clear-cut change occurred in commercial policy and policy affecting direct investment. The revised Treaty of Amity and Commerce, which came into effect in 1899, allowed the Japanese government to scrap the flat 5 percent duties on both exports and imports and to set its own duties. Export duties were abolished and import duties were generally raised except in the case of some raw materials for which domestic supply was inadequate. Even though the tariff rates were generally kept at moderate levels, seldom exceeding 50 percent, protection tended to encourage more capital-intensive industries and more capital-using technology. The new tariff, the new freedom for foreigners to live and travel

[11] Ohkawa *et al.* (1974), p. 16.

within Japan outside of the port areas, and their new right to purchase and own land encouraged direct investment in industries catering to domestic demand formerly supplied by imports. In many cases, direct investment went into heavy industries, creating demand for iron and steel, other metals, and energy.

Finally, the shift to militarism created a huge demand for weapons, including naval vessels. Military expenditure, which had not been a major component of the budget until the war with China, thereafter comprised a large portion of the total expenditure of the central and local governments, exceeding 40 percent in the war years and amounting to 20 to 30 percent in peacetime.[12] Actually, the impact of military expenditure on the economy in later years was even larger than these figures suggest, since government expenditure as a proportion of the GNP increased over time. Government military factories were greatly expanded, and the private sector also responded to the increase in demand. As shown in Table 3, two military-related industries, machinery and transport equipment, were unusually large for the per capita income and population of the time in 1914 as well as in 1935.

Optimism concerning the domestic supply of iron ore disappeared even before the end of the nineteenth century with the government-owned Yawata Iron Works producing iron and steel largely with imported ore and coking coal. However, Yawata's expansion and additional supply from private factories could not keep pace with the rapid growth of demand for iron and steel, and the import of pig iron, mainly from India, and steel, from Europe and the United States, increased. Neither could domestic supply of most other natural-resource commodities catch up with the growing demand. Table 4 compiled by Ippei Yamazawa documents the rapid transformation of Japan (present territory) from a resource-surplus country to a resource-deficit country. The transition from surplus to deficit took place in agricultural food products towards the end of the nineteenth century. About the same time, the originally small deficit in metal expanded greatly. Even in minerals, including coal, Japan became a deficit country in the first decade of this century. In all these categories, deficits expanded over time to reach large magnitudes in the 1930s.

By 1930, the composition of trade was quite different from that of 1880, reflecting a fundamental change that had taken place between these two years. As shown in Table 5, most of the major export items were manufactures, particularly textile products, while major imports were food, raw materials, and fuel, including cotton, rice, soybeans and soybean cake, lumber, and petroleum. Imports of coal exceeded exports, and the imports of steel, pig iron, and iron ore were considerable (not

[12] Koichi Emi and Yuichi Shionoya (1966), p. 22. The proportion was below 15 percent in the period of arms control in the 1920s and the early '30s.

TABLE 4

EXPORTS AND IMPORTS OF SELECTED NATURAL RESOURCE COMMODITIES, ANNUAL AVERAGE BY GROUP, 1877–1936

(1,000 yen)

	(1) *Metals*		*(2)* *Minerals*		*(3)* *Agricultural food commodities*	
	Exports	*Imports*	*Exports*	*Imports*	*Exports*	*Imports*
1877–1886	1,460	2,350	1,320	1,940	8,730	260
1887–1896	5,360	8,420	5,359	4,560	13,390	6,500
1897–1906	15,840	32,860	16,540	16,860	18,900	44,670
1907–1916	48,400	74,690	23,340	27,580	31,160	62,530
1917–1926	96,460	282,930	42,720	103,580	61,280	360,740
1927–1936	142,300	240,450	41,650	260,650	55,580	461,100

Source:
Yamazawa (1975a) presented in Appendix Tables of Ohkawa and Minami (1975), pp. 578–81.

shown in the table). The structure of imports resembled that of the period after World War II, except that, unlike in the latter period, really bulky commodities such as coal and iron ore came mostly from nearby areas of East and Southeast Asia.[13]

The search for strategic raw materials and fuel, population pressure, the distorted industrial structure, protectionism abroad and depression were some of the factors supporting the imperialist adventure in the 1930s and the 1940s. The validity of these economic arguments may be questioned, since output in real terms grew after 1931 at a rate exceeding 5 percent per year. The widely publicized deterioration in the terms of trade (31 percent deterioration between 1930 and 1938) meant a loss of only 0.5 percent of real income per annum, a small figure compared with the rate of growth of output.[14]

Yet, these arguments were in fact presented in support of the imperialist adventure. Since military buildup increased demand for natural resources, a vicious circle involving resources and imperialism

[13] The reduction of ocean transportation costs made possible the importation of bulky commodities, but the reduction was not enough to allow the importation of these commodities from countries far away as in the period after World War II. For example, the freight rate of iron ore from Dungun (Malay Peninsula) to Japan was as much as 38–40 percent of the c.i.f. price of ore in the middle of the 1930s (Yasuba, 1977a).

[14] Figures were taken from Ohkawa *et al.* (1974), *passim.*

TABLE 5

MAJOR EXPORTS AND IMPORTS OF JAPAN, 1930

(a) *Items*	*(b)* *Quantity* *(L. T.)*	*(c)* *Value* *(1,000 yen)*
EXPORTS:		
1. Raw silk	27,748	416,647
2. Cotton cloths[a]	—	316,993
3. Silk cloths[a]	—	79,343
4. Rayon cloths	—	34,934
5. Potteries	—	27,171
6. Sugar	214,788	26,735
7. Coal	2,472,550	26,200
8. Flour	180,469	22,704
9. Cotton yarn	11,785	15,032
Total exports		1,871,173
IMPORTS:		
1. Cotton	573,994	369,261
2. Sugar	1,009,617	169,873
3. Rice	1,158,168	167,785
4. Petroleum[b]	2,044,472kl	83,629
5. Soybean cake	893,247	58,960
6. Soybeans	620,707	54,153
7. Lumber	2,431,600M/T	53,058
8. Wheat	476,136	41,509
9. Coal	2,918,123	36,890
Total imports		2,005,399

Sources:

Nihon Teikoku Tokei Nenkan, Showa 6-nen (1932), pp. 171-80, and *Nihon Boeki Seiran* (1935), passim.

Notes:

[a] Exports of cotton and silk cloths from Taiwan are included under cotton cloths and not under silk cloths; [b] Crude and heavy oil, kerosene and gasoline.

was thus activated. The army and civilian migrants flowed out in large numbers to Korea, Manchuria, other parts of China, and eventually to Southeast Asia and the Pacific islands. At the end of World War II, some six million Japanese were stranded in these areas only to be evacuated later.

Great emphasis was put on reducing consumption of raw materials and fuel in industry. Here the success of the iron and steel industry was particularly noteworthy.[15] The expansion of small-scale industries using labor-intensive technology and electric motors must also have contributed significantly towards economy of capital and energy resources.[16] The switch from coal-burning steam engines to electric motors itself meant huge savings in resources because power supply depended on hydroelectricity to a large extent in the interwar period[17] and because energy loss was reduced by the elimination of the extremely wasteful intrafactory transmission of power through belts.

Technical progress was considerable in the manufacturing industry, with total factor productivity rising at an annual rate of 2.4 percent, 2.2 percent, and 5.7 percent, respectively, in 1908–17, 1918–31, and 1932–38.[18] Special attention should be paid to the fact that productivity growth was faster in the textile industry, an export industry, than in heavy industry, an import industry.[19] Such a pattern of productivity growth made it possible for Japan's exports to expand rapidly in the face of deteriorating terms of trade.

IV. FROM A WAR-TORN ECONOMY TO AN "OCEANIC STATE," 1945–72

The disastrous war ended with the loss of all the territories Japan had acquired through military expansion since the Sino-Japanese War, the repatriation of more than six million people from these territories and other occupied areas in Asia, and the destruction of approximately 1/3 of industrial capital in Japan proper.[20] The accommodation of six mil-

[15] At Yawata Iron Works, the consumption of cokes in the production of a ton of pig iron decreased from 1.6–1.7 tons in 1901–02 to 1.0 ton in 1930. Kojima (1945), p. 440, and Michio Kenmochi (1964), p. 612.

[16] The important role electric motors played in the rise of the small-scale industry was recently analyzed by Ryoshin Minami (1976).

[17] Sixty-five percent of power-generating capacity in 1930 was hydroelectricity (Kagaku Gijutsucho, 1962, p. 588).

[18] Ohkawa and Rosovsky (1973), p. 73.

[19] According to Hiroya Ueno and Sokichi Kinoshita (1968), pp. 14–48, the annual rate of technical progress between 1919 and 1936 was 4.4 percent for the textile industry and 1.6 percent for heavy industry (metals and machinery).

[20] The overall loss as a proportion of the total value of assets in 1935 was a somewhat lower 25 percent (Toyo Keizai Shinpo-sha, 1950, p. 789).

lion people from abroad within a few years was a particularly difficult task. Floods of people migrated to rural areas, increasing the labor force engaged in the primary sector (agriculture, forestry and fishery) from 44 percent of the total labor force in 1940 to 53 percent in 1947.

Yet, the population was increasing at an unprecedented rate exceeding 2 percent per annum due to the postwar "baby boom." Under the circumstances it was only natural that the legalization of abortion on economic grounds, a drastic measure by international standards of the time, was accepted without much resistance. Subsequently, the crude birthrate went down from 34 per 1000 in 1948 to 28 per 1000 in 1950 and then to 18 per 1000 in 1956; since 1956 the birthrate has remained at the same level, and the rate of increase of population has been kept at about 1 percent. The net reproduction ratio has been approximately one, a figure implying zero population growth in the long run.

The restriction of population growth helped, but it was basically economic growth and easier access to markets, raw materials, and fuel abroad that solved the population pressure on domestic resources. The major roles in growth were played by capital formation and technical progress. Capital formation as a proportion of GNP increased from 24 percent in 1952–59 to 36 percent in the 1960s, and in the first three years of the 1970s to 38 percent, twice as much as in 1934–36. A large proportion of new capital was channeled, through various policy measures, into manufacturing industry with the result that capital stock there increased at an annual rate of 11 percent between 1953 and 1963[21] and 14 percent between 1960 and 1971.[22]

The rate of growth of productivity has been estimated by a number of economists. For international comparison, it is best to cite the recent estimate by Edward F. Denison and William K. Chung. According to them, the growth rate of total factor productivity in Japan's private and public corporate sectors between 1953 and 1971 was 4.9 percent per year, as compared with 1.9 percent for the U.S. (1948–69) and 3.5 percent for West Germany (1950–62).[23] The rate of growth of productivity in manufacturing industry was higher, 6.0 percent in 1952–61[24] and 5.2 percent in 1960–71.[25] Unlike in prewar years, it was "heavy and chemical industries" such as metals, electric machinery, transport equipment, and

[21] Kunio Yoshihara (1973), p. 272.

[22] Miyohei Shinohara and Kiyoshi Asakawa (1974), p. 5.

[23] Edward F. Denison and William K. Chung (1976), pp. 98–99. A similar estimate by Kanamori puts the residual at 6.1 percent for 1955–68 (Kanamori, 1972, p. 159). Other recent estimates include 7.8 percent for total factor productivity in the manufacturing sector (1953–65) by Yoshihara (1973) and 4.4 percent for total factor productivity in the private and public corporate sectors (1952–68) by Ezaki and Jorgenson (1973).

[24] Tsunehiko Watanabe (1970), p. 120.

[25] Shinohara and Asakawa (1974), p. 23.

TABLE 6

THE PRICES OF ALL IMPORTS AND MAJOR RAW-MATERIAL IMPORTS RELATIVE TO THE AVERAGE PRICE OF EXPORTS, 1953–1972

(1953–57 = 100.0)

	All imports	*Metallic materials*	*Iron ore*	*Coal and petroleum*	*Textile materials*	*Food*
1953–57	100.0	100.0	100.0	100.0	100.0	100.0
1958–62	90.3	89.0	91.3	86.7	80.8	87.2
1963–67	96.9	91.4	87.5	75.5	85.7	100.0
1968–72	91.8	90.4	69.4	70.8	71.8	92.1

Source:
Gaikoku Boeki Gaikyo, various issues.

paper and pulp that made the fastest progress.[26]

Fast technical progress in these heavy and chemical industries occurred at the right time. By the late 1950s the demand for rapidly expanding capital formation and for personal consumption at the higher standard of living was more than offsetting the loss of military demand and shifting the demand pattern toward emphasis on these industries. In trade, even though lower barriers to imports in industrial countries helped the expansion of Japan's exports, it was clear that there was limit to the growth of traditional exports of Japan caused by the rise in wages, the emergence of substitutes (nylon for silk), and the rise of protectionism abroad (cotton textiles). A rapid technical progress in heavy and chemical industries proved to be timely in this regard as well.

After the middle of the 1950s, the terms of trade turned to Japan's favor (Table 6). The average price of imports relative to the average export price fell about 8 percent between 1953–57 and 1968–72. Particularly noteworthy was the decline in the relative prices of petroleum and coal and of textile materials. These prices declined by nearly 30 percent relative to the export price in the same period. The fall in the relative price of metallic materials as a whole was comparatively small (10 percent), but the decline for iron ore was as much as 31 percent.

Responding to the shift in the structure of demand, productivity and

[26] Sample rates of growth of productivity of more dynamic industries in 1952–61 are 11.1 percent for electric machinery, 10.6 percent for paper and pulp, 9.1 percent for metals, and 8.7 percent for transport equipments (Watanabe, 1970, p. 120). Comparable rates in 1960–71 are 10.1 percent for electric appliances, 10.6 percent for optical instruments, 7.3 percent for heavy electric machinery, and 7.1 percent for synthetic fibers (Shinohara and Asakawa, 1974, p. 23).

TABLE 7

OUTPUT OF VARIOUS INDUSTRIES AS A MULTIPLE OF THE "STANDARD OUTPUT," 1950, 1955, 1960, 1965 and 1970

	1950	*1955*	*1960*	*1965*	*1970*
Light industry	0.93	1.07	1.10	1.12	1.04
Food	0.63	0.90	0.88	0.93	0.91
Textiles	0.70	0.71	0.84	0.85	0.80
Heavy and chemical industry	1.51	1.42	2.04	2.11	2.02
Chemical	0.55	0.75	1.06	1.33	1.86
Primary metal	6.20	4.25	4.83	3.27	2.36
Machinery	3.70	2.64	3.52	2.39	2.03
Electric machinery	1.93	1.75	3.06	2.32	2.29
Transport equipment	2.43	1.65	3.32	2.97	2.33
Manufacturing (total)	1.05	1.16	1.44	1.53	1.53
Agriculture	1.69	1.36	1.19	0.95	0.78
Mining	8.68	5.11	3.23	1.67	0.73

Source:
Ministry of International Trade and Industry (1972), pp. 332–35.

Note:
The "standard output" (value added) is the hypothetical output of each industry corresponding to the actual real per capita GNP and population of the time. It is based on the regression equation derived from a 22-country cross-section analysis for 1963. The method is similar to that in Table 3.

costs, industrial structure changed toward an emphasis on heavy and chemical industries,[27] increasing the demand for natural resources. Table 7 shows that some of the heavy industries such as primary metals and machinery were already quite large for per capita income and population of the time as early as 1950. Later, the relative importance of these industries became somewhat smaller, while other heavy and chemical industries caught up with them. By 1970 the actual output was about twice as large, in each of the major heavy and chemical industries, compared with the "standard output" expected from per capita GNP and population. In contrast, light industry has been about as large as expected and agriculture and mining became substandard by 1970.

The shift of trade structure was somewhat delayed, but by the middle of the 1960s a lopsided structure of exports and imports emerged and has continued to the present day.[28] All the items in the list of 10 most important exports in 1972 (Table 8) were manufactured commodities, most-

[27] Heavy industrialization has been advocated most consistently by Miyohei Shinohara. *See,* for example, Shinohara (1968).

ly the products of heavy and chemical industries, and they were shipped not just to nearby countries but to the farthest corners of the world. All the items in the list of 10 major imports were natural-resource commodities including raw materials, fuel, and food, all of which except for cotton and wool can be considered bulky commodities.

The prominence of electronic appliances, motorcycles, synthetic fabrics, and scientific and optical instruments in exports should be readily understandable. Neither should it stretch the imagination to understand the performance of automobiles and ships. However, the emergence of iron and steel as the most important export item and the prominence of crudely-fabricated metal products and bulky chemicals such as fertilizer (not included in Table 8) may be more difficult to understand. Technical progress may have been particularly rapid in industries producing these commodities, but without the revolutionary change in ocean transportation and related activities, these crudely-fabricated resource-using commodities would not have become prominent in exports, nor would the super–long-distance transportation of large quantities of bulky import goods have been feasible.

Several related changes made the transport revolution possible. First, virtually all the new industrial complexes using natural-resource commodities were located facing the sea. Secondly, loading and unloading were made exceedingly efficient, inviting the emergence of specialized bulky-commodity carriers. Finally, the automation of the operation of ships was pushed to its limit, raising the efficiency of carriers of bulk cargoes enormously. A tentative estimate put the growth rate of total productivity of the newly-built bulk carriers at as much as 9 to 11 percent per year for the 1950s and the 1960s in the case of oil tankers (1950s and 1960s) and ore carriers (1960s).[29]

The rise in productivity induced the reduction of the freight rate and the freight factors for bulky commodities despite the rise in wages. Most spectacular are the cases of crude oil and iron ore, where average distance of transportation for Japan's imports became longer, and yet the average freight factors declined considerably. In the case of crude oil, the average distance of transportation increased from 6,100 to 6,400 miles between 1960 and 1972, while the average freight rate per metric ton decreased from $5.82 to $3.76, reducing the freight factor from 34 percent to 19 percent. In the same period, the average distance of transportation for iron ore increased from 4,400 miles to 6,000 miles, and yet the freight rate was reduced from $5.50 per ton to $3.66 per ton, and the freight factor from 37 percent to 32 percent. In the case of coal, the reduction in the freight rate was less impressive in the 1960s, but a substantial

[28] Japan's capacity to change the composition of exports was fully discussed in Kanamori (1968).

[29] Yasuba (1977b).

TABLE 8

MAJOR EXPORTS AND IMPORTS OF JAPAN, 1972

(a) *Items*	*(b)* *Quantity* *(1000 metric tons unless otherwise stated)*	*(c)* *Value* *(million yen)*
EXPORTS:		
1. Iron and steel	21,374	1,111,976
2. Automobiles	2,029[a]	913,069
3. Ships	9,887[b]	739,938
4. Radios	37,790[a]	318,195
5. Metal products	—	307,325
6. Motor cycles	2,207[a]	255,558
7. Synthetic fabrics	—	250,857
8. Scientific and optical instruments	—	237,673
9. Tape recorders	26,715[a]	202,942
10. Television sets	5,836	174,028
Total exports		8,806,072
IMPORTS:		
1. Crude oil	249,193[c]	1,209,669
2. Lumber	44,836[d]	531,916
3. Iron ore	111,520	392,685
4. Coal	49,278	332,043
5. Ores of non-ferrous metals	16,048	313,023
6. Non-ferrous metals	—	283,983
7. Cotton	867	190,846
8. Petroleum products	—	165,984
9. Soybeans	3,396	146,046
10. Wool	363	143,225
Total imports		7,228,979

Source:

Nihon Tokei Nenkan, 1973/74 (1974), pp. 292–95.

Notes:

[a] In thousand units; [b] In thousand gross tons; [c] In thousand kl; [d] In thousand cubic meters.

TABLE 9

PRICE OF IRON ORE AND COKING COAL JAPAN, U.S. AND WEST GERMANY, 1953–1972

(U.S. $ per metric ton)

	(1) Iron Ore			*(2) Coal*		
	(a) Japan	*(b) U.S.*	*(c) West Germany*	*(a) Japan*	*(b) U.S.*	*(c) West Germany*
	Average factory price, c.i.f.	*Bessemer-class, old area*	*Average factory price, c.i.f.*	*Average import price, c.i.f.*	*Consumer price*	*Ruhr at mines*
1953–57	16.50	10.65	14.18	21.80	10.37[a]	12.62
1958–62	14.57	11.52	11.70	18.11	10.84[b]	14.77
1963–67	13.26	10.82	9.86	15.80	10.71	16.32
1968–72	11.70	11.10	10.40	19.13	11.76[c]	19.34[d]

Source:

Tekko Tokei Yoran, various issues.

Notes:

[a] 1954 and 1957; [b] 1958, 60, 61, and 62; [c] 1968–69; [d] 1968–71.

economy was achieved in the 1950s.[30]

The prices of iron ore and coal in different countries shown in Table 9 are instructive. In the case of iron ore, the price in the 1950s was much higher in Japan than in the United States and in West Germany, the average price between 1953 and 1957 being $16.50 per metric ton in Japan and $10.65 and $14.18 for the latter two countries. By 1968–72, the price differentials narrowed significantly with the average price being $11.70, $11.10, and $10.40, respectively, for the three countries. The very substantial fall of the Japanese price is particularly noteworthy. A similar reduction in price differentials took place for coal. In this case, the reduction in Japanese price and the increase in German price were most pronounced.

The superior efficiency and, until recently, lower wages at Japanese mills were the decisive factors, but without the reduction of ocean transportation costs relative to land and lake transportation costs, such an oddity as the Japanese exports to the United States of a large amount of steel made with coking coal imported from Virginia and iron ore im-

[30] Yasuba (1977b).

TABLE 10

RESOURCE REQUIREMENT PER UNIT OF FINAL DEMAND, EXCESS OR SHORTFALL IN REQUIREMENT AS A PROPORTION OF JAPAN'S ACTUAL REQUIREMENT, 1970

(percent)

	(1) *Energy*	*(2)* *Agricultural and fishery products*	*(3)* *Minerals other than energy*
A. Total effect (actual requirement)			
U.S.A.[a]	10.9	−26.5	−59.0
United Kingdom[b]	2.2	−30.8	23.8
France[c]	−2.5	14.6	−41.2
Germany (West)	1.5	−34.2	−33.9
B. Effect on Japan's requirement if the demand structure of the specified country is used.			
U.S.A.[a]	64.8	−20.2	−34.8
United Kingdom[b]	101.7	−5.6	−17.4
France[c]	30.1	26.5	−16.0
Germany (West)	86.8	9.3	10.9
C. Effect on Japan's requirement if the input structure of the specified country is used.			
U.S.A.[a]	−5.1	−15.4	−28.9
United Kingdom[b]	−23.8	−31.0	30.5
France[c]	−6.2	−11.2	−34.3
Germany (West)	−28.0	−23.5	−53.1

Source:

Economic Planning Agency (1974), p. 291.

Notes:

[a] 1967; [b] 1968; [c] 1965.

A. 1930

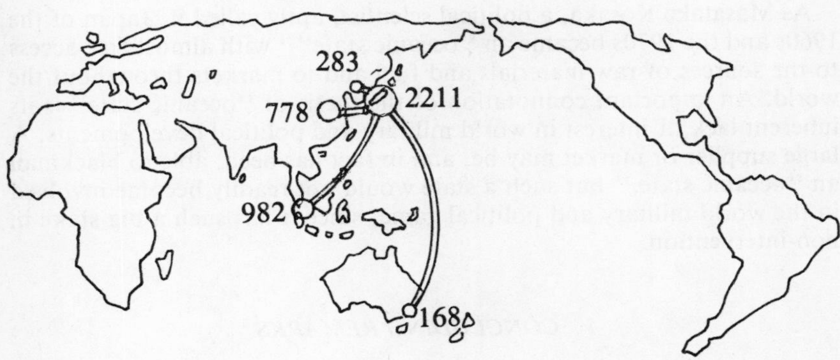

B. 1971

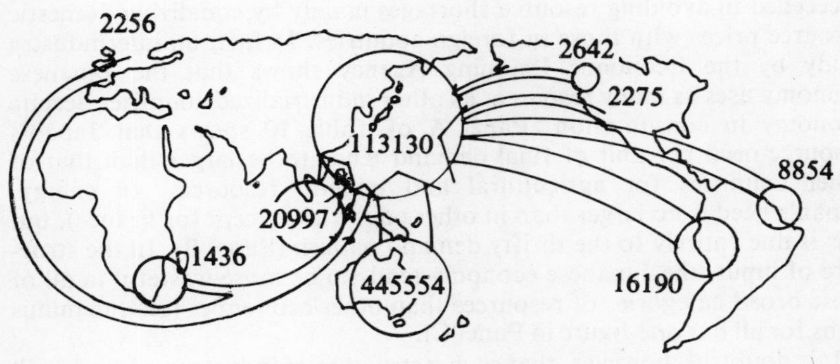

Fig. 1. Imports of Iron Ore to Japan, 1930 and 1971

Sources:

A: China, Straits Settlements and Australia; *Nihon Gaikoku Boeki Nempyo, Showa 5-nen*, Part II (1932), p. 337; Korea; *Hompo Kogyo no Susei, Showa 5-nen* (1931), p. 81.

B: OECD (1973), p. 107.

Note:

Figures are in units of thousand long tons.

ported from Brazil would not have occurred. Figure 1 shows for the case of iron ore the striking contrast in the geographical pattern of imports before and after World War II. Similar figures can be drawn for a number of other bulky commodities.

As Masataka Kosaka, a political scientist, aptly called it, Japan of the 1960s and the 1970s became an "oceanic state"[31] with almost free access to the sources of raw materials and fuel and to markets throughout the world. An important connotation of the thesis of "oceanic state" is its inherent lack of interest in world military and political developments. A large supplier or market may be, and in fact has been, able to blackmail an "oceanic state," but such a state would not readily become involved in the world military and political game, since it has such a big stake in non-intervention.

V. CONCLUDING REMARKS

The Japanese have contributed to conserving natural resources by consuming less food per capita, by living in smaller houses, and by using smaller consumer durables compared with the people in other industrial countries.[32] But, the major thesis of this paper is that Japan of today has succeeded in avoiding resource shortages mainly by equalizing domestic resource prices with those in foreign countries. In fact, an interindustry study by the Economic Planning Agency shows that the Japanese economy uses as many resources as other industrialized countries despite economy in consumption. Panel A of Table 10 shows that Japan's resource-need per unit of final demand tends to be larger than that of other countries for agricultural and mineral resources. In energy, Japan's need is no larger than in other countries (except for France), but this is due entirely to the thrifty demand pattern (Panel B). In the structure of input, the Japanese economy tends to be more wasteful in all of these broad categories of resources than other economies (see the minus signs for all but one figure in Panel C).

It is doubtful, however, that such a structure of industry and trade will continue into the future. Several external factors that may induce changes are already apparent. First, the high price of oil appears to be more or less permanent, and the cartel movement may spread to a few other commodities in the future as business conditions pick up. More important might be the change in the outlook of those people in resource-producing countries who are in a position to evaluate the intergeneration distribution of income. If they decide that the present rate of the depletion of underground resources cannot possibly be compensated

[31] Masataka Kosaka (1965).
[32] As emphasized by Ohkawa and Rosovsky (1973), pp. 157–66.

for by capital formation, output may be curtailed without resort to international cartels.

Secondly, the "protectionism" of resource-producing countries may become important in the future. As Katherine Saito found in her recent study, there was a systematic tendency for oil-producing countries to supply petroleum products domestically at prices far lower than in other countries. After the "Oil Crisis," such price differentials widened still further.[33] If such discrimination against nonresource countries continues into the future and spreads to other resources, the present patterns of international division of labor will be threatened.

Thirdly, the "Oil Crisis" brought an increase in ocean transportation costs by quadrupling the cost of bunker oil. Actually, the tramp shipping rates went down rather than up because of world recession and the decrease in the demand for oil tankers. However, it is expected that the increase in real costs will eventually push up the shipping rates.

Fourthly, protectionism may spread to industrialized countries, particularly if the world cannot solve the problem of unemployment. Since Japan's economic influence in the world is no longer insignificant, its cultural isolation and egocentric economic policy may well invite resentment and discrimination from the Western World. The so-called free ride on defense and less-than-enthusiastic participation in economic assistance may also provide excuses for discrimination.

The first factor will cause Japan's terms of trade to deteriorate. The second factor will not only cause the deterioration in the terms of trade but also adversely affect the competitive position of industry in Japan relative to that in resource-owning countries. The third factor will also cause the deterioration in the terms of trade and the decline in the competitive power of Japanese industry relative to that in resource-owning countries *and* in other countries closer to these countries. The fourth factor could be most damaging if it takes the form of discrimination against Japan.

These changes in external conditions certainly look ominous, particularly if it is remembered that some of them are future changes that will occur on top of the 35 percent deterioration in the terms of trade in the last five years. But then, history tells us that external threats have always been exaggerated. So, to strike a balance, I would like to conclude the paper by mentioning a few factors that tend to ease the tension.

First, Japan's propensity to import ore rather than metal is likely to be rectified partly by changes in external conditions but partly by the domestic demand for better environment. Secondly, the government will

[33] Katherine W. Saito (1975), p. 19. For instance, the price of regular gasoline in July 1973 was 24 cents per gallon in oil-producing countries and 68.9 cents in other countries. The prices after the Oil Crisis (July 1974) were respectively 26.1 cents and 107.4 cents. Similar patterns can be found for other petroleum products.

be more concerned with what is happening in the rest of the world. Military buildup is unlikely unless the foreign threat takes an extreme form. Instead, the government will be more eager to cooperate with foreign countries in economic, social, and cultural areas. Finally, growth, which has already slowed down considerably, is expected to be moderate, easing pressure on resources. After all, the Japanese have been working and saving too much for too long, assuming responsibility for a disproportionately large share of the polluting industries of the world.

REFERENCES

BOOKS AND ARTICLES:

Chenery, Hollis B.; Shishido, Shuntaro; and Watanabe, Tsunehiko. 1962. "The Pattern of Japanese Growth: 1914–1954," *Econometrica,* vol. 30, no. 1, January, pp. 98–139.

Denison, Edward F. and Chung, William K. 1976. "Economic Growth and Its Source," in *Asia's New Giant.* Edited by Hugh Patrick and Henry Rosovsky. Washington, D.C.: Brookings Institution.

Emi, Koichi and Shionoya, Yuichi. 1966. *Zaisei Shishutsu [Public Expenditure].* Long-Term Economic Statistics. vol. 7. Tokyo: Toyo Keizai Shinpo-sha.

Ezaki, Mitsuo and Jorgenson, Dale W. 1973. "The Measurement of Macroeconomic Performance in Japan, 1951–1968," in *Economic Growth: The Japanese Experience since the Meiji Era.* Edited by Kazushi Ohkawa and Yujiro Hayami. Tokyo: Japan Economic Research Center, pp. 286–361.

Hayami, Akira. 1973. *Nihon ni okeru Keizaishakai no Tenkai [Evolution of the Economic Society in Japan].* Tokyo: Keio Tsushin.

———. 1977. "Keizaishakai no Seiritsu to sono Tokushitsu" ["Formation of an Economic Society and Its Characteristics"] in *Atarashii Edojidai-Zo o Motomete.* Edited by Socioeconomic History Association of Japan. Tokyo: Toyo Keizai Shinpo-sha.

Ishii, Takashi. 1944. *Bakumatsu Boekishi no Kenkyu [A Study of Foreign Trade in the Last Part of the Tokugawa Period].* Tokyo: Economic Planning Agency.

Japan, Economic Planning Agency, ed. 1974. *Keizaihakusho, Showa 49-nenban [Economic White Paper, 1974].* Tokyo: Economic Planning Agency.

———, Ministry of International Trade and Industry, ed. 1972. *Tsusho Hakusho [White Paper on Industry].* Tokyo: Tsusho Sangyo Chosakai.

———, Science and Technology Agency, ed. 1962. *Nihon no Shigen [Resources in Japan].* Tokyo: Daiyamondo-sha.

Kanamori, Hisao. 1968. "Economic Growth and Exports," in *Economic Growth: The Japanese Experience since the Meiji Era.* Edited by Lawrence Klein and Kazushi Ohkawa. Homewood, Illinois: Richard D. Irwin, pp. 303–25.

———. 1972. "What Accounts for Japan's High Rate of Growth," *Review of Income and Wealth,* vol. 18, no. 2, June, pp. 155–71.

Kenmochi, Michio. 1964. *Nihon Tekkogyo no Hatten [Development of the Iron and Steel Industry in Japan].* Tokyo: Toyo Keizai Shinpo-sha.

Kojima, Seiichi. 1945. *Nihon Tekkoshi: Meiji-hen [History of the Iron and Steel Industry in Japan: Meiji Period].* Tokyo: Chikura-shobo.

Kosaka, Masataka. 1965. *Kaiyo Kokka Nihon no Koso [Outlook for Japan as an "Oceanic*

State"]. Tokyo: Chuokoron-sha.

Kuznets, Simon. 1971. *Economic Growth of Nations: Total Output and Production Structure.* Cambridge, Mass.: Harvard University, Belknap Press.

Minami, Ryoshin. 1976. "The Introduction of Electric Power and Its Impact on the Manufacturing Industries: With Special Reference to Smaller Scale Plants," in *Japanese Industrialization and Its Social Consequences.* Edited by Hugh Patrick. Berkeley: University of California Press.

Mitchell, Brian R. and Deane, Phyllis. 1962. *Abstract of British Historical Statistics.* Cambridge: Cambridge University Press.

Miyamoto, Mataji; Sakudo, Yotaro; and Yasuba, Yasukichi. 1965. "Economic Development in Preindustrial Japan, 1859-1894," *Journal of Economic History,* Vol. 25, no. 4, December, pp. 541–64.

O.E.C.D. 1973. *Maritime Transport, 1972.* Paris: O.E.C.D.

Ohkawa, Kazushi. 1976. "Initial Conditions: Measures of Economic Levels and Structure and Their Implications—Rough Notes," prepared for the Research Planning Conference on Japan's Historical Development Experience and Contemporary Developing Countries: Issues for Comparative Analysis, International Development Center of Japan. April 2–4. Mimeo.

——— and Minami, Ryoshin. 1975. *Kindai Nihon no Keizai Hatten [Economic Development of Modern Japan].* Tokyo: Toyo Keizai Shinpo-sha.

———; Noda, Osamu; Takamatsu, Nobukiyo; Yamada, Saburo; Kumazaki, Minoru; Shionoya, Yuichi; and Minami, Ryoshin. 1967. *Bukka [Prices]. Long-Term Economic Statistics.* Vol. 8. Tokyo: Toyo Keizai Shinpo-sha.

——— and Rosovsky, Henry. 1973. *Japanese Economic Growth: Trend Acceleration in the Twentieth Century.* Stanford, California: Stanford University Press.

———; Takamatsu, Nobukiyo and Yamamoto, Yuzo. 1974. *Kokumin Shotoku [National Income]. Long-Term Economic Statistics.* Vol. 1. Tokyo: Toyo Keizai Shinpo-sha.

Ranis, Gustav. 1957. "Factor Proportions in Japanese Economic Development," *American Economic Review,* vol. 47, no. 4, September, pp. 594–607.

Saito, Katherine W. 1975. "Petroleum Taxes: How High and Why?," *Finance and Development,* December; based on "An Examination of Changes in the Retail Price and Taxation of Petroleum Products," 1975, IBRD, Studies in Domestic Finance No. 9. Mimeo.

Shinohara, Miyohei. 1968. "Patterns and Some Structural Changes in Japan's Postwar Industrial Growth," in *Economic Growth: The Japanese Experience since the Meiji Era.* Edited by Lawrence Klein and Kazushi Ohkawa. Homewood, Illinois: Richard D. Irwin, pp. 278–302.

——— and Asakawa, Kiyoshi. 1974. "Gijutsu Shimpo no Sangyobetsu Keisoku" ["Measurement of Technical Progress by Industry"], *Keizai Bunseki,* no. 48, July.

Tashiro, Kazui. 1977. "17-Seiki Koki, 18-Seiki Nihongin no Kaigai Yushutsu" ["Exports of Japanese Silver in the Latter Half of the 17th Century and 18th Century"], in *Atarashii Edojidai-Zo o Motomete.* Edited by Socioeconomic History Association of Japan. Tokyo: Toyo Keizai Shinpo-sha.

———. 1976. "Tsushima han's Korean Trade, 1684–1710," *ACTA ASIATICA,* 30. Tokyo: The Toho Gakkai.

Toyo Keizai Shinpo-sha, ed. 1950, *Showa Sangyoshi [History of Industry in the Showa Period].* Vol. 3. Tokyo: Toyo Keizai Shinpo-sha.

Ueno, Hiroya and Kinoshita, Sokichi. 1968. "A Simulation Experiment for Growth with a Long-Term Model of Japan," *International Economic Review,* vol. 9, February, pp. 14–

48.

Watanabe, Tsunehiko. 1970. *Suryo Keizai Bunseki [Quantitative Economic Analysis].* Tokyo: Sobun-sha.

Yamamura, Kozo and Hanley, Susan B. 1971. "A Quiet Transformation in Tokugawa Economic History," *Journal of Asian Studies,* vol. 30, no. 2, February, pp. 373–84.

Yamazawa, Ippei. 1975a. "Boeki Kozo" ["Structure of Trade"], in *Kindai Nihon no Keizai Hatten.* Edited by Kazushi Ohkawa and Ryoshin Minami. Tokyo: Toyo Keizai Shinpo-sha.

———. 1975b. "Yushutsunyu Kakaku Shisu" ["Index of Prices of Exports and Imports"], in *Kindai Nihon no Keizai Hatten.* Edited by Kazushi Ohkawa and Ryoshin Minami. Tokyo: Toyo Keizai Shinpo-sha.

Yasuba, Yasukichi. 1977a. "Freight Rates and Productivity in Ocean Transportation for Japan, 1875–1939," *Explorations in Economic History.* Forthcoming.

———. 1977b. "Gaikokaiun to Keizai Hatten" ["Ocean Transportation and Economic Growth"]. Paper presented to the Annual Meeting of the Socioeconomic History Association of Japan.

Yoshihara, Kunio. 1973. "Productivity Change in the Manufacturing Sector, 1906–1965," in *Economic Growth: The Japanese Experience since the Meiji Era.* Edited by Kazushi Ohkawa and Yujiro Hayami. Tokyo: Japan Economic Research Center.

STATISTICAL SOURCES:

Gaikoku Boeki Gaikyo [The Summary Report: Trade of Japan]. Edited by Okurasho Zeikankyoku. Various issues.

Hompo Kogyo no Susei, Showa 5-nen. 1931. Edited by Shokosho.

Nihon Boeki Seiran [Foreign Trade of Japan: Statistical Survey]. 1935. Tokyo: Toyo Keizai Shinpo-sha.

Nihon Gaikoku Boeki Nempyo, Showa 5-nen [Annual Returns of the Foreign Trade of Japan]. 1930. Part II. Edited by Okurasho.

Nihon Keizai Tokei Sokan [Statistical Overview of the Japanese Economy]. 1930. Edited by Asahi Shimbun-sha. Osaka.

Nihon Teikoku Tokei Nenkan, Showa 6-nen [Statistical Yearbook of the Empire of Japan]. 1932. Edited by Naikaku Tokeikyoku.

Nihon Tokei Nenkan [Japan Statistical Yearbook, 1973/74]. 1974. Edited by Sorifu Tokeikyoku. Tokyo: Nihon Statistical Association.

Tekko Tokei Yoran [Statistical Yearbook on Iron and Steel]. Edited by Nihon Tekko Renmei. Various issues.

Comments

Akrasanee Narongchai (Thailand)

First I would like to offer my general comment about the paper. I think the paper was very well written. In only about twenty pages Professor Yasuba gave us the whole history of resources problems in Japan economic development. The methodology applied was extremely useful. These include the division of history into four periods, the comparison of Japanese output with normal output, and various calculations presented. Several important works of other prominent scholars were properly cited. It is only in the concluding remarks that Professor Yasuba did not fully use his preceding analysis, and many were speculative in nature, based on rather unconvincing evidence available elsewhere. (I will come back to this point when I make my specific comment.)

At this point I would like to make an observation. I think the history of Japanese economic development can be explained by the Japanese attitude towards natural resources. The Japanese have what I call resource phobia, or resource paranoid. This is the fear of having no resources, which I think basically explained why Japan adopted militarism before the Second World War, and commercial expansionism since then. In what follows I shall argue that this fear was too much exaggerated, and actions taken to overcome the problem of resources actually led to more problems. I think it is also the message implied by the author in the paper.

Let me now turn to specific issues in the paper. On the whole I am in agreement with the analysis presented. I would like to add a few more interpretations which may be drawn from the paper.

First it is important to recognize that resource is a *relative* problem, rather than absolute. Japan used to be net exporters of resources earlier in her history, and she became net importers after about 1900. This is a normal development. Korea became net importers of resources in 1975, and will soon compete with Japan in resources purchases. I think this is nothing but a historical fact. When we seriously think of it, what do we mean by resources? Analytically resources should be treated as raw and intermediate materials. In this sense petrochemical products are raw materials. My country Thailand as well as most other Eastern Asian countries *depend* on Japan for these raw materials. So Japan is perhaps still the net exporter of raw materials. And we are just as fearful of Japan's cutting down our supply as much as Japan is fearful of the Arab's cutting down her oil supply. Therefore a resource-deficit country in the old sense does not imply a disadvantage.

Another conclusion one can draw from the paper is that the production pattern of Japan was very much lopsided towards heavy and chemical industry (Table 6). This would be all right if a country produced heavy and chemical industry products and traded them for light industry. But when we look at imports, Japan hardly imported products of light industries (Table 8). This fact has very important implications. One, Japanese people do not have access to cheap consumer goods from overseas. Two, Japan will tend to have problems with countries, especially LDCs, from which she buys natural resources. The reason is obvious. A country like Thailand sells natural raw materials to Japan, and buys synthetic (or processed) raw materials and machinery from Japan. If trade is limited to these, it is likely that Thailand cannot afford to pay for her imports, unless she could sell light manufactured goods to Japan as well. Japan has for a long time used non-tariff barriers to prevent light manufactured goods from entering into Japan. Consequently, the ordinary Japanese have to pay exorbitant prices for consumer goods, whereas the country accumulates more and more reserves, and the value of the yen continues to rise. What kind of economy is this? On the one hand the country of Japan is rich, while the ordinary people have to continue to sacrifice. This point was also clearly stated in the last sentence of Professor Yasuba's paper.

I would now like to comment on Professor Yasuba's concluding remarks, which I consider speculative. First are the resource scarcity and the fear of resource cartels, which I touched on a little earlier.

On resource scarcity, there has not been enough evidence that the world will reach a "doomsday" in the near future. This is especially true when we understand the nature of resource development. In a recent paper Professors Stephen McGee and Norman Robins explain that a resource has a product cycle.[1] First, certain inventions create a boom in demand for resources. Second, the boom in demand leads to higher prices, which leads to two things—new supply and research and development to substitute or to save the use of that resource. Then comes the third stage when a synthetic is produced, and the resource has a perfect substitute. The synthetic uses other kinds of resources, thus creating a new cycle of demand. The implications of this theory are several. One, the world will always have something to use as raw materials. Two, the market mechanism will work in such a way that a substitute will be ready to be developed, by an advanced country. Three, the long-term trade of natural resources will remain constant. Therefore, it is not the advanced country like Japan which should worry about the problem of resources. It is LDCs which rely on exports of natural resources that should worry. This leads me to the question of cartel. Experiences have told us that except for oil, there has not been a successful cartel. The tin agreement, which has the longest life, has not been found to be beneficial to the world as a whole because of its tendency to raise prices. Even with the oil cartel, in the long run the non-oil exporting LDCs will be hurt much more by the actions of OPEC. Thus in relative terms the poor countries gain less and lose more in the development of world trade. So whenever I hear rich countries complaining about having no resources, I think it is rather unfortunate.

I also do not share the author's view on "protectionism" of resource-producing countries, especially LDCs. Even though there may be discrimination, that will be a minor problem. Resource-producing countries will always give the best treatment to the biggest buyer like Japan, no matter what they say in the press. And a country like Japan can enter into an arrangement which Professor Kiyoshi Kojima calls *"import development cum long-term contract"* such as those Japan has with Australia on iron ore, with Papua New Guinea on copper, with Indo-

[1] Stephen McGee and Norman Robins, "Raw Materials Product Cycle," paper presented at the Ninth Pacific Trade and Development Conference, August, 1977.

nesia on oil, and with even Vietnam on cattle. The long-term contract will assure Japan of a long-term supply at stable prices.

The last comment I have is that I am surprised to see Professor Yasuba write that "protectionism may spread to industrialized countries." In fact this has been happening for more than a decade, as evidenced by the fact that while tariffs were being reduced, all sorts of non-tariff barriers were used in their places. People like myself have been screaming about this for years. I do, however, understand that political reality makes it difficult for DC's to comply. But I will continue to make the point. 

I would like to end my remarks by saying that resources are not a serious problem in Japanese economic development unless Japan makes it so. And it is no one else but Japan who is capable of creating a long run serious resource problem for herself. The militarism before the Second World War, and the commercial expansionism since then, both of which created resource problems, were invented by Japan herself. I hope this "resource phobia" will be lessened, and Japan will think of other aspects of life rather than only business of securing resources. Japan can continue to grow at a moderate rate of 5–6 percent, gradually raising the standard of living of the mass of Japanese people. The growth at this rate will also be much more accommodatable to Japan's neighboring Pacific countries whose trade and economic development are so interdependent with Japan now. It is a welcoming sign that Japan is now more ready and more willing to have other forms of relationships in addition to the commercial one with her neighboring Asian countries. As Prime Minister Fukuda said in his Manila speech at the conclusion of his visit to ASEAN countries, "Japan will do its best for consolidating the relationship of mutual confidence and trust based on 'heart to heart' understanding with these countries, in wide ranging fields covering not only political and economic areas but also social and cultural areas." I am sure that Professor Yasuba shares this view, and I hope the attitude is also shared by the majority of the Japanese people.

Resource Potentials of Continental East Asia and Japan's Material Needs

E. Stuart Kirby *University of Aston and St. Antony's College, University of Oxford*

This paper stresses Japan's international economic relations, as no less determinant than the internal dynamics with which the Congress has almost exclusively dealt. Japan, though a world power economically, is positioned in Asia, where macropolitical trends necessitate consideration in terms of political economy rather than econometrics. The dominant Asian forces are the U.S.S.R. and China, which are also major future suppliers and markets. In fundamental conflict, each will strive to draw Japan and Asia to its side, utilizing the prevalent amalgam of nationalism and communism. Current models and projections of Japan's economic future are criticized on that background.

I. WORLD ECONOMY AND POLITICS

This paper is a summary discussion of some aspects of Japan's place in the world economy. The whole life of Japan—in all spheres, economic, social, and cultural—has always been primarily conditioned by its international relations. The nature, forms and directions of what happens in Japan, and the prospects for Japan as a whole, depend very much on the external, exogenous, and extrinsic circumstances and trends as well as (or even more than) the internal, endogenous, and intrinsic character and initiatives of the Japanese themselves. This was true in the past from the earliest times and was greatly accentuated in the modern period in which Japan became modernized and industrialized—thus involved in the world economy and international politics in broadly the same way as the United Kingdom, by dependence on the importation of materials and ideas and the exportation of products and services. It will remain fundamentally true in the future.

This point apparently needs some emphasis, as practically all the papers in this section of the present Congress approach Japan mainly from the aspect of Japan's internal dynamics, and in historical rather than prognostic terms. It is necessary to consider the issue more dialectically; at the present day and in future, Japan's livelihood, ideas and practices are and will be very much shaped by the international setting, but reciprocally Japan has and will have great formative influence on the world economy. The present paper stresses two further points in this connection. One is that, however much Japan may now and henceforward be in the category of the industrially and technologically "advanced" nations, it is still by location and by character very much an Asian entity, with very basic affinities and responsibilities in Asia. The other point is that an econometric approach, though certainly essential and illuminating, is not enough; the problems and the outcomes are to a great extent political and psychological. This paper will therefore be largely in terms of Political Economy rather than "purely economic" reasoning.

The present writer had hoped that other papers would chart out the worldwide and Asian situation, with special reference to Japan's present and prospective relation to that macrocosm; then this paper would concentrate on the two great Asian neighbors of Japan, China, and the U.S.S.R., especially as potential elements in the supply of and demand for resources. Since—judging at least *a priori* from the explicit titles listed—the other papers take mainly an "internal Japanese" point of view (in the terminology used above), the present essay must discuss the overall framework and place the great communist economies within it, rather than surveying the latter in detail as such.

Asia generally is strongly dominated today by nationalism, so powerfully as to make what remains of *yamato damashii* (the Japanese spirit) look comparatively cosmopolitan and accommodating, as a mentality.

There are many varieties of Asian nationalism: in fact more than the number of countries concerned, since fission continues, with local separatist movements within most of the countries. Asian nationalism is distinctly suffused, at the same time, with the other powerful movement in Asia, communism or Marxism—which has a very considerable influence within Japan also—or at least the two have widely and quite effectively formed a "united front" spanning the whole spectrum of nationalism/Marxism/socialism/anti-capitalism/anti-foreignism. To a great extent, political considerations are ruling over economic ones.

There is some convergence, really, in this: if Asian nationalisms are suffused with Marxism, *vice versa*, the great Marxisms are suffused with nationalism. The Soviet Union is probably now the most patriotic nation on earth, though offering a high-technology base. China has tremendous nationalistic confidence in itself as a cultural model for solution of the problems of the Asian peoples in particular and of poor and oppressed peoples in general. In principle, Marxism is supposed to take economic considerations as basic, the rest as superstructural (even if the two are interdependent); nationalists are supposed to seek national ways of solving heterogeneous local problems. In practice, the former are more active ideologically than economically in the international arena; the latter seek primarily economic national advantage and subscribe to economic determinism, postulating that economic development is the key to all other kinds of development.

In this welter of mixed and conflicting trends, Japan is in an invidious position: drawn to ever-increasing identification and association with the advanced capitalist group (OECD), yet suffering strongly adverse internal effects from hyperindustrialization and high technology (pollution and social stresses) and finding that its feet and its heart are in Asia though its head is in the Western world. The econometricians had gotten everything worked out very nicely, futurology even predicting that Japan would be the world's top nation by the year 2000, when the realities of Japan's dependence were brutally emphasized by the "oil shock" of 1973. It is not improbable that some more directly political shock may come if some event occurs to emphasize another brutal reality; namely that Japan is equally and deeply vulnerable to the tides of nationalism and communism in Asia.

Militarily and strategically, an "ultimate deterrent" is held objectively by the Soviet Union in particular; Japan is very vulnerable to the Russian power, not only from propinquity for bombing but from possible blockading of the sea traffic on which Japan depends and other possible pressures. China wields no such strategic menace, but the whole Marxist-nationalist complex has enormous and widespread power in Asia to raise practical, as well as militant, resistances, which could wreck any peaceable plans for general expansion of the international and the Japanese economy. These are extreme possibilities; but even short of them on some lesser plane of threat and resistance, the prospects for Japan could become somewhat uncomfortable. The corollary is that Japan must go far actually to placate, or at least not distinctly offend, the Asian communist countries (ACCs—principally the U.S.S.R. and Mainland China, but also North Korea and Vietnam), similarly to cooperate with the nationalistic Asian developing countries (ADCs).

In recent years, leading "think-tanks" in Japan have advanced two international types of initiative for Japan. One is to rise high within the group of advanced (capitalist, Western) countries. This is hardly to the Communist liking; if Japan raises its head too high in the Western orbit, it will become a giraffe, whose body and legs in Asia (or "soft underbelly," in an immortal phrase) will be viciously attacked. The implied further hyperindustrialization will also raise cardiac and digestive trouble at home—Narita, Kashima, Lockheed, etc., on a much greater scale. The other line of advance is proposals for economic unions, especially of the Free Trade Area type, *e.g.*, West Pacific, North Pacific, West Asia, South Asia, etc., Free Trade Areas—PAFTA, WAFTA, SAFTA, NAFTA ... NANJA MONJA. These are intrinsically rational, and well worked out econometrically, but again the implications, such as the opening of many doors to Japanese investment, multinational firms' activities, etc., will raise bitter opposition and antagonism. It has already been seen how much resistance there is to Japanese economic penetration on the more *ad hoc* present basis (agitation, boycott, burning of Japanese goods, etc.); it is easy to anticipate further development of the cry that such proposals represent a new Japanese economic imperialism, a new peacetime capitalist version of the former wartime militarist Greater East Asia Co-Prosperity Sphere.

Another root error is the assumption that foreigners, even fellow Asians, will react in either the form or the spirit that the Japanese themselves show. The logical processes and the customary criteria differ. There was a curious public resentment in Japan a few years ago to Western use of the expression "economic animal." This was actually founded on etymological ignorance—"animal" simply means anything having life or a spirit (as the Japanese certainly have) and even the Japanese word *dobutsu* simply means a thing that moves (though there is a linked word, *dojiru*, meaning to be perturbed; one sees that the Japanese are both mobile and prone to worry). However, foreigners may dislike the assumption that their responses are going to be those of an "economic Japanese."

China and Russia have enormous resources, potentially complementary to Japan's needs. These resources are extremely varied: this could be a point of some significance, since the progress of technology and society is quite kaleidoscopic, and flexibility is a great advantage. These two countries have the largest long-term industrialization plans in the world. This situation offers great challenges to Japan, which Japan is especially suited to meet. The present writer offers the suggestion that this too could be an important factor. The best performances (as the lady said) come out of challenge, not out of satiety. If Japan passes easily into super-affluence, it may pass also into decadence.

Obviously, China contains a quarter of the world's population and is as large as the United States. The U.S.S.R. has one-sixth of the world's land area. Siberia, the whole expanse east of the Urals and north of Soviet Central Asia, has relatively high resource-potentials over and above its own local needs. Aspects like the following are often underestimated. Siberia straddles the top of Asia. People are commonly talking about Asia and thinking only of the southern half of it. Archaic also is the Europocentric view that Siberia is remote; in fact, apart from satellite and electronic surveillance, Russians are in telescope sight of Japanese at the southern Kuriles (the return of which to Japan they refuse even to discuss) and of Americans in the Bering Strait. There are dramatic conflicts of interest about such matters as fisheries. Nor is Siberia an empty, useless quarter. It has a population half that of Britain or one-quarter that of Japan, not evenly dispersed at the average of one person per square kilometer, but actually more than 75 percent urban, in large modern cities pursuing modern occupations. Siberia has been the scene of one of the largest and technologically most sophisticated industrializations in Asia, thoroughly integrated into that of the Soviet Union as a whole.

Siberia is the world's largest remaining storehouse of untapped or undertapped natural resources. A simple illustration is that Siberia is similar geographically and geologically to Canada, with closely comparable developmental possibilities, but nearly 30 percent larger. Many of these resources are already important to the world, and to Japan in particular; given the increasing usage by the latter of resources, this great reserve will be of major importance in the future, with sharper competition for such resources in the looming sellers' market. Shorter-run calculations are based on the items that have been hitherto the staples, such as timber, coal, and other bulk minerals, with which the U.S.S.R. is well endowed. Petroleum and natural gas are now added, as acute urgencies; Siberia has the world's largest reserves of these. In the next phase, for the kind of new technology-based industries that are in view for Japan, a number of other minerals, hitherto of minor significance, may spring into key importance—somewhat as uranium, previously rather useless, did when nuclear energy came on the scene. Siberia is rich in these.

China also has variety and high potential geologically, sub-tropical as well as northern resources, offering important plant products too, but under different conditions. China has to cater, in its ambitions for a high level of industrialization, to its own huge population, on a somewhat less favorable resource base. Siberia's economic future lies in extensive, land- and capital-using development; China's in intensive, labor-using development. The forégoing considers both countries primarily as potential suppliers to Japan and the world; of course they are equally to be considered as potential markets. At this point, the argument is concerned with identifying the general perspectives; only if these are clear can detailed analysis be effective. China does not have the strategic grip on Japan that Russia has; it has however some psychological hold, a supposed cultural affinity plus some feelings of guilt about past Japanese actions in China, but these sentiments are not altogether convincing and are rapidly disappearing in succeeding generations. Nevertheless Chinese sensitivities have to be placated. China still presents a picture of instability, if not of uncertainty—great crises, drives, and fluctuations of policy in which cryptic utterances replace systematic information, and politics is conducted by graffiti (wall posters) rather than by state pronouncements. The Soviet progress seems steady, continuous, and systematically predictable, in comparison.

Both countries are however extremely secretive. The windows of Russia are made mostly of one-way glass, but some regular and systematic information is divulged. In respect to Siberia, there is one particular form of obscuration; statistical and other information on Siberia is included in the reporting for the whole Russian Soviet Federal Socialist Republic (RSFSR), of which Siberia is a part, but this RSFSR includes everything from the borders of Poland, White Russia, the Ukraine, and Central Asia to the Pacific coasts, lumping everything together from the metropolitan activities to the most distant. Specific information on China is however much scarcer; very little is known of what exactly is going on; Mainland China has supplied to the United Nations (for instance) neither its quota of staff personnel nor the normal flow of national information to that organization. This is no exaggeration: the Chinese desks in the international organization remain largely unmanned since the expulsion of Taiwan; China's trade statistics have to be compiled from the trade returns of other countries; Mainland China's national income, population, and other indicators are matters of conjecture. Any Japanese pressure for improvement in this respect, as a condition of cooperation and peaceful coexistence, would be a major contribution to international understanding.

To complete the preliminary survey of perspectives, one other aspect of the scenario is crucial. That is, the great struggle and rivalry between Russia and China, which is deep and lasting. The intensity and incompatibility of the Sino-Soviet dispute is often underestimated; the question is pursued below. It is important to the present topic because it is actually the main positive force in Asian relations at present, where all the other influences are weak and ill-defined. It was stressed above that Japan must necessarily refrain from doing anything that is too distinctly to the disapproval of either of the Asian communist great powers. What this means becomes primarily now a question of not doing anything that will benefit one of the rivals more than the other; for the great priority of each, in any transaction or set of transactions, is to disadvantage the other; above the specific gain inherent in the transaction as such, each assesses the side-effects on its antagonist. The nature and implications of the Great Schism must be considered further, as follows.

It is not merely a territorial dispute about some territory in Siberia taken in the past from China—though irredentism is still in itself cause enough for war. This is an ideological war, with all the fury and irreconcilability of a religious war, within the communist faith. There are two popes in Marxism today, or rather an established sumptuous Vatican in Moscow and a Protestant puritan poor man's Asian do-it-yourself communism centered in Peking—with other sects between and around them. Both consider that the heretic is for burning or liquidation; mere pagans can be nurtured, pending their conversion, and prevented from aiding or comforting the opponent. The two sides have excommunicated, anathematized, and unforgivably abused each other. They are embattled all along Asia's northern frontier, a much longer, more complicated line than the "Iron Curtain" in Europe, an equally heavily armed area, in which actual bloodshed has occurred. The Soviets also encircle China with a marked naval predominance. The rivals contend for influence, by all possible means, all over the world and especially in the developing countries and not least in Asia. They have thus vied, with varying success, for a whole generation now in the Indian sub-continent and in Japan, the principal pillars of the Asian economic structure at its western and eastern ends.

Asia presents a parallelogram of forces—forces conjointly economic, technological, organizational, cultural, political, and military—with the strongest at the northeastern corner. Its northern side is strong, its western one less so, the southern line faint. The center, Southeast Asia, is comparatively void, the European political and military presence there having been largely withdrawn some time ago, the American one more recently expelled; the consequent local vacuum is filled only by a heterogeneity of local nationalisms, which could do little to counter materially any advance from their north and northeast. The practical question is where the line of balance between Soviet and Chinese influence will be drawn; it is already emerging in Vietnam, a country of proven courage, which is now well armed with captured American equipment. Thus, Vietnam is able to balance between the Soviet appeal, offering mainly sophisticated technology, and the Chinese offers of methods ostensibly more Asian, yet (to keep other doors open) more on the Yugoslav pattern, *e.g.*, to draw in Japanese and other participations in the work of reconstruction and development.

Triangular relationships are now the rule. The biggest triangle is America-Russia-China, but others are possible. The communist superpowers are interested in keeping the relationships angular—and pointing all the existing and possible triangles against each other. If Japan seeks better relations with China, it must oppose Soviet "hegemonism": and vice versa. If Japan elects mainly for partnership with the West or with Southeast and South Asia, the Asian continental powers will separately seek to point the situation against any success of such plans. This propensity probably renders infeasible, in practice, any of the visionary constructs for major regional or super-regional compacts, Asian or Pacific Communities and the like, or for New World Orders more generally, that are being canvassed. Because of the inability or unwillingness, also, of the weaker Asian countries to operate such understandings, such plans are politically unrealistic, however well worked out economically.

II. JAPAN'S REQUIREMENTS

The above is the setting in which the alternatives for Japan must practically be considered. Within that external framework, what is the prospective structure of Japan's needs? And what part can Russia and China play in meeting such needs? These questions could be asked in either order; if the answer to the latter were established first, theoretically the former could be shaped to that answer. The supply possibilities of the ACCs are, however, varied and diffuse; they are known to be large, but very difficult to quantify. Whereas there is more relevant and specific evidence on the likely structure of Japan, the options from an internal Japanese point of view fall within closer and more definable limits than do the supply possibilities from the ACCs. This second part of the present paper therefore takes the former question, the next part goes into the less determinate question of the Chinese and Russian potentialities.

Out of an extensive literature, it may be useful to select one clear text at this point for that purpose: the cogent work of Professor Kiyoshi Kojima, which contains a very good exposition and criticism, *Japan and a New World Economic Order*, extant in Japanese since 1975, but just recently made available in English.[1] The first two-thirds of this book (pp. 1–119) analyze the world-economy setting as the formative matrix shaping the present and the future Japan. The exogenous problems are in effect viewed as the determining variables: stressed, in a chapter each, are the need for world solutions of trade problems, currency problems and investment problems, amounting altogether to a quest for a "new economic order," virtually prerequisite for a satisfactory solution by Japan itself of its own problems. Japan must seek a global, not merely national solution; one promoting overall economic development in "interdependence" (as stressed in this paper) "especially with neighboring Asian-

[1] Kiyoshi Kojima, *Japan and a New World Economic Order*, English translation (Boulder, Colorado, and London: Croom and Helm, 1977).

Pacific countries'' (p. 10). In that context, writes Professor Kojima, ''today Japan faces an important turning point'' (p. 9). Hitherto it ''could clearly imitate the development pattern of more advanced American and European economies,'' which it successfully did by 1965 with a rapid development of chemical and heavy industries and by 1973 with a ''phenomenal'' growth of exports. This was through a period of world-wide growth, strongly led by the United States and by international arrangements for the liberalization of trade, exchange, and investment; which however came into diminishing returns by the 1970s and was further wrecked by the oil crisis. The emergency is particularly acute for Japan, which ''depends on imports of almost every kind of primary commodity'' and must restructure its economy for five reasons that are conjointly ''both internal and external'' (p. 121). These reasons are given as follows:

(1) ''Shortage of coastal sites'' in mountainous Japan ''suitable for the location of new heavy and chemical industries.''
(2) Growth of the ''pollution problem'' and ''opposition from local residents.''
(3) The needs of ''heavy and chemical industries'' for ''raw materials and fuels'' are ''getting beyond the level at which supply from overseas is economically manageable.'' Thus Japan is ''compelled to move [to] industries less dependent on raw materials and fuels and to transfer production of intermediate goods to sites overseas.''
(4) ''Labour-intensive goods'' of ''traditional light industries such as textiles are losing their comparative advantage through labour-scarcity and rising wages'' so Japan will be ''forced to move to new types of industry.''
(5) ''For similar reasons it has become imperative to revolutionise the structure of agriculture and medium- to small-scale business so as to concentrate production exclusively in efficient large-scale firms.''

Thus, he concludes, ''through the 1970s [already] the Japanese economy must carry through a structural transformation toward 'knowledge-intensive' industries'' such as the following:

(a) ''R and D–intensive'' ones like ''electronic computers, aircraft, electric cars, industrial robots, atomic-power related industries, integrated circuits, fine chemicals, new synthetics, new metals, special ceramics and the development of the oceans.''
(b) ''Sophisticated assembly industries'' like ''communications equipment, business equipment, pollution control devices, large household coolers, teaching equipment, industrially produced housing, automated warehouses, large construction machinery

and high-grade plant."

(c) "Fashion industries" like "high quality clothing, furniture, household fittings, electric sound equipment and electronic musical instruments" and

(d) "Knowledge industries such as information processing systems, education-related video industries, software, systems engineering and consulting."

Many such developments were already foreshadowed by 1970, and the conclusion was sharpened by the oil crisis. The implications are much worked out and discussed in various "visions," "outlooks," etc., in Japan. The following comments are the present writer's own reactions. Taking first the five reasons (1–5 above):

(1) The Asian developing countries (ADCs) will say the shortage of sites is Japan's own problem, to be solved at the cost of Japan, not that of the ADCs. They are willing to host such activities on partnership terms and with heavy outlay (investment) by Japan; thus Japan would still bear a heavy charge, even if less than that of finding internal Japanese solutions. The joint nationalist and Communist opposition will say land shortage in Japan is a pretext; the real motive is a neo-imperialist one of taking advantage of cheap labor in the ADCs. Asia is not just an empty quarter into which Japanese enterprise can automatically expand, in some "industry-cycle" like that of the well-known "product-cycle" by which advanced countries shed obsolescent activities and "transfer" them to less advanced countries. Nor is the receiving-ground so efficient as the home ground in Japan; the ADCs require a great deal of preliminary infrastructural work before they can effectively host transferred "heavy and chemical" industries. They have been trying to do that infrastructural preparation ever since they became independent. In fact, they have become to a notable extent disillusioned with the "heavy industry first" approach; more "balanced" development is everywhere in vogue now; and "heavy" proposals are hardly as welcome as they were a generation ago. (Incidentally this militates also against the Marxist approach, which remains committed to the view that development must begin with iron and steel. Especially the Soviet approach; the Chinese one may be preferred in this respect, as it purportedly is more suited to the Asian grass-root conditions).

(2) The same considerations apply to the notion of "exporting" Japan's problem of "pollution," similarly opposed on mixed anticapitalist, nationalist, and antiforeign grounds. The ADCs hardly recognize that they have a pollution problem, turning a blind eye to it (for a good survey of the actualities, see the *Far Eastern Economic Review*, Hong Kong). The ADCs, with an equal lack of veracity, claim that this and other problems do not arise under socialism and that the problem would be eliminated if Japan were a socialist country, and never arise if the ADCs were socialist.

(3) Are the raw material and fuel needs really "getting beyond" manageable procurement? If so, this problem must surely be still more "unmanageable" for economically small ADCs than for industrialized Japan, which has great market (monopsony) power as a bulk purchaser and other ways of counter-pressuring the sellers of these things. If this is a factor powerful enough to "force" the emigration of mighty industries from Japan, it may be strong enough to inhibit the reincarnation of such industries in ADCs, except under heavy subsidy. Will Japan offer such subsidy? Rather than support the same industries at home, *e.g.*, by devoting the same outlay to reclaiming "coastal" or "clean" sites in Japan? The ACCs might offer such subsidies because they have political aims in ADC-Asia in a way Japan has not. The Chinese are pleased to do things like building the Tanzania railway, on the one hand; on the other they advocate a decentralized pattern of generalized development. The Soviets can point to the very considerable development of heavy and medium industry in Siberia and elsewhere in Russia, in more sophisticated style and allegedly without pollution or social stress.

The Soviets have another possible trump card if they are able to offer to Japan or others some of the rich resources of Siberia for the needs of heavy industries in those countries, and/or new markets for their products. The Japanese think-tanks' thesis on this point may, moreover, be somewhat more dubious on intrinsic grounds. There are some reserves of the key materials for traditional types of industries; doomsday had not yet arrived in that respect, and prophecies of its imminence are constantly being postponed. Japanese ingenuity might profitably be devoted to transforming existing types of activities as well as to transferring them to less well-prepared situations elsewhere. Might not reducing—for instance at random—the inputs required to produce a ton of steel be a better "knowledge-industry" than creating things like "industrial robots"? And the results be far more marketable in Asia and elsewhere, where there is not much demand for robots, but a great eagerness to reduce costs within the existing structure?

(4) In contrast, the proposal that Japan should cease competing in labor-intensive activities generally would be welcome to the ADCs, whose problem is to give full employment to their relatively "unlimited" supplies of labor. They would however prefer Marxian or Arthur Lewisian solutions to that of becoming annexes of the economy of Japan. It may be doubted, nevertheless, whether in the real world there is such a stark dichotomy between labor-intensive and capital-intensive as existed in the days of Marx and Lenin or in early-Showa Japan. Even the comparatively labor-intensive activities nowadays require considerable inputs of sophisticated techniques and capital equipment. Professor Kojima, as quoted above, identifies "labor-intensive goods" as those "represented by traditional light industries such as textiles." The modern textile industry in the ADCs is by no means "traditional" in the sense of primitive. It

is not a question of Mr. Gandhi's spinning wheel, but of the things happening in such centers as Hong Kong—where it has actually been said "we establish the finest modern mills, with all the labor-saving equipment, then use our plentiful labor to run them very intensively." This was said in jest, but with some epigrammatic point. Many of the other objections and queries mentioned above apply on this point also.

(5) "To concentrate production exclusively in efficient large-scale firms" may raise the biggest objections of all. Are the largest firms *ipso facto* the efficient ones? Diseconomies of scale arise, in different ways at different levels of development. Giant firms in foreign countries or in the home country, especially foreign ones or multinationals, are the subjects of the greatest hostility and suspicion. A development of this kind would alienate Japan to a considerable extent from ADC-Asia, and drastically from the ACCs. Is this not a recipe for proliferating Narita and Kashima type conflicts in Japan and exporting them also to ADCs?

Next, the suggested product-list for the industries of Japan's third industrial revolution, (a) to (d) above, may be briefly considered. Many of these require inputs which are scarcer or more difficult to procure than those for the present basic industries. The impression is conveyed that the new industries require less fuel input. This might be true of coal; but electricity? For the rest, many new items are brought on to the shopping list, which were unknown or of little interest to previous generations. In such categories as the rare earths and metals, semiconductor materials, etc., great and difficult requirements arise; meanwhile dependence on the old staples by no means disappears, as also seems to be suggested, because the demand for such things as nickel, copper, industrial diamonds, etc., will continue. Of new key items, in a long list that cannot be detailed here but covers such things as beryllium, columbium, lithium, molybdenum, tantalite, vanadium, rubidium, etc., the U.S.S.R and Siberia in particular are diversely rich. China is not so liberally and miscellaneously endowed in this field as in others and will require a larger proportion for its own use.

Innovations like those scheduled in (a) to (d) above, the "twenty-first century spectaculars," are in many cases a gamble rather than a certainty. With swiftly changing technology and swiftly changing social structures, changes are more severe and more stochastic on both the supply side and the demand side. As emphasized above, the political risks may be greater than the economic ones. On balance, it seems that Japan certainly can and should evolve many of the "futurological" or "futuristic" activities in question; but might be well advised not to abandon the old "staples," which are not without worth and merit, being still open to evolution, innovation, and improvement. A diversified economy (diversified both in the range of industries and in terms of stages of the product- and industry-cycles) is a better "hedge" for the national investment of the developed country; and finally, such a multifarious and multi-level structure can do much more to help the developing countries. Last but perhaps not least, it is the supply and procurement problems from Japan's point of view as a wealthy industrial purchaser that tend to dominate the thinking in this issue; the converse and counterpart must be studied at least as much, *i.e.*, the demand side, the problems of markets, and outlets for Japan's production. Market analysis is more difficult and hypothetical than procurement analysis and cannot be gone into in this short space. This second section of the present paper has discussed the possible nature of Japan's future desiderata; the third part gives some summary perspectives of the industrial scene, with reference especially to the significance of the U.S.S.R. and China in the East Asian framework.

III. REGIONAL PROPORTIONALITIES

The projected pattern of Japan's output structure to 1985 is summarized in Table 1. By broad categories (a), all types of activities are to increase their outputs, even primary industries; though the share of these decreases percentagewise, it is already only a small part of the total. It is striking that secondary industry (manufacturing and others—the latter, oddly, includes mining in this projection) and tertiary (service) industries only maintain their percentage share, at around 50 percent and 35 percent, respectively. Tertiary industry and manufacturing show double the growth indices of primary industry, but other secondary industries (*i.e.*, mining and construction) show the highest index. (This is in value terms here.) The second part of Table 1 (b) details the industries much more fully, making clear where the large increases are expected, both quantitatively and in the indices. All these will require imports that are either quantitatively or qualitatively important; for many of which continental Asia is an important potential area of procurement.

Table 2 summarizes the expected implications in terms of Japan's international trade. Table 2 (a) shows the enormous growth in import requirements, in every single category; Table 2 (b), the enormous export effort that must be made, also in all the categories, including food. In this table also, it is easy to see where the big figures emerge; yet the picture is one of an overall enlargement rather than a complete change of Japan's industrial and trade structure—at any rate, not till after 1985—and Continental Asia is clearly of special importance both as a potential supplier of many of the essential inputs and as a market for many of the anticipated end-products.

Table 3 completes this set of projections by showing the distribution between regions. Note, again, that imports and exports are expected to increase greatly for all parts of the world. The Communist Bloc actually shows the largest index of increase of all with respect to the imports Japan will need and one of the highest with respect to the required ex-

TABLE 1

JAPAN'S PROJECTED INDUSTRIAL PATTERN, 1970–85

	Output			*%*			*Index*	
	1970	*1980*	*1985*	*1970*	*1980*	*1985*	*1980*	*1985*
(a) Summary								
Primary industry	7	9	10	4	3	2	120	133
Manufacturing	80	157	213	50	51	51	196	266
Other secondary	13	33	47	10	11	11	254	362
Total, secondary	93	190	260	60	62	62	210	278
Tertiary industry	57	109	149	35	35	36	192	261
(b) Industries								
Agriculture, forestry, fishery	7	9	10	4	3	2	120	139
Mining	1	1	2	1	0	0	139	166
Food manufacture	10	16	22	6	5	5	169	230
Textiles	5	7	8	3	2	2	135	155
Paper pulp	3	5	6	2	2	2	185	238
Chemicals	5	11	15	3	4	4	207	286
Oil and coal processing	3	6	8	2	2	2	193	256
Ceramics	3	5	7	2	2	2	193	270
Iron and steel	11	21	24	7	7	6	187	217
Non-ferrous metals	2	4	5	1	1	1	234	286
Metal products	4	9	12	2	3	3	235	330
Machinery:								
general	8	18	26	5	6	6	215	314
electrical	8	19	29	5	6	7	252	373
transport	8	14	18	5	5	4	185	235
precision instruments	1	2	3	1	1	1	199	290
Total, machinery	25	53	76	16	18	18	212	304
Other manufactures	10	20	29	6	6	7	194	283
Building	16	31	45	10	10	11	191	279
Electricity, city gas	2	4	6	1	1	1	190	264
Transport and communication	7	15	20	5	5	5	195	269
Commerce	14	27	36	9	9	9	186	245
Finance and insurance	5	8	10	3	3	3	165	210
Services	28	56	77	18	18	19	199	273
Total	162	308	418	100	100	100	190	259

Source:

International Trade and Industry Research Association, MITI, *A Long-Term Vision for Industrial Structure,* November 1974.

Notes:

Output in billion yen, at 1970 prices. % = percentage of total output. Index: growth in output, 1970 = 100.

TABLE 2

JAPAN'S IMPORTS AND EXPORTS: PROJECTION BY COMMODITY GROUPS

	Amounts			*%*			*Index*	
	1970	*1980*	*1985*	*1970*	*1980*	*1985*	*1980*	*1985*
(a) Imports								
Food	3	16	24	14	12	10	233	918
Raw materials: textile	1	3	3	5	2	1	294	339
metal	3	9	18	14	7	8	348	653
others	3	11	17	16	9	7	372	566
Total, raw materials	7	23	38	35	28	16	351	569
Mineral fuels	4	55	91	21	43	39	1,427	2,330
Chemicals	1	6	14	5	5	6	619	1,445
Machinery: general	1	6	13	7	5	5	509	1,014
electrical	0.5	3	8	2	2	3	696	1,573
transport	0.5	1.5	3	2	1	1	374	727
precision	0.2	1	3	1	1	1	783	1,885
Machinery, total	2	5	26	12	9	11	542	1,137
Other manufactures	2	17	43	13	13	18	696	1,752
Total	19	131	236	100	100	100	692	1,236
(b) Exports								
Food	0.7	2	2.5	3	1	1	263	383
Machinery: general	2	24	52	10	18	22	1,183	2,578
electrical	3	25	50	15	18	21	859	1,759
transport	3	28	49	23	21	20	825	1,409
Total, machinery	9	82	160	46	68	68	915	1,794
Automobiles	1	14	23	1	11	10	1,083	1,732
Ships	1	10	14	7	7	6	693	977
Precision instruments	1	5	10	3	4	4	812	1,565
Metal products	4	19	22	20	14	9	501	578
Iron and steel	3	14	14	15	10	6	482	475
Chemical products	1	12	22	6	9	10	959	1,815
Textiles	2	7	8	12	5	3	279	315
Nonmetal mineral products	0.4	2	2	2	1	1	444	665
Others	2	11	20	10	8	8	575	1,032
Total	19	134	237	100	100	100	692	1,227

Notes:
In billion current \$. % = percentage of total imports/exports. Index: growth, 1970 = 100.

TABLE 3

REGIONAL DISTRIBUTION OF JAPAN'S TRADE

	1960			*1970*		*1980*			*1985*		
	V	%	*I*	*V*	%	*V*	%	*I*	*V*	%	*I*
(a) Imports *from:*											
U.S.A.	1,554	35	27	5,551	29	27,056	21	487	51,370	22	925
Canada	193	4	19	944	5	5,620	4	595	9,661	4	1,023
Europe	395	9	20	1,963	10	11,633	10	593	27,806	12	1,417
Oceania	337	8	20	1,662	9	11,240	9	676	19,794	8	1,191
South Africa	58	1	18	321	2	1,569	1	489	2,592	1	807
Southeast Asia	916	20	30	3,021	16	21,567	17	714	44,065	19	1,459
Middle and Near East	449	14	19	2,341	12	31,370	24	1,340	42,416	18	1,812
Africa, others	135	3	13	1,001	5	4,575	4	459	10,604	5	1,009
Central and South America	310	9	22	1,378	7	7,320	6	531	13,903	6	1,009
Communist Bloc	124	7	14	887	5	7,320	6	825	16,495	7	1,860
Total	4,472	100	23	18,881	100	129,270	100	678	238,706	100	1,264
(b) Exports *to:*											
U.S.A.	1,102	27	19	5,931	31	32,370	24	546	59,029	25	995
Canada	118	3	20	580	3	4,949	4	853	9,483	4	1,635
Europe	478	12	16	2,898	15	24,746	19	854	45,043	19	1,554
Oceania	166	4	24	695	4	6,019	5	866	9,957	4	1,433
South Africa	58	1	18	328	2	1,605	1	489	2,608	1	795
Southeast Asia	1,009	25	21	4,907	25	25,816	19	526	43,620	18	889
Middle and Near East	178	4	28	637	3	12,975	10	2,037	23,707	10	3,722

Africa, others	251	6	25	1,005	5	7,490	6	745	13,039	6	1,297
Central and South America	304	8	26	1,178	6	9,898	7	840	17,543	7	1,489
Communist Bloc	73	2	7	1,043	5	8,026	6	770	13,987	6	1,341
Total	4,005	100	21	19,318	100	133,760	100	692	237,066	100	1,772

	1960	1970	1980	1985
(c) Trade balance *with:*				
U.S.A.	−657	+380	+5,317	+7,644
Canada	−75	−364	−671	−198
Europe	+84	+943	+11,872	+17,213
Oceania	−208	−963	−5,246	−9,989
South Africa	0	0	+36	+16
Southeast Asia	+93	+1,889	+4,158	+2,585
Near and Middle East	−272	−1,703	−18,582	−19,267
Central and South America	+7	−200	+2,627	+3,625
Communist Bloc	−52	+158	+671	−2,659
World	−417	+437	+4,490	−1,640

Source:

MITI, *A Long-Term Vision for Industrial Structure,* 1974.

Notes:

V = value in million $. % = percentage of total. *I* = index, 1970 = 100.

(c) Trade balance; customs clearances, million $.

TABLE 4

JAPANESE INVESTMENTS OVERSEAS

	1960			1970		1980			1985		
	A	%	*I*	*A*	%	*A*	%	*I*	*A*	%	*I*
North America	88	31	10	912	26	9,181	20	1,007	18,103	18	1,985
Europe	3	1	0.5	639	18	6,440	14	1,008	10,923	11	1,709
Oceania	2	1	0.7	281	8	4,385	10	1,560	11,668	12	4,152
Total, developed countries	93	33	5	1,833	52	20,006	44	1,091	40,694	40	2,220
Latin America	85	30	15	567	16	8,261	18	1,457	18,785	19	3,313
Asia	49	17	7	751	21	10,671	23	1,421	24,014	24	3,198
Middle and Near East	56	20	17	334	9	3,072	7	920	6,909	7	2,069
Africa	1	0	1	92	3	1,978	4	2,150	7,312	7	7,948
Total, less developed countries	190	67	11	1,744	49	23,982	53	1,375	57,020	56	3,269
Centrally planned economies	0	0	0	0	0	132	1	—	1,454	3	1,102[a]
Grand Total	283	100	8	3,577	100	44,120	100	12,334	99,168	100	27,724

Sources:

Fiscal and Financial Statistics Monthly, Ministry of Finance, and *The Future of World Economy and Japan*, Japan Economic Research Center, 1975.

Notes:

A = amount: authorizations in million \$. % = percentage of total. *I* = index, growth: 1970 = 100.

[a] 1980 = 100.

ports. The highest index number of all is for the Middle East, on account of fuel; if the communist countries can supply fuels to Japan, this would be one of their most helpful contributions. Table 3 (c) notes the consequent pattern of trade balances, which also tends to emphasize the same points; Japan expects to import substantially, on balance, from the communist countries, especially after 1980. In that period the colossal deficit arises with the Middle East, apparently to be offset by equally colossal gains in the trade with Europe (principally) and the United States. But Europe is protesting bitterly against the "unbalanced" trade with Japan; the United States and Southeast Asia, even South America, may be inclined to do the same. These formerly "soft" areas of trade expansion are likely to harden to some extent (on political and general grounds as well as economic), making it necessary to try new directions such as the communist countries, difficult as it is to deal with the latter.

This proposed structure would be buttressed by a tremendous expansion of Japanese investment overseas, which also displays peculiar features (Table 4). By 1985 there is to be more Japanese investment in Latin America than in North America, more in Africa than in the Middle East. How realistic are such prognostications? Investment in the "centrally-planned economies" will take a great leap in the period 1980–85, according to this prognosis, though remaining still a tiny percentage; it is to increase ten times (in current values) in that short period, while Japanese investment in the developed countries is "merely" to double, that in the less developed countries to increase 140 percent. On current prospects, the "Communist Bloc" looks as if it might be much less monolithic by then; there will probably be large entries in it for Vietnam, and perhaps others, as well as the U.S.S.R. and China.

In conclusion, there should be as definite as possible a calculation of the potential contribution of the communist countries to Japan's needs, both as suppliers of inputs and as market outlets. Unfortunately, this cannot be done with great accuracy because (a) of the lack of information (myriads of scraps of partial information are available, but it would take volumes, rather than a short paper, to marshal and appraise them); (b) because of the extreme variability of the technology in the modern world; and also (c) the equally extreme variability of policy and of international alignments. It is necessary to have at least the perspectives and the proportions of actual industrial progress in the relevant areas well in mind. This is attempted in Table 5, which summarizes the relativities in the present phase and the trends in the most recent decade for which full and comparable data are available. Table 5 uses United Nations returns, which are broad and approximate, but ostensibly "official," being accepted for publication by the governments concerned. These show vividly the rather low level and pace of progress in China, as compared to the U.S.S.R., and the larger contribution the latter could make if the basis continues much as at present—which surely it will—till the end of this

TABLE 5

SOME COMPARISONS OF INDUSTRIAL PROGRESS

	1964		*1973*		
	O	*%*	*O*	*%*	*I*
Bauxite					
China	400	1	600	1	150
U.S.S.R.	4,300	12	4,300	6	100
World	34,600	100	67,400	100	195
Antimony (Sb content)					
China	15,000	24	12,000	17	80
U.S.S.R.	6,100	10	7,100	10	116
Japan	503	1	0	0	—
World	63,700	100	69,400	100	109
Cement					
China	10,500	3	23,500	4	219
U.S.S.R.	32,895	8	78,118	15	237
World	413,000	100	512,000	100	124
Copper (smelter)					
China	100	2	118	1	118
U.S.S.R.	820	14	1,300	15	159
Japan	342	6	950	11	278
World	5,850	100	8,490	100	145
Aluminum					
China	100	2	150	1	150
U.S.S.R.	800	14	1,360	11	170
Japan	374	7	1,639	14	438
Canada	764	13	930	8	122
U.S.A.	2,816	50	5,033	42	179
World	5,680	100	12,020	100	122
Asbestos					
China	120	4	210	4	175
U.S.S.R.	735	22	1,280	26	174
Japan	16	0	14	0	88
World	3,390	100	4,980	100	147

TABLE 5 *(Continued)*

	1964		*1973*		
	0	*%*	*0*	*%*	*I*
Chrome ore (CrO_3 content)					
U.S.S.R.	550	31	800	24	145
World	1,800	100	3,290	100	183
Diamonds (000 metric carats)					
U.S.S.R.	4,000	11	9,500	21	238
World	37,480	100	44,630	100	119

Notes:

O = output: 000 m.t., unless otherwise stated. % = percentage of world output.
I = index, 1964 = 100.

	1970		*1973*		
	O	*%*	*O*	*%*	*I*
ENERGY:					
Production (mn. m.t. of coal equivalent)					
U.S.S.R.	1,189	17	1,374	17	116
Japan	55	1	40	0	73
World	6,989	100	8,026	100	115
Consumption per capita of population (kg. of coal equivalent; % of world average)					
U.S.S.R.	4,345	230	4,927	240	113
Japan	3,215	170	3,601	176	112
World	1,892	100	2,050	100	108
Consumption (mn. m.t. of coal equivalent)					
U.S.S.R.	1,055	15	1,230	16	117
Japan	332	5	390	5	117
World	6,820	100	7,796	100	114

TABLE 5 *(Continued)*

			No.	*L*		*W*
Power industry (electricity, gas and water)						
U.S.S.R.	1968		1,611	628	R	939
	1972		1,383	657		1,148
	1972	I	86	105		122
	1973		1,333	661		1,198
	1973	I	83	105		128
Japan	1968		234	200	Y	196
	1972		252	203		314
	1972	I	108	102		160

Notes:
No. = number of units reporting. *L* = persons employed, 000s. *W* = wage bill: mns. of yen and rubles.

	1964		*1973*		
	O	*%*	*O*	*%*	*I*
IRON AND STEEL:					
Iron ore production					
China	20,350	7	39,000	8	192
U.S.S.R.	76,129	25	118,151	24	155
Japan	1,432	0	588	0	41
World	304,516	100	492,300	100	162
Crude steel production					
China	14,000	3	25,000	4	179
U.S.S.R.	85,038	19	131,481	19	155
Japan	39,799	9	119,322	17	300
World	445,000	100	684,000	100	154
Pig-iron production					
China	18,000	6	33,000	5	183
U.S.S.R.	62,377	20	95,933	14	154
Japan	24,449	8	92,043	14	376
World	437,800	100	683,800	100	156
Lead (Pb content)					
China	100	4	100	3	100
U.S.S.R.	330	13	470	14	142
Japan	54	2	53	2	100
World	2,540	100	3,410	100	134

TABLE 5 *(Continued)*

	Reserves		*Output*				
	1973		*1969*		*1973*		
	R	*%*	*O*	*%*	*O*	*%*	*I*
FUELS:							
Crude petroleum							
China	2,026	3	14,600	0	50,000	4	342
U.S.S.R.	6,464	9	328,373	16	429,037	15	131
World	74,280	100	2,071,100	100	2,774,600	100	134
Natural gas							
China	165	0	1,275	0	2,832	0	222
U.S.S.R.	18,400	3	32,601	18	44,901	19	138
World	74,280	100	181,121	100	236,326	100	130

	1961/2–1965/6		*1973/4*		
	O	*%*	*O*	*%*	*I*
FERTILIZER CONSUMPTION:					
China	1,491	4	1,918	2	129
U.S.S.R.	3,704	10	12,560	15	339
Japan	1,799	5	2,375	3	132
World	38,000	100	83,600	100	220

Notes:

Crude petroleum: Reserves, bn. m.t., Output, mn. m.t.

Natural gas: Reserves, bn. cu.m., Output, mn. cu.m.

Fertilizer consumption: 100 m.t. Annual average, 1961/2–65/6 and annual 1973/4; percentage of world. Totals of nitrogen (N content) + phosphate (P205 content) + potash (K20 content).

TABLE 5 *(Continued)*

	1964		*1973*		
	O	*%*	*O*	*%*	*I*
Magnesium (m.t.)					
China	1,000	0	1,000	0	100
U.S.S.R.	32,000	22	57,000	23	178
Japan	5,185	4	19,310	8	372
World	148,000	100	244,000	100	165
Magnesite (000 m.t.)					
China	1,000	16	1,000	11	100
U.S.S.R.	n.a.	n.a.	1,550	17	—
World	6,240	100	9,040	100	145
Manganese (Mn content; m.t.)					
China	300	5	300	3	100
U.S.S.R.	2,272	37	2,839	31	125
Japan	93	2	53	1	57
World	6,090	100	9,200	100	151
Meat production					
China	10,068	15	11,965	14	188
U.S.S.R.	n.a.		n.a.		
Japan	501[a]	1	1,161[a]	1	232
Asia	14,472	21	18,185	21	126
World	69,102	100	88,203	100	128
Mercury (m.t.)					
China	900	11	900	11	100
U.S.S.R.	1,210	15	1,729	25	143
Japan	338	4	129	2	36
World	8,000	100	7,970	100	100
Molybdenum (Mo content; m.t.)					
China	1,500	3	1,500	2	100
U.S.S.R.	6,000	14	8,200[b]	12	137[b]
World	43,090	100	83,040	100	193
Nickel (Ni content; m.t.)					
U.S.S.R.	75,000	19	136,000	20	181
Canada	207,288	52	243,949	35	118
World	395,700	100	692,400	100	175

TABLE 5 *(Continued)*

	1964		*1973*		
	O	*%*	*O*	*%*	*I*
Silver (m.t.)					
China	25	0	25	0	100
U.S.S.R.	1,110	15	1,177	13	104
Japan	271	4	356	4	131
World	7,690	100	9,180	100	119
Sulphur (native, free mineral; m.t.)					
China	130	—	130	—	100
U.S.S.R.	950	—	2,300	—	242
TEXTILES:					
Cotton yarn (pure and mixed; 000 m.t.)					
China	1,155	—	1,450	—	126
U.S.S.R.	1,274	—	1,535	—	120
Japan	501	—	510	—	101
Cotton fabrics (mn. m²)					
China	n.a.		n.a.		n.a.
U.S.S.R.	5,814	—	7,137	—	123
Japan	2,965	—	2,380	—	80
Rayon fibers (mn. m²)					
China	33	0	69	0	209
U.S.S.R.	322	4	654	9	203
Japan	672	9	1,255	17	187
World	4,020	100	7,500	100	187
Timber: sawnwood (bn. m³)					
China	14	3	15	4	118
U.S.S.R.	110	28	119	27	108
Japan	40	10	43	10	107
World	393	100	436	100	111
Phosphate rock					
China	800	1	3,000	3	375
U.S.S.R.	10,660	19	21,230	22	199
World	55,800	100	98,500	100	177
Tungsten (concentrates: WO_2 content; m.t.)					
China	12,200	34	10,600	22	87
U.S.S.R.	6,600	18	9,300	19	141
World	36,050	100	48,230	100	134

TABLE 5 *(Continued)*

	1964		*1973*		
	0	*%*	*0*	*%*	*I*
Zinc					
China	90	2	100	1	111
U.S.S.R.	445	12	670	14	151
Japan	338	9	869	18	257
World	3,690	100	4,910	100	133
WOOD PULP (air dry):					
Chemical					
China	512	1	700	1	137
U.S.S.R.	2,688	5	5,260	6	196
Japan	4,064	7	8,720	10	215
World	54,750	100	87,811	100	160
Mechanical					
China	312	1	550	2	176
U.S.S.R.	1,168	6	1,697	6	145
Japan	956	5	1,400	5	146
World	20,920	100	28,486	100	136
Newsprint					
China	366	2	700	3	191
U.S.S.R.	633	4	1,212	6	191
Japan	1,137	7	2,106	10	185
World	16,190	100	22,003	100	136
Other paper and board					
China	2,721	4	3,450	3	127
U.S.S.R.	3,487	5	6,213	5	178
Japan	6,229	8	13,868	12	223
World	75,895	100	115,555	100	152

Source: United Nations Statistics.
[a] inspected.
[b] 1972.

century when it can hardly be changed as drastically as some people seem to expect.

The array of figures in Table 5 gives an overall topography of the industrial map in terms of the relative place of these areas in the world-industrial economy and their relative progress in the pre–oil crisis decade. The facts do not sustain the popular impression that China is a particularly rich storehouse of industrial supplies in what are still the dominant requirements, whether in absolute quantities, percentage of the world supply, or growth in either of these respects. The U.S.S.R. appears considerably more important in practically every way. This broad conclusion is reinforced when divisors are introduced for the respective populations of these countries (800 million to one billion for China and one-fourth of those magnitudes for the Soviet Union) and what they must absorb out of their currently realized outputs for their own national development programs (both civil and military). In the latter respect, China is in the early and more primary phase of industrialization and will be required to devote much to that process; while the U.S.S.R. has passed into the "industrialized" and "high technology" categories, with quite a different relationship between availabilities and needs.

The two great continental countries of Asia are of considerable importance to Japan's plans and prospects from an economic point of view. From the political, human, and cultural points of view, any "models" that leave them out of account are dangerously unrealistic. By the yardsticks of industrialization and technological progress, Japan can be considered to have moved into a different category, to have become a hyperindustrialized and super-technological unit. Some models suggest—logically from that premise—that Japan should be grouped with the Western world, the United States and Europe, and form combinations (economic associations or communities) under its leadership in its own hemisphere. Japan is however firmly located in Asia and remains an Asian nation and an Asian people in many fundamental ways and through many linkages. The economic and technological ties with Asia are not weakening in absolute terms, they are weakening only in relative terms, *i.e.*, those with the advanced capitalist countries are developing more than those with Asia.

The political situation is such that these considerations are absolutely compelling. There are several lines of force working on and in Asia, and on and in Japan; the two greatest of these are the communist powers. By an irony of history, these two are in deadly enmity with each other: an uneasy situation for Asia in general and Japan in particular. In a world of economic logic, the East Asian nations would be highly complementary to each other, whether by the principle of comparative advantage or on a Heckscher-Ohlin basis or by other kinds of reasoning; maximization demands full economic cooperation between the Soviet Union, China, and Japan. The two former are however committed to autarchic development and are in full hostility to each other. This is an unprecedented situation for Japan (and indeed for the world at large). Historical models' introspective or retrospective analyses, it is submitted here, however valid in themselves, are of limited use in this situation, which requires a multilateral response on the part of Japan, maximizing or optimizing both the possibilities of joining with the West, which are so attractive to Japan, and the possibilities in the Asian situation. This is a much more complicated task than the classic responses of ancient Yamato, Meiji, and early Showa, which were responses to one principal pressure at a time. It is a problem of Political Economy rather than of econometric projection. Among the "resources" that are the topic of this IEA World Congress Section—"Growth and Resources Problems Related to Japan"—the most important remains the resourcefulness of the Japanese people to meet international as well as national responsibilities.

Economic Growth in Japan and Energy

Masao Sakisaka *National Institute for Research Advancement*

Rapid energy consumption increases have accompanied Japan's rapid economic growth since World War II. During this period, the type of energy consumed shifted from hydraulic power and coal to petroleum. Energy consumption became particularly high in the 1960s due to the development of the heavy and chemical industries. In the 1970s, the growth pattern of the economy changed, with personal consumption and commodity exportation the driving forces of economic growth; this change, combined with higher petroleum prices and economic stagnation, has led to a decline in the elasticity of demand for energy in the 1970s. Even so, to sustain necessary economic growth in the future, Japan should not rely solely on Middle Eastern petroleum sources, but should obtain more diversified energy sources and new energy sources. Conservation of energy is also an important target for Japan.

I. FORMATION OF A HIGH-ENERGY-CONSUMPTION TYPE ECONOMY

It was after the 1960s that the Japanese economy became a high-energy-consumption type. The volume of energy consumed during fifteen years up to 1975 far exceeded the volume consumed from the beginning of Meiji Era up to 1960. Average annual per capita energy consumption during the last five years was about 4.5 times that in the period prior to World War II. The transition of the economy to a high-energy-consumption type coincided with the shift of energy from hydraulic power and coal to petroleum.

After World War II, rapid energy consumption increases occurred along with rapid economic growth, and the increase rate was particularly high in the 1960s because the elasticity of energy consumption to GNP became greater due to the rapid development of heavy and chemical industries. The elasticity of energy shifted from the 1.0 of the 1950s to an average of 1.2 in the 1960s and further to an average 0.75 during the initial six years of the 1970s (the decline since the turn of the decade will be explained later).

Energy consumption grew during the 1960s by an average annual rate of 13 percent. This is explained as follows. In the first place, the economic growth rate was high at an annual average of 11 percent. Secondly, industrial production, particularly in the area of materials industry (such as metals, chemicals, and ceramics), which use a large volume of energy, grew rapidly, causing a change in the industrial structure. The secondary industry represented 51 percent of total industrial production in 1960. This ratio grew to 60 percent in 1970, while, likewise, the representation of manufacturing industry by heavy and chemical industries grew from 47 percent in 1960 to 63 percent in 1970. This was the major reason for the increased elasticity of demand for energy. Thirdly, the pattern of transportation also went through a changeover, and the burden on modes of mass transportation such as railroad and ships reduced while the burden on individual transportation with a low energy efficiency such as trucks, personal cars, and airplanes—particularly automobiles—increased rapidly. Automobiles represent 78 percent of energy consumed for transportation purposes. Fourthly, energy consumption at home increased because of the spread of various home-use electric appliances and heating and hot-water supply equipment. Fifth, energy consumption has consistently increased in the service industry—hotels, department stores, and other stores.

In Japan technological innovation bloomed in the 1960s. Materials industries, such as metals, chemicals, and synthetic textiles, developed modern mass production systems. And mass assembling industries, such as automobiles and home-use electric appliances, developed based on the supply of such materials of a good quality for a low price. Consumer durables rapidly saturated homes, and "throw-away" became a commonplace word. Facility investments increased in industries, and public investments such as roads and highways and ports and harbors also increased. Heavy and chemical industries developed extensively, and the economy grew rapidly. An "affluent society of mass production and mass consumption" was formed.

II. FRAGILITY OF ENERGY SUPPLY BASE

This "affluent society" is a grand house built on petroleum. The Japanese economy had come to consume a huge quantity of energy as a result of its rapid growth. But what furnished the increasing energy consumption was imported petroleum. Demand for primary energy expanded from the 95 million tons of 1960 to 284 million tons in 1970, and 87 percent of this increase was supplied by petroleum—almost all imported.

The composition of primary energy supply largely changed in the 1960s. In 1960 coal represented 52 percent of total energy supply, while petroleum represented only 32 percent, and domestic energy supply—coal and hydraulic power—represented 56 percent. But in 1970, the ratios of coal and petroleum reversed to 23 percent and 69 percent, respectively, and the ratio of domestic supply declined to 16 percent. Thus, reliance on petroleum and reliance on imported energy increased as energy consumption expanded.

When rapid economic growth began in Japan under technological innovation, rapid increases in petroleum production began in the Middle East and Africa. International oil companies raced to increase their market shares and reduce the price of petroleum. OPEC could not stop it. The world entered into an oil age. The Japanese economy took full advantage of low-priced petroleum supply. For Japan, which has only meager energy resources, both the solution of employment problem and the accomplishment of a high consumption level would have been impossible without the importation of a large quantity of petroleum.

But since the beginning of 1970, the international petroleum situation has largely changed. Because of worldwide expansion in petroleum consumption, the demand and supply of petroleum became tight. The position of OPEC was strengthened, and nationalism rose in oil-producing countries where the governments intensified controls over their petroleum resources. The petroleum strategy of OPEC in the fall of 1973 at the time of the Fourth Middle East War caused a decisive impact on the oil situation of the world. During the war, OPEC increased the price of crude oil by a substantial margin, and the government of oil-producing countries acquired powers to determine the price and the volume of production of crude oil. The era of abundant supply of low-priced oil ended, and the world entered the era of high-priced oil. The Summit Conference of OPEC Countries, which was held in Algiers, the capital of Algeria, in March 1975, stated in the proclamation that it adopted that "we will act jointly in order not to allow only advanced nations to continue to enjoy economic prosperity at the sacrifice of interests of oil-producing countries." The political situation is unstable in the Middle East. The Third World, with OPEC in the lead, is trying to have advanced nations accede to their various economic demands. The "affluent society" in Japan, which was built on petroleum, is now being shaken from the ground.

III. CHANGE IN GROWTH PATTERN AND THE DECLINE OF ENERGY ELASTICITY

At the time the world's oil situation changed, the pattern of Japanese economic growth also changed. Since the beginning of the 1970s, increases in private equipment investment have slowed down because new technology had been introduced into all sectors, environmental destruction had advanced, and the spread of consumer durables had slowed down. In lieu of private equipment investment, increases in personal consumption and commodity exportation became the driving force of economic growth. Metals, chemicals, and other materials production increases slowed down, and the ratio of secondary industry production to GNP started to drop. Such growth pattern change and industrial structure change naturally affected energy consumption. And this tendency was accelerated by the energy price hike and economic stagnation since the oil crisis of the fall of 1973.

A review of energy consumption since the beginning of the 1970s reveals that the average annual consumption increase during the period from 1970 through 1973 was 7 percent—a fair decline from such average during the 1960s—and elasticity to GNP also declined to 0.99. Then in 1974 and 1975, energy consumption shrank because of the economic stagnation after the oil crisis, but it expanded in 1976 only to resume the level of 1973. If discounted for the prolonged economic stagnation, the elasticity of demand for energy must be said to have lowered to a fair degree in the 1970s in comparison to the 1960s. The elasticity in 1970 through 1976 is estimated at 0.75.

A review of the composition of consumption shows that this decline was attributable to the fact that increase in energy consumption of the industrial sector has become relatively small. Also a review of energy consumption in relation to GNP indicates that consumption for industrial purposes has been on the decline, while that for household purposes has consistently increased. (Energy consumption for transportation purposes remained little changed since the 1960s.) The ratio of household consumption of energy to total consumption has been still too small to cover in full the decline in industrial consumption, and energy elasticity to

GNP has declined.

IV. FUTURE ECONOMIC GROWTH AND ENERGY CONSUMPTION

The Advisory Committee for Energy, an advisory organ of the Government of Japan, compiled a report last June under the title of "Preliminary Prospect of Energy Demand and Supply up to 1985." This report contemplates an average annual economic growth of 6.1 percent for the 10 years up to 1985 and, based on it, estimates the demand for energy in 1985 at 740 million kl (in terms of petroleum oil). And this report sets an objective of conserving 5.5 percent, or hopefully 10.8 percent, of said demand by 1985 through the accomplishment of higher efficiency and the reduction of waste in energy uses under the Save-Energy policy. If this objective is achieved, 1985 energy demand will be 700 million kl and 660 million kl, respectively, and energy consumption elasticity will become 0.95 and 0.89, respectively.

Under this estimate, the economy will need to grow by about 6 percent annually in order to sustain full employment, to accomplish fuller social security, and to develop public facilities. And it is believed that the economy will have that much potentiality provided that there is no limit on energy supply.

On the other hand, it is possible that energy demand will fall short of the above estimate in view of stagnant industrial equipment investment and slowdowns in the production of steel, chemicals, and other high-energy-consumption type industries in 1976 and 1977 (and perhaps 1978). There is no telling what level energy consumption elasticity will be. Assuming that it will be 0.9 in consideration of the elasticity during the period from 1970 through 1976, future recovery of the equipment investment rate, and the estimate (0.91) under the New Energy Plan of the Government of West Germany, energy demand under economic growth of about 6 percent will be 660 million kl in 1985. Conservation of energy can depress the demand to less than that indicated. If 5.5 percent saving is achieved, demand will be 625 million kl in 1985.

V. CONTINUOUSLY HIGH FUTURE RELIANCE ON PETROLEUM

A future energy demand forecast involves numerous factors of uncertainty. But if the super economy of a high-energy-consumption type is to continue to grow at a certain rate, the satisfaction of its ever-increasing energy demand will not be an easy task. The higher the economic growth rate, the greater the problem in securing a stable supply of energy. But if the rate of economic growth is low, the economy will have less problem in securing energy supply, but will face an employment problem and other economic difficulties.

Policy for stable energy supply is conceived of in the directions as indicated below.

(1) Domestic resources will be exploited as much as possible.
(2) Development of nuclear power generation will be propelled at social consensus, while making efforts to increase its safety and reliability.
(3) Of imported energy, the utilization of natural gas and coal will be expanded.
(4) While reducing reliance on imported petroleum through the above listed measures, petroleum supply sources will be diversified.

Potential energy supply volumes estimated by the Advisory Committee for Energy under the above policy are presented in Table 1.

TABLE 1

LONG-TERM PROJECTION OF ENERGY DEMAND AND SUPPLY BALANCE

Fiscal Year	*Unit (millions)*	*F.Y. 1975 Actual*		*F.Y. 1985 Projection*			
				(MITI)		*(Sakisaka)*	
Demand for Energy before Conservation	kl			740		660	
% of Reduction of Energy Demand by Conservation	%			5.5		5.5	
Energy Demand after Conservation	kl	390		700		625	
Type of Primary Energy		*Actual*	*%*	*Estimate*	*%*	*Estimate*	*%*
Hydro—General	kw	17.80	5.7	19.50	3.3	19.50	3.7
Electric—Pumped storage	kw	7.10		19.50		19.50	
Geothermal	kw	0.05	0.0	0.50	0.1	0.50	0.1
Domestic Oil—Natural Gas	kl	3.50	0.9	8.00	1.2	8.00	1.3
Domestic Coal	tons	18.60	3.4	20.00	2.0	20.00	2.2
Nuclear	kw	6.62	1.7	26.00	5.4	26.00	6.1
Imported LNG	tons	5.06	1.8	24.00	4.9	24.00	5.5
Imported Coal (steam coal)	tons	62.34 (0.50)	13.1	93.00 (6.00)	10.7	93.00 (6.00)	12.0
Sub Total	kl	104.00	26.7	195.00	27.8	195.00	31.2
Imported Oil	kl	286.00	73.3	505.00	72.2	430.00	68.8
Grand Total	kl	390	100	700	100	625	100

The possibility of exploitation of Japan's domestic resources is low. Also, the development of nuclear power generation will proceed only slowly due to citizens' movement against nuclear power generation. The government inevitably will lower the previous development target of 49,000 MWe by 1985 to 26,000 MWe. Even if smooth operation of these nuclear power plants is assumed, the total volume of energy supplied from domestic sources, including such plants, will be only about 84 million kl (in terms of petroleum).

Therefore, an overwhelming portion of energy demand will have to depend on imports. In order to diversify imported energy, the potential was estimated under the policy that the importation of natural gas (liquefied) and coal will be increased as much as possible. Southeast Asian and Middle Eastern regions have unexploited natural gas resources at numerous locations. These countries are looking forward to the exploitation of such resources. In view of a high possibility in Japan of market acceptance and developmental importation of natural gas, due to the fact that it has few elements of atmospheric air pollution, a maximum importation volume was estimated. As for coal, importation of hard coal for thermal power generation was estimated in addition to material coal for steel mills. This is to open ways for the expansion of power generation by coal in preparation for a chronic oil shortage.

The total potential supply volume of energy sources other than petroleum, as discussed in the above, is only 195 million kl (in terms of petroleum), and the balance must depend on imported petroleum. Assuming that 5.5 percent conservation will be accomplished by 1985, said energy demand will be from 625 to 700 million kl, which means that from 430 to 505 million kl (or from 7.4 to 8.7 million barrels per day) of petroleum will have to be imported. This will be from a 55 to 77 percent (or from 4.5 to 5.9 percent annual) increase during the next 10 years from the 286 million kl (4.93 million barrels per day) in 1975. Reliance on imported petroleum will remain at the level of 70 percent.

If economic growth of about 6 percent is to be sustained, continuously greater amounts of petroleum will have to be imported. Even if the source is diversified to China and Indonesia, an overwhelmingly large portion of importation will have to come from the Middle East. In view of the political instability in the Middle East, the energy base of the Japanese economy is extremely fragile.

The petroleum price will continue to rise in the future. Can Japan import so much petroleum and still maintain its international balance of payments? (If the price is to rise annually by 5 percent, petroleum imports in 1985 will be from 60 to 70 billion dollars; in 1975, it was 22 billion dollars). If the balance can be maintained, it means that trade deficit with oil-producing countries will be filled with a huge amount of trade surplus with the United States and European countries. Will they acquiesce to such a trade surplus for Japan? Also, will they not blame

Japan for causing oil price hikes by the purchase of a large volume of petroleum? In order for the Japanese economy to continue to grow in the future, it will face a number of problems with regard to energy.

VI. WORLD LIMIT TO OIL PRODUCTION

The seriousness of energy problem to Japan becomes even greater when the demand and supply of energy in post-1985 period is taken into consideration. If demand for primary energy is to increase by the annual rate of from 2.5 to 3.0 percent during the period from 1985 to 2000, the demand in 2000 (working from the median of energy demand of 1985) will be from 950 million to one billion kl. If domestic resources are exploited and the utilization of nuclear power generation expanded (100 million MWe) and if efforts are exerted for the expansion of importation of natural gas (60 million tons) and hard coal (35 million tons), the total of these supplies will satisfy only about 40 percent of the demand, and if the remainder is to be satisfied by imported petroleum, the volume of petroleum importation will be about 65 billion kl (10 million barrels per day). It is doubtful if this much petroleum can be imported. The skepticism has the following base.

Petroleum experts believe that the petroleum production of the world (excluding communist countries) will reach its peak in the latter half of the 1980s or the first half of the 1990s and will subsequently continue to decline. The main reasons for this are the deterioration of the discovery rate (new petroleum fields) and the preservation policy of oil-producing country governments.

(a) The discovery rate will drop because of the worsening of natural conditions (deep seabed, extremely remote locations, etc.), rise in development costs, decline in investment potentials of international oil companies and national companies, and the shrinkage of profits of oil companies (expansion of tax revenue of oil-producing country governments), and (b) the gap between increasing production volume and the discovery rate will open in the shape of scissors, causing reduction in the volume of proven deposits. (c) When production reserve ratio *(R/T)* declines to a certain level (10–15), production increase reaches its limit. This is the physical limit of petroleum production increase.

When this limit is reached will depend on relationship between the volume of petroleum consumption and discovery rate. It is highly probable that this limit will be reached in the middle of the 1990s.

On the other hand, it is doubtful if all oil countries will continue to produce petroleum at the maximum production rate. While countries whose developmental potentials are high (Iran, Iraq, Algeria, etc.) will need to increase their revenue by increasing oil production, oil countries of the Arabian Peninsula will adopt policies to restrict the volume of oil production due to their excessive revenue from oil, as has already been done by Kuwait and the United Arab Republic, and will eventually be done by Saudi Arabia, the country owning the world's greatest petroleum resource. When this will take place will depend on a number of factors, such as the political situation in the Middle East, the influence of the United States, and the value guarantee of oil money. The government of Saudi Arabia might consider that the limit of production is about 15 million barrels per day (currently 9 million barrels per day), and there is little possibility of Saudi Arabia continuing to increase oil production up to the level of 20 million barrels per day. When Saudi Arabia curbs its production, world oil production will begin to shrink. It is highly possible that the transition from production increase to production decline will take place in the latter half of the 1980s.

If OECD countries fail to conserve on energy and to develop and expand the utilization of a substitute energy by the time of this transition, but increase demand for OPEC oil, the world will suffer from an oil shortage. Industrial nations will fight each other to secure oil (in which the U.S.S.R. will join), and the price of oil will surely rise and the world politics and economy will be hurt substantially.

The lead time for energy conservation and the successful development of a substitute energy is from five to 10 years. Unless OECD countries establish a goal for reduced petroleum consumption in the future and immediately intensify their policy efforts for energy conservation and development of substitutes for petroleum, a chronic shortage of petroleum will be inevitable. And if this situation is not avoided, non-oil-producing developing nations will be hurt seriously.

VII. ENVIRONMENTAL PRESERVATION AND ENERGY CONSERVATION

Atmospheric air has been greatly polluted in Japan since the 1960s because of rapid increases in the consumption of fossil fuels. While pollution by SO_x has been mitigated by emission controls aimed against the sources of pollutants, no noteworthy achievement has been seen with regard to NO_x. If a large portion of increased energy consumption in the future will be satisfied by nuclear, geothermal, and hydraulic power generation, atmospheric air pollution will be mitigated. But if such a portion is to be satisfied by fossil fuel, the emission volume of air pollutants will increase and emission control will have to be strengthened. Purification of atmospheric air will make no advancement unless all methods—direct desulphurization of heavy oil, desulphurization and denitration of stack gas, modification of combustion systems, the use of fuel with little N content—are utilized. Investments for pollution control therefore will reach a huge amount. Also, desulphurization of smoke will accompany the site problem for the desulphurization and denitration facilities.

Under this situation, energy conservation increases its significance. A large amount of investment for higher efficiency in the utilization of energy will be no loss to the nation's economy. It is necessary that a policy to encourage investments for the conservation of energy by enterprises and households be strengthened. Also, the reduction of materials per unit of production, the discontinuation of "throw-away" in homes, the recycling of waste materials, and the elongation of useful life of consumer durables are important methods of energy conservation from the view of the nation as a whole. Also, the growth of industries with higher value-added than that of the materials industry is important, needless to say.

More than for any other country, energy conservation is an important target for Japan, which has formed a high-energy-consumption economy in its narrow national land, in view of environmental preservation and the unstable supply of petroleum.

VIII. RESEARCH AND DEVELOPMENT OF NEW ENERGY

The petroleum age will come to an end eventually. The era of high consumption of renewable and inexhaustible energy such as nuclear fusion, solar energy, and geothermal energy will not come until several decades after the next turn of the century. Meanwhile, what sort of energy composition will satisfy Japanese energy demand? Along with the importation of petroleum and natural gas, expansion of coal utilization will be important. But coal, too, will have to depend on imports and, therefore, there will be a limit to the expansion of coal utilization. Then, atomic power by nuclear fission will have to be depended upon to a fair degree.

The basic question is how much risk of radiation can be tolerated. But what is essential is to improve the light water reactor, thereby increasing safety and reliability and establishing the recycling of nuclear fuel. Also, new type reactors will have to be developed, such as a breeder reactor and a high-temperature gas reactor. Otherwise, a limit to nuclear power utilization will be reached if full attention is not paid to the prevention of proliferation of nuclear weapons. Non-proliferation may not be accomplished through technical methods alone; it must depend on a solution by international politics. The next 10 years or so should be deemed as the period in which the utilization of nuclear power will ripen, and preparation should be made now for the 1990s and subsequent periods when greater reliance will have to be placed on the utilization of nuclear power.

The Government of Japan should make greater research and development efforts for future energy, such as nuclear fusion, solar energy, geothermal energy, synthetic coal, heavy oil, sand oil, and shale oil, while at the same time making positive contributions to international cooperation.

"Resource Constraints": A Problem of the Japanese Economy

Jun Nishikawa *Waseda University*

The influence of the resource constraints problem on the growth trend of the Japanese economy as it was manifested at the time of the "oil crisis" has been very important. However, this was not the first time the "constraints" problem was manifested; it had been known since the end of the 1960s. The "oil crisis" radically accelerated this problem, which originated, not in the physical bottleneck of production, but in resource distribution and in problems closely connected to the worsening of the "North-South issues." Domestically, the destruction of the environment that accompanied high economic growth intensified the resource constraints problem. A necessary measure for the Japanese economy at this time of change in the worldwide structure of resources supply is a lowering of economic growth targets to a moderate degree. However, a quantitatively lower rate of growth will not solve the resource constraints problem facing developed countries if these constraints originate in the high growth and mammothization of big enterprises, in the related worsening of the North-South problem (deterioration of the terms of trade of developing countries), and policies supporting overseas investment and securing raw materials at any cost. The answer to the "resource constraints" problem facing the Japanese economy is: realizing the control of citizens over big enterprises that are responsible for the resource constraints problem and concentrating people's efforts to resolve the environmental problem as well as the North-South issues.

INTRODUCTION

Since the middle of the 1950s, the Japanese economy has experienced a very high rate of growth, approximately 10 percent per annum. It is a unique experience in the world's economic history to see a country whose national income was around $200 per capita reach the stage of $3,500 per capita within a quarter of a century. But, around 1974–75, this high economic growth stopped; since then the policymakers in the government began to aim for a moderate 6 percent of growth rate, which is called "the road of stabilized growth." At the same time, they enumerate several conditions that affect this "diminishment of the speed" of economic growth, among which one of the most important reasons seems to be the "resource constraints" problem.[1] We shall examine in this paper the following two questions:

(1) What are the elements that affected the trend of growth in the Japanese economy and how far have the so-called "resource constraints" contributed to it?
(2) With the slowdown of the economic growth rate to the level of 6 percent as intended by the policymakers, could the Japanese economy absorb and overcome the crisis manifested in the middle of the 1970s?

To examine these problems, we shall see first the apparition of the "resource constraints" problem in the Japanese economy in the 1970s, then analyze the nature of these "resource constraints," and finally inquire into the future problems of resources relating to the predicted image of the Japanese industrial structure. This kind of examination will bring us to some policy recommendations that are not necessarily coincident with those prevailing in official circles.

The resources can be broadly defined as material factors that are necessary for the production of goods and services to satisfy human needs (raw materials, fuels, and environmental resources such as land, water, or climate) or that which are consumed by the human being in order to reproduce himself (food resources) and also as nonmaterial factors that organize this production and consumption process, such as human resources and cultural resources. But, here we shall limit our analysis only to natural resources.

[1] Economic Planning Agency, *Keizai Hakusho [White Paper on Economy]* (1975), Part II, Chap. 1.

I.

People have discussed extensively the reasons that led to the high economic growth or "economic miracle" in the postwar Japanese economy: democratic reforms made after the Second World War, technological innovations, the high rate of education and the high work morale among the Japanese people, the role of government intervention, etc. But undoubtedly, the abundant supply of natural resources that Japan benefited from during these two decades constitutes one of the pillars of this growth process. This supply of natural resources had been assured mainly by the outer world, especially by developing countries.

It is not the first time that the Japanese economy has encountered the "resource constraints" problem. Already, before the First World War, in the 1890s, when the Japanese economy proceeded to heavy industrialization, the Japanese entrepreneurs experienced difficulty in the supply of raw material, and they sought iron ore in China. We know the consequence of this road was the Japanese imperialist expansion and the war with neighboring countries. Many Japanese remember even today that one of the official motives for that war was the claimed "encirclement by ABCD" (encirclement for natural resources by the Americans, British, Chinese, and Dutch). We shall not discuss now the accuracy of this slogan (it could be judged through the analysis of the nature of "constraints" problem in this paper), but we should point out that since then, after the Second World War, there have occurred some major changes concerning supply of raw materials and fuels in the international market: a boom of industrialization that led to a considerable expansion of the world's mineral production and dissolution of the colonialist economic "bloc" motivated by independence of the former colonized Third World countries. These phenomena assured Japan of a necessary supply of raw materials, fuels, and food through the process of postwar economic growth. This supply of resources had been so smoothly met that the majority of the Japanese, including experts, did not notice the existence of the "resource problem" until the burst of the "oil crisis." In fact, Japanese economic growth in the postwar period has been accompanied by a huge increase in resources consumption.

From 1960 to 1970, consumption of copper increased by 10.0 percent per year; that of zinc, 12.3 percent; nickel, 17.0 percent; aluminium, 19.0 percent; iron ore, 18.1 percent; coal, 13.0 percent; and petroleum, 20.4 percent; each exceeding by two or three times the world average rate of increase. Thus, the consumption of iron ore was 21.1 million tons in 1960, and it reached to 111.0 million tons in 1970. In the same period, the consumption of petroleum had increased from 29.5 million kiloliters to 185.5 million kiloliters.[2] This huge increase in resources consumption

[2] Ministry of International Trade and Industry (MITI), *Shigen Mondai no Tenbo [The Perspective of the Resource Problem]* (1971), Table I-1-1 and I-1-2.

TABLE 1

JAPAN: OVERSEAS DEPENDENCE RATE FOR MAIN RESOURCES (1960–1970)

	1960 (%)	*1970 (%)*
Copper	50.6	75.6
Lead	54.6	54.6
Zinc	26.3	54.5
Aluminium	100.0	100.0
Nickel	100.0	100.0
Iron ore	68.0	87.9
Coal	35.8	78.5
Petroleum	98.6	99.7
Natural gas	0	34.8
Uranium	—	100.0
10 resources average	70.8	90.4

Source:
Ministry of International Trade and Industry, *Shigen Mondai no Tenbo [The Perspective of the Resource Problem]* (1971).

was met principally by overseas supply.

Between 1960 and 1970, the overseas dependency rate of copper increased from 50.6 to 75.6 percent; of zinc, from 26.3 to 54.5 percent; of iron ore, from 68.0 to 87.9 percent; of coal, from 35.8 to 78.5 percent; of petroleum, from 98.6 to 99.7 percent. For the ten principal resources, the average dependency rate increased from 70.8 to 90.4 percent in this decade. (*See* Table 1).

As for the imports of these resources, the share of the developing countries is important: in 1975, 55 percent of foods, 54 percent of raw materials (79 percent of metal material), and 86 percent of mineral fuels came from the nonindustrial world and, on average, 63 percent of total imports originated in the developing countries.[3] The role of the developing countries in resource provision to the Japanese economy, where the rate of resource consumption surpasses by two or three times the world average, has been, we have to affirm, very important.

According to the forecast made by the Economic Planning Agency in 1975, the Japanese share of the world consumption of petroleum should become 20.1 percent in 1985, when we assume that the actual trend of growth would continue, following just after the United States (22.4 per-

[3] MITI, *Tsusho Hakusho [White Paper on International Trade]* (1977), Statistical Annexes, Table 1.

cent) and the European Community (20.7 percent): actually their shares were each 7.8, 32.2, and 15.3 percent in 1972. In the consumption of iron and steel, aluminium, and copper, Japan would become in 10 years the first consumer in the world: its shares have each become 25.8, 28.7, and 27.3 percent, surpassing by far those of the United States (15.1, 20.8, and 17.4 percent) and of the E.E.C. (14.8, 11.2, and 18.2 percent). This increase of the Japanese share, it must be realized, is at the cost of the other developed countries because their share must be decreased considerably.[4]

Already, in 1971, the Ministry of International Trade and Industry made the following forecast for Japan's absorption of resources in the world market. In 1980, Japan's share of the world's import of petroleum should be increased to 24 percent (in 1969, 10 percent), and of raw materials to 39 percent (in 1969, 14 percent).[5]

It is in a sense normal that this accelerated increase would cause problems in securing the sources of resources supply. That is why the main governmental reports concerning the natural resources problem began to mention in the late 1960s the necessity of "autonomous development" or "cooperative development" of natural resources abroad. These slogans mean that Japan has relied on resources supply made by "international oligopolistic firms," but this supply source has become endangered from the point of view of prices as well as quantity, precisely because of the behavior of this "oligopolistic system" or rise of economic nationalism in the supplying countries.[6] That is why the Japanese government has to promote overseas investments in mining and energy sectors for the sake of securing natural resources.

But exactly when an awareness arose concerning the necessity of "autonomous development" of natural resources, in the beginning of the 1970s, there appeared the "resource constraints" problem in the Japanese economy.[7] Before entering into the analysis of this problem, we have to note two new aspects of the resources problem that emerged ac-

[4] Economic Planning Agency (1975), Figure 74.

[5] MITI (1971), Table I-1-3.

[6] In 1969, the Committee on Resources Study installed in the Economic Council published a report entitled *Kokusaika Jidai no Shigen Mondai [Resources Problem in the Age of Internationalization]*, in which was pointed out for the first time the necessity of "autonomous development." This report stressed the impact of the "oligopolistic system," but did not pay attention to the problems caused by nationalism in the producting countries. But, as the international environment had been rapidly changing, the same committee was obliged to publish in 1972 the second report called *Henka no naka no Shigen Mondai [Resources Problem in a Time of Change]* in which they replaced the expression "autonomous development" by "cooperative development."

[7] Since 1974, the imports of crude oil and iron ore have been stagnated, and we see here a significant correlation between the supply of raw materials and diminishment of economic growth.

companying the high economic growth realized in the postwar period.

First, the economic growth has been achieved by industrialization in the heavy and chemical industries: in 1960, the share of the products of the heavy and chemical industries in the total of industrial products accounted for 55.4 percent, but in 1970 this share increased to 68.5 percent. This promotion of heavy and chemical industries followed the traditional doctrine of international division of labor: Japan must have comparative advantage in heavy and chemical industries, and consequently it specialized in these sectors. This specialization in the heavy and chemical industries meant that other sectors of economy had to be discouraged or even destroyed, especially the sectors of agriculture and domestic development of natural resources. For example, the domestic supply accounted still for 76 percent of the primary energy supply to the economy in 1955, but this share decreased in 1970 to only 15 percent. Instead, the share of imported energy (mainly petroleum) increased from 24 to 85 percent in these 15 years (Table 2). This substitution of energy sources was accomplished under the successive "Measures for the Coal Problem" promulgated by the government from 1959 to 1966. One of the consequences of these measures was the considerable decline of domestic reserves of coal: in 1955, the exploitable reserves of coal were estimated as 20,246 million tons (possible exploitable years: 64), but, in 1966, these reserves declined to only 7,075 million tons (possible exploitable years: 25)[8] because of the shutting down of the coal mines.

In the agriculture plan, the self-sufficiency ratio of food decreased considerably. In 1960, the average self-sufficiency ratio of food was 90 percent, but in 1973, it dropped to 71 percent.[9] Especially for cereals, the self-sufficiency ratio declined from 83 to 41 percent in this period, mainly because of massive imports of cereals from abroad. Domestic agriculture was discouraged under the "Basic Agrarian Law" applied since 1961.

Certainly imports of cheaper overseas resources contributed to maintaining the strong competitiveness of Japanese industry, and especially imports of cheaper food guaranteed the low wage as well as the supply of labor power from the rural areas to industry in the decade of the 1960s, but these were achieved at the cost of discouragement or destruction of domestic resources.

Second, the heavy and chemical industrialization has led to destruction of environmental resources. The heavy and chemical industries consume a lot of energy and raw materials and are heavy polluters. In Japan, around 1970, the density of raw steel production per km² of habitable

[8] Statistical Division of the Prime Minister's Cabinet, *Nihon no Tokei [Statistics of Japan]* (1976), p. 78.

[9] Ministry of Agriculture and Forestry, *Norin Suisan Tokei [Statistics of Agriculture and Fishing]* (1975), p. 71.

TABLE 2

JAPAN: COMPOSITION OF PRIMARY ENERGY SUPPLY

	1955 (%)	*1960* (%)	*1970* (%)	*1985* (%)
Water power	21.2	15.3	6.7	3.7
Atomic power	—	—	0.6	9.6
Coal	50.2	42.1	17.5	13.1
Petroleum	20.2	37.7	73.5	65.1
Natural gas	0.4	1.0	0.9	—
LNG	—	—	0.4	7.9
Thermal energy	—	—	—	0.5
Woods	8.0	3.9	0.4	—
Domestic energy	76.0	55.8	15.1	17.6
Imported energy	24.0	44.2	84.9	82.4
Total	100.0	100.0	100.0	100.0

Source:
Resources and Energy Agency, *Sogo Enerugi Tokei [Synthetic Statistics of Energy]*. The estimates for 1985 are quoted from *Chukan Toshin [Intermediate Report]* of the Research Committee on General Energy Problems (1975).

area was 28.6 tons, which is 14 times higher than the U.S. level and six times higher than the French level. Energy consumption in the heavy and chemical industries amounts to 2,223 tons (in coal equivalent) per km² habitable area, which greatly surpasses the figures of the United States (325 tons), the United Kingdom (1,230 tons), France (393 tons), or West Germany (1,531 tons).

Today, we can affirm that Japan is one of the most polluted countries in the world. As for total wastes, in 1971, these were estimated as 820 million tons (of which, 85 percent is of industrial origin) and, in 1975, it reached to 1,250 million tons (of which 88 percent is of industrial origin).[10] As is well known, with the progress of industrialization and the consumption civilization, the share of industrial wastes accelerates. As

[10] Environment Agency, *Kankyo Hakusho [White Paper on Environment]* (1974), pp. 47–48.

for water pollution, if we take for example the Chemical Oxygen Demand in the evacuated waters from mining and industry, it amounts to 5,860 thousand tons in 1969, a figure that is also increasing rapidly.[11] In the lagoon Kasumigaura (Ibaragi Prefecture), the degree of transparency had decreased 40 percent in the decade of the 1960s. As for air pollution, the oxides of nitrogen and of sulfur (NO_x and SO_2) evacuated in 1971 in the Tokyo metropolitan district amount to 460 tons per km^2 of habitable area, which is 7.3 times higher than the 1955 level[12] and which is ranked as one of the highest in the world.

In this way, the rapid industrialization has caused a heavy burden on the environment, and the nation is losing one of the most precious natural resources in this country.

II.

We shall examine next the nature of the so-called "resource constraints" caused both in the international and domestic plans.

The famous Club of Rome report entitled *The Limits to Growth* (1972) predicted that in the very near future the world reserve of natural resources would be exhausted because, according to the report, the "geometrical" exploitable years of these resources are limited: aluminium, 31 years; iron ore, 93 years; natural gas, 22 years; petroleum, 20 years; copper, 21 years; etc., and since these reserves are diminishing year to year, it is urgent to realize "equilibrium state" through population control and restraint of industrialization.

It is true that the nonrenewable natural resources have some physical limit, but it is nonsense to talk about the "physical depletion" of resources starting from the comparison between the actual proved reserves to the actual production (*R/P*— the so-called exploitable years) because the "reserves" could be reevaluated year by year according to new discovery, technological change, technological innovations, change in costs, or changes in demand or market prices. The "resources" are by their very nature a very dynamic notion. If, to take an example, by some technological innovation, the exploitation of laterite or oil-shale became economically feasible, the "reserves" of iron ore or petroleum would jump to an enormous degree. And it is normal that, when one resource becomes exhaustible in some exploited area, people begin to invest in exploration of other areas or in development of substitute goods. Also, the price rise that should accompany the exhaustion of one resource should justify these efforts, and new mines or oil fields or new substitution

[11] Sangyo Keizai Kondan-kai, *Sangyo Kozo no Kaikaku [Reforms of Industrial Structure]* (1973).

[12] Environment Agency (1974), p. 127.

goods could always be developed.

That is why we cannot conclude that the "resource constraints" that the Japanese economy has faced have their origin in any physical depletion of resources.

The main problems of resource constraints in the Japanese economy can be analyzed in two plans.

First, in the international plan, we see that it has been caused by a dynamic change in the supply structure of principal natural resources.

This change had been manifested already in the 1960s when some developing peripheral countries asked for nationalization, participation, indigenization, or nationalization of foreign mining companies. The clear expression of this movement was seen at the time of the so-called "oil crisis." The unity of Arab oil-producing countries at the moment of the Fourth Middle East War created a psychological panic phenomenon concerning the oil supply among the developed countries, and the high oil prices set by the producing countries have been maintained through their cartelization. This move was followed, as we know, by many producing countries of raw materials or agricultural products such as copper, bauxite, mercury, iron ore, tungsten, banana, sugar, coffee, and others. Of course, depending on the products, the bargaining power of trade partners and market conditions are very different, and we could consider that the case of oil, where perfect monopoly was once established, is rather exceptional. Nevertheless, we have to see the fact that peripheral producing countries, which produce actually 40 percent of the world's iron ore, 69 percent of tin ore, 70 percent of bauxite, 46 percent of copper, 46 percent of manganese and export the majority of these raw materials to the international market,[13] have begun to gather themselves, for the first time in the world's modern economic history, to exercise control over their own resources.

Also, in the area of ocean law, the installation of the exclusive 200 sea miles' economic zone, which originated from the demand of the developing countries, became a common practice. This affected 3/4 of Japanese ocean fishing and 45 percent of total fishing production.[14]

These movements took their ideological formulation in the demand of

[13] The share in 1970 of the developing countries in the world's exports of iron ore was, according to the survey of UNCTAD (A19544/Add. 1), 42 percent; tin, 77 percent; tin ore, 64 percent; bauxite, 88 percent; copper, 42 percent; and manganese ore, 51 percent. As for the other raw materials and foods, crude oil (89 percent), coffee (94 percent), cocoa (85 percent), tea (80 percent), cotton (58 percent), sugar (69 percent), crude fertilizer (43 percent), ground nuts (79 percent), palm oil (94 percent), jute (95 percent), sisal (97 percent), lumber (41 percent), and rubber (98 percent) are the commodities for which the export share of the developing countries is high.

[14] Ministry of Foreign Affairs, *Daisanji Kaiyoho Kaigi [The Third Conference of Maritime Law]* (1975), pp. 100–103.

the developing countries for the foundation of the New International Economic Order, which was crystallized in the Charter of Rights and Duties among the States adopted in the General Assembly of the United Nations in 1975 and which aims essentially for the establishment of permanent sovereignty over natural resources and economic activities in general and the stabilization of the prices of primary goods.[15]

Examining the evolution of this movement to control their own resources, we have to recognize that the developing peripheral countries are proceeding irreversibly in the way to use and transform their own resources in their own countries. This is perhaps the most important and profound meaning of the "resource constraints" problem that the developed countries are facing.

In the 1970s, the United States experienced the famous "energy crisis." This is the consequence of various conjunctures: difficulty in increase of imports of fuels and raw materials from Latin America because of rising nationalism, lag of construction of refineries to meet increasing demands, delay of development of new energy resources, the strategy of transnational corporations to increase the oil prices, and some climatic conditions. But we see essentially that the energy crisis that the United States is experiencing has its origin in the delayed and inadequate policy response to the rapidly changing conditions of the world supply of fuels.[16]

If the United States, which has rich fuel reserves, is experiencing the energy constraints and importing nearly half of its supply of oil for energy consumption from abroad, how could the other developed countries, especially Japan, which are not necessarily favored in natural resources endowment, avoid the coming constraints problem in resources supply?

The resource constraints problem that Japan is facing today derives not from the physical limit of world resources, or the strategy of "international oligopoly," or the energy crisis of the United States, but essentially from the changing international conjuncture concerning the supply of natural resources.

But there is another aspect of resource constraints that is eminently domestic, and this is the second point: people's minds are changing in Japan.

The big enterprises, some of the principal promoters of economic growth, internalizing the external economies and externalizing their diseconomies, have realized at low cost the rapid expansion of production. This is one of the principal reasons for the high degree of international competitiveness of Japanese enterprises and undoubtedly a reason for the high economic growth realized recently in Japan. Nevertheless this

[15] For the analysis of the New International Economic Order, *see* my *Daisan Sekai no Kozo to Dotai [The Structure and Dynamics of the Third World]* (1977), Chap. 1.

[16] *See* my *Daisan Sekai to Nihon [The Third World and Japan]* (1974), Part 1, Chap. 1.

TABLE 3

JAPAN: ASSESSMENT OF CITIZENS ON ECONOMIC GROWTH
(1971–1974)

	1971 *(%)*	*1974* *(%)*
Consider good	27	18
Part good/part bad	29	25
Cannot decide	18	24
Consider bad	14	24
Do not know	12	9

Source:
Inquiry made by the Prime Minister's Cabinet, quoted by Agency of Environment, *Kankyo Hakusho [White Paper on Environment]* (1975).

behavior by big enterprises resulted in a high degree of pollution in every part of the country, as we have noted already.

With the progress of pollution and the destruction of the environment, people's view of economic growth began to change. According to an interesting inquiry made by the Cabinet of the Prime Minister in 1971 and in 1974, the Japanese citizens' assessment of economic growth has been modified considerably in these several years.

In 1971, the percentage of citizens who considered economic growth good amounted to 27 percent of total replies, and the percentage of citizens who replied that economic growth was bad was 14 percent. Three years later, this proportion was reversed, and the citizens who appreciated economic growth now accounted for only 18 percent, while the citizens who judged economic growth as bad increased to 24 percent. If we include the intermediate replies, which said that the consequences of economic growth were partly good and partly bad, or which said they could not decide, 3/4 of the citizens now have doubt about economic growth. (*See* Table 3.)

Throughout this period, according to the change observed in people's attitudes, we saw the rise and development of a citizens' movement against pollution and for preservation of the environment. In 1972 and 1973, the victims and citizens' movements achieved victories in the so-called Four Big Pollution Suits: the cases of the diseases Itai-Itai, Minamata, Niigata-Minamata, and Yokkaichi; this fact demonstrates well the people's changing mentality regarding the behavior of enterprises.

In addition to these cases, where the enterprises responsible were ordered by the tribunal to compensate the victims, private enterprises became more and more obliged to pay for the prevention of pollution.

For example, in mining and industrial sectors, in 1965, investment for the prevention of pollution amounted to 3.1 percent of total investment; in 1975 it increased to 18.6 percent.[17] Also, the taxes levied by the local communities from the corporations had a tendency to be increased. The strengthening of antimonopoly law adopted in the Diet in May 1977 must be understood in the context of these movements.

Thus the enterprises, especially big enterprises, which have long benefited from the free use of, or the use at extremely low cost of, environmental resources such as land, water, or air began to feel constraints on the expansion of their production. In other words, there arises the "location problem" of their plants. This constitutes one of the reasons, which is not yet principal but whose importance is growing, for vertically increasing overseas investment of private enterprises. We have to see that, behind the cry of "resource constraints," also lies this move by enterprises to look abroad for locations where new resources will be at their disposal.

Until now, we have seen the nature of the "resource constraints," which are loudly discussed among the Japanese opinion leaders. We shall now examine the problem of choice for a possible industrial structure in the near future and of the supply of necessary resources for it.

III.

Since 1974, the Ministry of International Trade and Industry began to publish yearly the forecast of the future Japanese industrial structure, presented by the Council on Industrial Structure attached to the Ministry.[18]

According to this report, a 6 percent rate of economic growth is presented as optimum for the Japanese economy in the medium run. Also, Japanese industry must, according to the report, increase technological intensification and increase the amount of value added, responding to the "resource constraints" problem.

In this view, as well as policy, of the evolution of the Japanese industrial structure, the specialization of the manufacturing industry is further increased: in 1960, the proportion of production by secondary industry to total domestic production was 36.4 percent; in 1970 it was 60.3 percent; and in 1985 it is forecast to be 63.4 percent. Instead, the share of primary industry (which decreased from 14.8 percent in 1960 to 4.4 percent in 1970) is further predicted to diminish and to account for only 2.4

[17] MITI, *Minkan ni okeru Sangyo Kogai Boshi Setsubi Toshi no Doko [The State of Investment for the Prevention of Industrial Pollution in the Private Sector]* (1975).

[18] Council on Industrial Structure, *Sangyo Kozo no Choki Bizion [The Long-Term View of Industrial Structure]* (1974, 1975, 1976).

TABLE 4

JAPAN: EVOLUTION OF INDUSTRIAL STRUCTURE

	Production (%)			*Employment (%)*		
	1960	*1970*	*1985*	*1960*	*1970*	*1985*
Primary industry	14.8	4.4	2.4	30.2	17.4	8.9
Secondary industry	36.4	60.3	63.4	28.0	35.2	37.5
Service sector	49.0	32.3	31.6	41.8	47.4	53.7

Source:
Ministry of International Trade and Industry, *Sangyo Kozo no Choki Bizion [The Long-Term View of Industrial Structure]* (1976).

percent in 1985 (Table 4). This trend in the industrial structure, which shows further the drive of the secondary sector, is also observed in the plan of employment; from 1970 to 1985 the share of employment in secondary industry would increase from 35.2 percent to 37.5 percent, but that of the primary industry would decrease from 17.4 to 8.9 percent.

In this way, the push for further industrialization does not give up the traditional international division of labor doctrine, which contributed to creating conflicts with raw material–producing countries that had suffered the deterioration of terms of trade and which desire to promote their own industries. Nevertheless, we have to remark that, taking advantage of these conditions, the report intends to develop the traditional division of labor doctrine in an original way: it recommends that the Japanese economy move into a "higher state of international division of labor," in other words, to invest massively abroad. This promotion of private overseas investment is justified mainly for three reasons, each connected very closely with the proposed reply to the "resource constraints" problem: first, to secure the supply of raw materials;[19] second, to conserve domestically the limited natural environment necessary for citizens' life; and, finally, to maintain the "vitality" of Japanese enterprises in an era of low economic growth (we recall that the annual growth rate is forecast as 6 percent).[20]

Thus Japanese overseas investment (permission base) remained in 1967 at the level of $1,458 million (per capita of the nation, $15), increased to $19,405 million (per capita, $176) in 1976, and is predicted to reach more

[19] Thus, the Japanese government gives a special taxation deduction of 100 percent to investments in the resources development sector (in general, 50 percent of the taxation deduction is given to investment in developing countries and 10 percent of the deduction to those in the developed countries); the investing firms can calculate the sum of investment as "reserve for losses on overseas investments."

[20] Council on Industrial Structure (1976), pp. 144–45.

than $80,000 million (per capita, $800) in 1985. This evolution of overseas investment reflects, of course, economic tendencies such as market competition, shortage of labor force, location problems, increasing profit opportunities in the developing countries, and so on. But it is also supported and pushed firmly by the government's special treatment measures: the automatic permission system, a favorable tax system (recognition of foreign investment as reserve), the insurance for overseas investment, and loans of foreign currency (actually a loan of 5 dollars is given to 5 yen).[21]

Can this promotion of overseas investment, based on a philosophy of ever-continuing economic growth and conceived as a reply to the "resource constraints" problem,[22] resolve fundamentally the problem of the Japanese economy?

According to our analysis, this does not constitute an authentic reply to the actual "resource constraints" that the Japanese economy is facing but rather will worsen this problem.

The "resource constraints" problem in the Japanese economy did not originate in any sense in an absolute resource depletion problem, but originated in the worsening of "North-South" issues, which in turn resulted in the change in supply structure of the world's raw materials. The problem also resulted from the spectacular development of domestic heavy and chemical industries based on the doctrine of international division of labor, which has destroyed and deteriorated domestic resources.

If our analysis is valid, the policy to push foreign investment is not the basic solution to the "resource constraints" problem because foreign investment is precisely one of the most controversial issues in "North-South" relations: many developing countries, which control in large part the world's exports of raw materials and fuels, are intending to establish permanent sovereignty over economic activities in their territories (and the Japanese government is opposing the articles concerning permanent sovereignty in the Charter of Rights and Duties among States). If one does not take into account this fundamental fact and only pushes overseas investment "to secure" the raw materials or to find suitable plant location sites "to preserve" one's own environmental resources, this behavior will result only in further worsening of the North-South relation-

[21] The system of loans of foreign currency began in September 1972 with the ratio of 7 dollars to 3 yen. It was modified in May 1973 and became the ratio of 9 dollars to 1 yen, but after the "oil crisis," since January 1974, the actual ratio was adopted.

[22] After the "oil crisis," the main pillars of the Japanese resource policy were conceived as follows: (1) development of domestic energy sources, (2) saving of energy consumption, (3) overseas investment to secure the resources, and (4) stockpile of resources (oil, lumber, soybeans, barley, feedstuffs, and nonferrous metals such as copper, lead, and zinc). These policies indicate the confrontation attitude of the Japanese government *vis-à-vis* the resource producing countries. In fact, the Japanese are hurrying to develop nuclear energy.

ship and accelerate the seriousness of the "resource constraints" problem. On the other hand, to push the foreign investment of heavy and chemical industries for the sake of "maintenance of their vitality" without making reassessments of economic growth policy would only mean the export of pollution to the world and the consequent deterioration of the world's environmental resources.

To make a fundamental response to the "resource constraints" problem in the Japanese economy, in our view, it is necessary to proceed to modification of the "economic growth" philosophy, which has for over a century, and especially since the end of the Second World War, influenced the Japanese mentality as well as the evolution of Japanese society.

Concretely, three aspects of this modification should be developed:

First, it is necessary to abandon the development philosophy based on the theory of the international division of labor and specialization in industrialization and to reestablish the balanced structure of the economy. This means the reestablishment of the primary industry sector, such as agriculture and fishing, in the domestic plan.

Second, it would be better to leave the flow of foreign investment to the market mechanism; governmental support for overseas investments could be the origin of fears concerning domination, or hegemony, on the part of the developing countries (which receive 60 percent of the Japanese overseas investment). Rather, the government should make efforts to ameliorate the actual relations between the North and the South, respecting and recognizing the economic sovereignty of the developing countries, guaranteeing the long-term contract of raw material imports, increasing official development aid, and trying to realize an effective transfer of technology.

Third, the conventional beliefs concerning economic growth and the proliferation of mass consumption have led the peoples of the developed countries, including Japan, to be chronic consumers and wasters of natural resources. Actually, one U.S. citizen is more wasteful in resource terms than 50 Indians, and one Japanese is more wasteful than 30 Southeast Asians. If the peoples of the developed countries really feel the "resource constraints," they have to contain their consumption of resources at home rather than recommend population control to the peoples of the developing countries. Resource saving efforts are very necessary in the developed world in the age when the peoples of the world are becoming aware of the need for a more equitable and rational distribution of resources.

For the Japanese economy, the true solution of the "resource constraints" problem is to modify the actual road of economic growth, and proceed to construct an economic structure based on people's welfare, social justice, and harmony with nature. It should not move toward a huge and unbalanced industrial country, which invests massively abroad to secure at all costs the natural resources needed for domestic industry. This way only means the worsening of "resource constraints" both in domestic and global plans. But, for the Japanese economy to progress in the first direction, the citizens' control of the economy and of resources must be the primordial element, and here we see the genesis of a society where the resources problem is resolved by the efforts of citizens themselves.

IV.

We can now respond to the two questions that were posed at the beginning of this report.

First, the influence on the growth of the Japanese economy of the resource constraints problem, as it was manifested at the time of "oil shock," has been very important. However, this "constraints" problem was not a new problem brought about for the first time by the "oil shock." It was already seen as a problem at the end of the 1960s with the progress of the Japanese economy. The "oil shock" accelerated the growing perception of a resource constraints problem in a radical way. On the international scene these resource constraints did not originate in physical bottlenecks of production, but rather in the resource distribution problem, and it was closely connected with the worsening of the North-South issues. Also, domestically speaking, the destruction of the environment, which accompanied high economic growth, has intensified the resource constraints problem.

Second, the lowering, or the lowering out of necessity, of economic growth targets to a moderate level constitutes, together with efforts to reduce the elasticity of resource consumption to GNP, a necessary measure for the Japanese economy at a time of change in the worldwide structure of resources supply. In spite of this, the mere agreement on a quantitatively lower rate of growth does not in any sense solve the resource constraints problem that the developed countries are facing. If these constraints originate in high growth, in the increasing dominance of big enterprises, and in the related worsening of the North-South problem (deterioration of the terms of trade of the developing countries, which is derived partly from the administered price inflation of the developed countries; conflicts; and accumulation of debts in the developing countries associated with the progress of transnational corporations), then one cannot expect beneficial results from a policy of supporting overseas investment and securing raw materials at all cost to maintain growth rates and promote industrialization led by the big enterprises. It will only intensify further the environmental problem as well as the North-South conflicts on a worldwide scale. Under these conditions, the Japanese economy will inevitably experience a second and a third "oil shock" in every field of natural resources supply. The authentic reply to the "resource constraints" problem that the Japanese economy is facing is the following: establishing citizen control over the big enterprises whose resource demands gave rise to talk of "resource constraints" and concentrating the efforts of the people to resolve the environmental problem as well as North-South issues.

REFERENCES

Japan, Economic Council, Committee on Resources Study. 1969. *Kokusaika Jidai no Shigen Mondai [Resources Problem in the Age of Internationalization]*. Tokyo.

———. 1972. *Henka no naka no Shigen Mondai [Resources Problem in a Time of Change]*. Tokyo.

———, Economic Planning Agency. 1975. *Keizai Hakusho [White Paper on Economy]*. Tokyo.

———, Environmental Agency. 1974. *Kankyo Hakusho [White Paper on Environment]*. Tokyo.

———, Ministry of Agriculture and Forestry. 1975. *Norin Suisan Tokei [Statistics of Agriculture and Fishing]*. Tokyo.

———, Ministry of Foreign Affairs. 1975. *Daisanji Kaiyoho Kaigi [The Third Conference of Maritime Law]*. Tokyo.

———, Ministry of International Trade and Industry (MITI). 1971. *Shigen Mondai no Tenbo [The Perspective of the Resource Problem]*. Tokyo.

———. 1975. *Minkan ni okeru Sangyo Kogai Boshi Setsubi Toshi no Doko [The State of Investment for the Prevention of Industrial Pollution in the Private Sector]*. Tokyo.

———. 1977. *Tsusho Hakusho [White Paper on International Trade]*. Tokyo.

———, MITI, Council on Industrial Structure. 1974, 1975, 1976. *Sangyo Kozo no Choki Bizion [The Long-Term View of Industrial Structure]*. Tokyo: MITI.

———, Statistical Division of the Prime Minister's Cabinet. 1976. *Nihon no Tokei [Statistics of Japan]*. Tokyo.

Meadows, Donella H., *et al.* 1972. *The Limits to Growth: A Report for the Club of Rome's Project on the Predicament of Mankind*. Washington, D.C.: Potomac Associates.

Nishikawa, Jun. 1974. *Daisan Sekai to Nihon [The Third World and Japan]*. Tokyo: Ushio Shuppan-sha.

———. 1977. *Daisan Sekai no Kozo to Dotai [The Structure and Dynamics of the Third World]*.

Sangyo Keizai Kondan-kai. 1973. *Sangyo Kozo no Kaikaku [Reforms of Industrial Structure]*. Tokyo: Taisei Shuppan-sha.

Japanese Economic Developments, 1970–1976*

D. J. Daly *York University*

The magnitude of the price increases in raw materials and the extent of the recession were unusually severe in Japan from 1970 to 1976, in comparison with postwar experience in Japan and the other major industrialized countries. This paper examines these two developments in the perspective of historical experience in comparable recessions and considers some associated macro policies. Special emphasis is given to the cyclical changes in productivity and the associated volume of exports and imports.

* This study was financed by a research grant from the Canadian Federal Department of Energy, Mines and Resources. Research assistance has been provided by Mrs. Fumiko Yamada. The seasonal adjustments were done by Census Method X-11, and the charts were done on a Hewlett-Packard Model 9825A computer by J. J. Singer Associates of Toronto. Helpful comments on an earlier version have been provided by Martin Bronfenbrenner, Sueo Sekiguchi, and Ippei Yamazawa.

INTRODUCTION

The period 1970 to 1976 witnessed a major upheaval in the world economies, including the most widespread and severe peacetime inflation in history, the most widespread recession in many of the major industrialized countries occurring roughly simultaneously since the 1930s, the collapse of the IMF system of fixed exchange rates and the widespread adoption of managed floating exchange rates by many industrialized countries, and a major increase in world petroleum prices initiated by the OPEC countries.

The aim of this paper is to assess these developments in Japan. The paper will essentially consist of four parts, dealing initially with the role of imports of natural resource products in the Japanese economy. Secondly, the effects of higher materials prices on the terms of trade will be explored, and the options open to deal with the large change that developed will be considered. The third part will discuss domestic economic developments within Japan where the recession was the most severe experienced by any industrialized country since the 1930s. The fourth part will discuss international trade developments—considering imports, the competitive position of Japanese manufacturing, and the renewed growth of exports. The paper will use some new seasonally adjusted series, which will identify the timing and magnitude of the marked changes in domestic activity and international trade that took place after late 1973.

I. THE ROLE OF NATURAL RESOURCE IMPORTS IN JAPAN

Japan consists of a series of islands with a high proportion of the land area being mountainous and rugged, with only a small proportion of the land available for agriculture, for business purposes, and for dwellings for the more than 100 million people. Japan has the lowest levels of land and mineral production in relation to population of any of the major industrialized countries. Some evidence of the magnitudes of the differences can be seen in the accompanying Table 1. Relative to the size of the total labor force, Japan is clearly the least well endowed by all three measures. In terms of arable land per person employed, Japan is only 5 percent of the U.S. level and 2 percent of the Canadian level, and lower than any of the individual European countries,[1] and the differences are even greater for total land area. Mineral production per person is less than 4 percent of the U.S. level and only 2 percent of the Canadian level. Both employment and production in mining in Japan were lower in 1975

[1] Edward F. Denison and William K. Chung, *How Japan's Economy Grew So Fast: The Sources of Postwar Expansion* (1976), Appendix 0, p. 258.

TABLE 1

LAND AREA AND MINERAL PRODUCTION PER PERSON EMPLOYED, 1960 AND 1970

(Relatives, U.S. = 100)

	Land Area Per Person Employed		*Value in $ U.S. of Mineral Production Per Person Employed*	
	All Land	*Arable Land*	*Denison List*	*Expanded List*
Japan (1970)	6.4	5.2	3.7	
Northwest Europe (1960)	13	20	26	26
United States	100	100	100	100
Canada (1970)	1036	218	181	222

Sources:

United States and Europe: Denison and Poullier, *Why Growth Rates Differ* (1967), Table 14-2, p. 184; Japan: Denison and Chung, *How Japan's Economy Grew So Fast* (1976), Appendix 0, p. 258; and D. Walters, *Canadian Income Levels and Growth: An International Perspective* (1968), Table 64, updated to 1970 with sources as described on pp. 233-34.

TABLE 2

INTERNATIONAL COMPARISONS OF GROSS DOMESTIC PRODUCT, 1970, PERCENTAGES OF U.S. VALUES

Price Weights	*U.S.*	*Japan*	*France*	*West Germany*	*U.K.*	*Italy*
GDP per person employed						
United States	100.0	55.2	79.1	73.3	60.4	61.6
Other country	100.0	44.3	65.6	61.5	51.2	49.2
GDP per capita						
United States	100.0	68.1	81.8	80.3	67.9	53.5
Other country	100.0	54.6	67.8	67.3	57.5	42.7

Source:

Denison and Chung (1976), p. 5, drawing on results from Irving B. Kravis and others, *A System of International Comparisons of Gross Product and Purchasing Power* (1975).

than at the start of the decade, continuing a relative decline that started much earlier.

By 1970 Japan had a level of real gross domestic product per capita above Italy, and close to the United Kingdom, but still below France and West Germany. The Japanese levels are not relatively as high on a per person employed basis, as the Japanese have a larger proportion of the total population employed than any of the other countries. Results for 1970 are shown in Table 2.

Since 1970, increases in real GNP per capita and per person employed have been more rapid in Japan than in the other industrialized countries, in spite of the severity of the recession in the Japanese commodity-producing industries from late 1973 to early 1975. This would have moved the Japanese position up relatively further in relation to the European countries, but still well below the levels in North America for the economy as a whole.

A growing Japanese economy involves significant increases in the volume of imports. This is reflected in Table 3, where the volume of imports is compared to real GNE and the index of industrial production. The volume of imports has increased at a significant rate of growth (13.1 compounded from 1955 to 1973). This is more rapid than the rate of growth in real GDP, but about the same as the index of industrial production over this period. Both imports and industrial production dropped during the recession.

It might be noted that Japan had a very high effective tariff rate as recently as the early 1960s, but these rates were reduced sharply as part of the Kennedy Round reductions, and Japan made a further unilateral tariff reduction across the board in late 1972 of 20 percent. By 1972 the nominal tariff rate was 2.3 for raw materials, and the effective tariff rate was under 1 percent, well below the effective tariff rates on manufactured products. In addition, import quota restrictions were widespread early in the 1960s, but the number of items under quota restriction had been reduced sharply by late 1973.[2] Imports were much less restricted by Japanese tariff and nontariff barriers than in earlier years when the major price increases in oil and other primary products took place.

The limited availability of primary products domestically, the high rates of domestic growth, and the reduced tariff and nontariff barriers on imports of primary products have been reflected in a high degree of Japanese dependence on imports. The Krause-Sekiguchi chapter highlights this point saying "The most striking characteristic of Japan that influences its economic relations with other countries is its almost total

[2] Bela A. Balassa, "Tariff Protection in Industrial Countries: An Evaluation" (1965), p. 591; Thomas Hout, *Japan's Trade Policy and the U.S. Trade Performance* (1973), pp. 9–10; and Lawrence B. Krause and Sueo Sekiguchi, "Japan and the World Economy" (1976), pp. 426, 428, and 457–58.

TABLE 3

PHYSICAL MEASURES OF IMPORTS AND DOMESTIC ACTIVITY, JAPAN, 1954 TO 1975, 1970 = 100

	Imports (1970 prices)	*Industrial Production*	*GDP*
1954	12.3	11.6	22.0
1955	12.5	12.7	23.9
1956	15.6	15.6	25.7
1957	19.3	18.5	27.6
1958	16.3	18.1	29.1
1959	20.3	21.9	31.7
1960	25.1	27.2	35.9
1961	32.8	32.4	41.2
1962	32.6	35.1	44.0
1963	38.5	38.6	48.7
1964	44.0	44.6	55.0
1965	44.7	46.3	57.8
1966	51.7	52.5	63.5
1967	63.5	62.6	71.8
1968	71.7	73.7	81.4
1969	83.9	87.9	90.2
1970	100.0	100.0	100.0
1971	99.9	102.6	107.4
1972	112.9	110.1	117.0
1973	144.5	127.3	128.6
1974	141.2	123.3	126.9
1975	123.1	109.7	130.0

Sources:

International Financial Statistics for the data on imports and industrial production; Annual Report on National Income Statistics 1977 by Economic Planning Agency; Government of Japan for the data on gross domestic products.

lack of domestic sources of raw materials."[3] Illustrations of the high dependence on imports of primary products are shown in Table 4. In 1971,

[3] Krause and Sekiguchi (1976), p. 386. *See* also the papers in this volume by Yasukichi Yasuba ("Resources in Japan's Development," which provides a longer-term historical background and some intercountry comparisons on the use of natural resources) and by Jun Nishikawa ("'Resource Constraints': A Problem of the Japanese Economy," which contains data in Table 1 similar to Table 4 of this paper). The Nishikawa paper tends to be critical of the MITI proposal to ensure a more adequate supply of natural resource materials by investment in foreign countries and to favor slower growth in the Japanese economy instead.

TABLE 4

JAPAN'S DEPENDENCE ON IMPORTS OF PRIMARY PRODUCTS

	1974
Beef	15.7
Wheat	95.9
Maize	99.6
Greasy Wool	100.0
Raw Cotton	100.0
Lumber	41.2
Copper	90.1
Lead	80.3
Zinc	64.5
Bauxite	100.0
Tin	98.5
Nickel	100.0
Iron Ore	99.4
Coal (including coking coal)	72.2
Crude Oil	99.7
Natural Gas	34.8
Uranium	100.0

Sources:

JETRO, *White Paper on International Trade Japan* (1976), p. 10. The data for natural gas and uranium relate to the Fiscal Year 1970, from MITI, *White Paper on Prospect of Natural Resource Problems in Japan,* English summary in *Trade and Industry of Japan,* No. 167, 1970, Table 4, p. 37.

Japan was the largest importer of all the OECD countries of such raw materials as coal, crude petroleum, iron ore, manganese ore, copper ore, zinc ore, lead ore, wool, and cotton and was the second largest importer of rubber, and the third largest of bauxite.[4]

II. JAPAN'S TERMS OF TRADE AND POLICY OPTIONS

Under the circumstances, the increase in primary product prices during the worldwide inflationary period beginning about 1973 had a major impact on Japan. The Arab oil embargo in late 1973 and the sharp increase in the price of imported crude oil in 1974 were a further shock in light of the almost complete dependence on imports shown in Table 4. The price increases in primary products were reflected in an increase of about 140

[4] Krause and Sekiguchi (1976), p. 386.

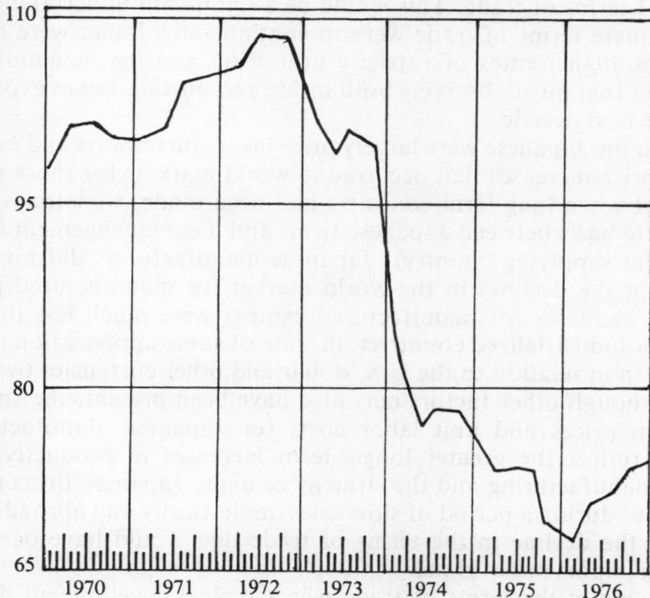

Fig. 1. Terms of trade, Japan (1970 = 100)
Quarterly averages

percent in the import price index from the summer of 1971 to a high early in 1974, and a tripling of the London Economist Sterling price index from late 1971 to a 1974 high. These external shocks coming so soon after the changes in U.S. policy initiated by President Nixon in August 1971 on tariff surcharges on manufactured products, the breakdown of the gold-exchange standard, and the resulting readjustments in exchange rates created real concern in Japan about their increased interdependencies with the world economy.

The increases in prices of imports to Japan after the summer of 1972 were much sharper than the price changes of exports, which largely consist of a wide range of manufactured products. The commodity terms of trade are shown in Figure 1; these fell from a peak of about 108 (1970 = 100) to under 70 in late 1975.[5] This would imply that in late 1975, Japan would have to export about 50 percent more in physical quantity to pay for the same quantity of imports as would have been required with

[5] Ministry of International Trade and Industry, *Statistics on Japanese Industries 1976* (1977), pp. 102–04.

the 1970–73 terms of trade. This would be a significant undertaking, especially if these terms of trade were to continue and Japan were to recover to the higher rates of capacity utilization and the medium-term growth rates that most observers both inside and outside Japan expect to see over the next decade.

Although the Japanese were largely price takers on imports and experienced the price increases that occurred in world markets for those products (except when long-term contracts had been made, sometimes on a joint venture basis between Japanese firms and local management in the raw material supplying country), Japanese manufacturers did not passively accept the declines in the world market for manufactured products. Price increases for manufactured exports were much less than in other major industrialized countries, in spite of some appreciation of the Japanese yen in relation to the U.S. dollar and other currencies over the period. Although other factors may also have been present, the smaller increases in prices and unit labor costs for Japanese manufacturing appear to reflect the greater longer-term increases in productivity in Japanese manufacturing and the strategy of many Japanese firms to cut export prices during a period of slow sales domestically and abroad. This intensified the decline in the terms of trade that would have occurred from rising import prices alone.

This decline in the terms of trade was a typical development during previous periods of low or declining economic activity. Miyohei Shinohara emphasizes this development in both the pre–World War I period and the interwar period and discusses the influence of the deterioration in the terms of trade during the downward phases of long cycles.[6] Leon Hollerman discusses the same phenomenon and finds that increased exports frequently occurred during periods of excess capacity.[7] Hugh Patrick also notes that the major improvements in export competitiveness have come in periods of recession.[8] Kazushi Ohkawa and Henry Rosovsky also explore this area and consider the rate of growth in productivity largely oriented to the domestic market as a factor in the declines in relative prices and unit labor costs in the modern manufacturing industries.[9] This decline in the terms of trade has been a typical phenom-

[6] Miyohei Shinohara, *Growth and Cycles in the Japanese Economy* (1962), pp. 63–71 and 98–106. Martin Bronfenbrenner suggested this point in commenting on an earlier version of this paper and this and the preceding paragraph in the text have been added in response to his comment and to those by Ippei Yamazawa.

[7] Leon Hollerman, *Japan's Dependence on the World Economy: The Approach Toward Economic Liberalization* (1967), pp. 43–49.

[8] Hugh J. Patrick, "Cyclical Instability and Fiscal Monetary Policy in Post-war Japan" (1965), p. 573.

[9] Kazushi Ohkawa and Henry Rosovsky, *Japanese Economic Growth: Trend Acceleration in the Twentieth Century* (1973).

enon during a period of underutilization during the current century. This development has occurred in spite of the declines in primary product prices during such periods, with different exchange rate environments and different movements of the Japanese yen in relation to major foreign currencies. The implications of these developments for the volume of exports will be considered later.

In the calendar year before the sharp decline began in the terms of trade, the current account of the balance of payments of Japan was in surplus to the extent of $6.6 billion U.S., and a basic balance of about $2.1 billion U.S. This was a sizeable current account balance, but a number of factors were already underway to reduce its size, including the revaluation of the yen in December 1971 and its further appreciation in early 1973 when the Bank of Japan no longer attempted to peg the exchange rate, and the changes in September and October 1972 to free up imports, lower tariff rates, and liberalize international capital restrictions. In addition, Japan was in the process of enlarging investment elsewhere, frequently in the form of joint ventures to provide imports of labor intensive products from lower wage countries in Asia, or investments in raw material sources in other countries, themes that have been emphasized in subsequent White Papers, such as on future industrial structures.

A continuation of a high level of economic activity in Japan and the associated continuance of a high level of imports in volume would clearly have been associated with a dramatic basic balance-of-payments deficit in Japan. The deficits would have been larger during the 1974–75 period of lower domestic activity if imports had not dropped more than 20 percent from their peak in volume in November 1973 to a low point a year-and-a-half later in May 1975 (based on a 3-month moving average of a quantum measure of imports, seasonally adjusted). If Japanese imports had grown at the same rate of increase from 1973 to 1975 as they had from 1963 to 1973, the 1975 level of merchandise imports would have been almost 50 percent higher in 1975, instead of falling 14 percent. As imports were $49.7 billion U.S. in 1975, the 1973–75 rate of increase would have implied a level of imports about $24 billion higher than actually recorded. This difference would have been important in relation to the 1975 level of GNP, but would have been even more significant in relation to net national product in manufacturing.

Japanese policymakers had three possible policy options open to them as a response to this marked deterioration in their terms of trade: exchange depreciation; selective taxes and controls to curtail imports or encourage exports; or domestic demand deflation.

Exchange depreciation or devaluation was one possibility that would further discourage imports by higher prices and make exports relatively more profitable. However, this would increase the prices of primary products in Japan even further when this was already regarded as a factor in the rapid domestic price increases. Furthermore, Japan had been criticized for having an undervalued yen during the late 1960s when her balance of payments was strong, and the yen had appreciated about 25 percent from early 1971 to March 1973. Depreciation or devaluation of the yen would have been inappropriate in that situation.

A second possibility was to reverse the tendency towards lessened tariff and nontariff barriers to trade that had been underway since early in the 1960s. Japan had made a unilateral tariff reduction of 20 percent across the board in late 1972 and had sharply reduced the range of items subject to global import quotas by 1973. It is of some significance that these reductions in barriers to trade were not reversed.

The third possibility was a restriction in domestic demand, and this was the process that has in fact emerged. The previous rate of increase in prices and costs domestically was bound to have been an important factor in the decision-making process. From 1973 to 1974, for example, the increase in the cash earnings index of regular workers was 26.4 percent, while the wholesale price index increased 31.3 percent and the consumer price index went up 24.2 percent.[10]

It would divert us from our major theme to explore the developments in Japanese monetary and fiscal policy and their interrelations with international financial and exchange rate policy, partly as the postwar background on these topics is covered in the Patrick-Rosovsky Brookings volume.[11] However, these studies did not explore developments in the money supply from a monetarist point of view, an omission raised in the Bronfenbrenner-Minabe review.[12] Figure 2 shows the annual rates of change in the money supply (M_2—the wide definition). It indicates a much lower rate of change after about 1973 and reduced short-term variation. When the slower rate of growth in the money supply is combined with the sharp rates of price increases referred to above, it is clear that a significant decline in the money supply has occurred in real terms. As loans from banks and other financial institutions are an important source of funds (38 percent of total liabilities are in the form of short-term and long-term liabilities), the slower growth in total bank assets and deposit liabilities and a significant decline in real terms can have significant repercussions on business spending on inventories and fixed assets. The change in loans and discounts of all banks was still down more than 30 percent in 1974 and 1975 than in 1972, and changes in other sources of

[10] Sueo Sekiguchi, "Recent Development in Japan's Foreign Trade and Investment" (1976), pp. 18–25, esp. pp. 19–20.

[11] Patrick and Rosovsky, *Asia's New Giant. How the Japanese Economy Works* (1976), especially the chapters by Gardner Ackley and Hiromitsu Isi; Henry C. Wallich and Mable I. Wallich; Lawrence B. Krause and Sueo Sekiguchi. *See,* however, the reviews by Martin Bronfenbrenner and Shigeo Minabe (1977), pp. 145–77.

[12] Martin Bronfenbrenner and Shigeo Minabe (1977), pp. 147 and 155.

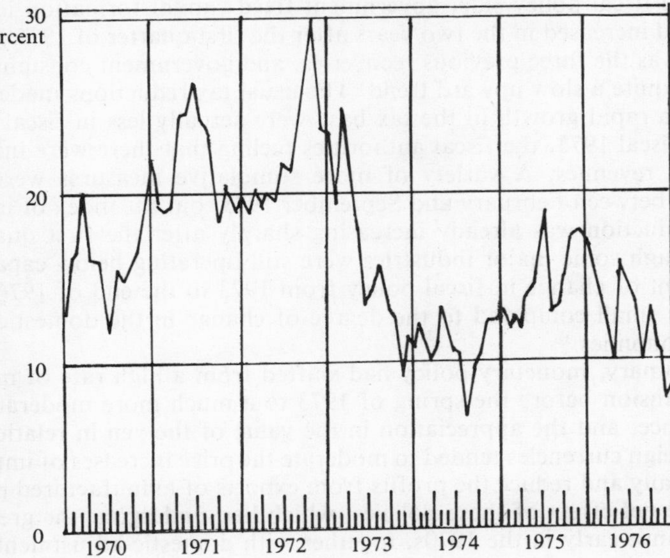

Fig. 2. Rate of change in the money supply (M2)
Annual rate, S.A., 3 month moving avg.-Japan

supply of industrial funds were insufficient to offset the lower levels of loans and discounts of all banks.[13] It is interesting that the Bank of Japan had been testing monetarist models during this period. A study published in its Monthly Report on Research for July 1975 (*Chosa Geppo*) gives results for several definitions of money supply and regression results for later changes in prices, using Almon distributed lags.[14]

The sharp declines in net profits for the principal enterprises in nominal terms between 1973 and 1975, and the even larger decline in real terms (if depreciation of physical assets were valued at replacement cost), would be further depressing elements in business spending, although profits were recovering by 1976.[15]

[13] MITI (1977), p. 97.

[14] Bank of Japan, "The Importance of Money Supply in the Japanese Economy" (1975). An English translation of this important article is available in limited quantities from the author.

[15] Further discussion is contained in Hisao Kanamori and Sueo Sekiguchi, *The 1974–75 Recession and Economic Stabilization Policy—Japanese Experience* (1977); Hisao Kanamori, ed., *Recent Developments of Japanese Economy and its Differences from Western Advanced Economies* (1976), especially Chapters 13 and 16 by Hugh Patrick; Kunio Saito, "The Japanese Economy in Transition" (1977); and other studies in Japanese.

On the fiscal policy side, government fixed capital formation in real terms had increased in the two years after the first quarter of 1975 about the same as the three previous recoveries, and government consumption grew on quite a slow upward trend. The usual tax reductions made possible by a rapid growth in the tax base were actually less in fiscal 1975 than in fiscal 1973, the fiscal authorities feeling that there were insufficient tax revenues. A variety of more stimulative measures were announced between February and September 1975, but the index of industrial production was already increasing sharply after the first quarter, even though some major industries were still operating below capacity. The extent of change in fiscal policy from 1973 to the end of 1976 was relatively small compared to the degree of change in the domestic and world economies.[16]

In summary, monetary policy had shifted from a high rate of monetary expansion before the spring of 1973 to a much more moderate increase since, and the appreciation in the value of the yen in relation to most foreign currencies tended to moderate the price increases of imports domestically and reduce the profits from exports of manufactured products. The net effect of these policies, which had undergone the greatest change since early in the 1970s, together with domestic adjustments on inventories, capital spending, and other areas of domestic demand, tended to depress the level of imports. Thus, this brief summary of government policies at the macro level provides evidence that the government had really followed the third option outlined earlier.

III. DOMESTIC ECONOMIC ACTIVITY IN RECESSION AND RECOVERY

Initially, the typical pattern of developments in recession and recovery in the labor market and physical activity in North America will be sketched to provide perspective on whether the Japanese recession of 1973 to 1975 was similar. The weaknesses in final demand and inventory investment associated with such periods of demand weakness are reflected in more marked declines in physical output in the commodity producing industries than in GNP. The increases in employment that have been experienced over the postwar years slow down and even decline in absolute terms if the recession is moderately severe, and increases in unemploy-

[16] Wesley Clare Mitchell, *What Happens During Business Cycles: A Progress Report* (1951); Thor Hultgren, *Changes in Labor Cost during Cycles in Production and Business* (1960); Geoffrey H. Moore, "Business Cycles and the Labor Market" (1961), pp. 505–13; Julius Shiskin, *Signals of Recession and Recovery* (1961); Thor Hultgren, *Cost, Prices and Profits: Their Cyclical Relations* (1965); Geoffrey H. Moore and Julius Shiskin, *Indicators of Business Expansions and Contractions* (1967).

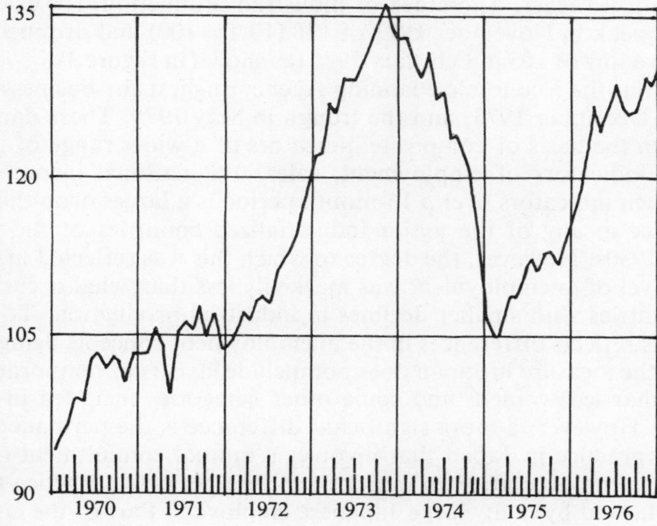

Fig. 3. Index of industrial production, Japan (1970 = 100)
All industries—seasonally adjusted

ment develop. However, part of the cutbacks in production is reflected in reductions in hours worked and declines in output per man hour, rather than in the increases that have tended to take place on a longer-term basis. The interconnections of these changes in physical output and productivity with the developments in wages, prices, and per unit costs are reflected in sharp declines in corporate profits and profit margins.

Once a lower turning point occurs, the subsequent rebound in physical activity leads to the sharpest increases in employment and physical activity of the whole business cycle, made possible by the degree of slack present during the early stages of the expansion. The increases in output per man hour are typically above the long-term average and can partially or fully offset increases in wage rates and other costs, so costs per unit frequently increase more slowly than during the later stages of the previous expansion. The increases in corporate profits can frequently be very dramatic.[17]

Many of these developments that have been the historical experience of other industrial countries have been apparent in the 1974–75 recession and recovery in Japan, with the magnitudes sometimes being significant-

[17] Geoffrey H. Moore and Philip A. Klein, "Recovery and Then?" (1976), p. 55.

ly different, however. The index of industrial production, for example, reached a peak in November 1973 of 136 (1970 = 100) and dropped 22.8 percent to a low of 105 in February 1975 (as shown in Figure 3).

Data from the Economic Planning Agency suggest the business cycle peak was December 1973, and the trough in May 1975. These dates are selected on the basis of composite indicators of a wider range of industries and indicators of employment, sales, and national income. The drop in such indicators over a 15-month period is a larger drop than has taken place in any of the major industrialized countries of the world since the 1930s. However, the degree to which this was reflected in an increased level of unemployment was markedly less than what occurred in other countries with smaller declines in industrial production. To some degree this reflects differences in the unemployment concepts being measured, as the measure in Japan does not include layoffs of temporary employees, married women, and some other categories included in other countries. However, a more significant difference is the permanent employment practice in Japan that involves a mutual commitment of employer and employee after an initial three-year period, a practice that is widely followed by many large Japanese employers. Part of the cutback in production was associated with a reduction in hours worked, which dropped more than 10.8 percent from early 1973 to early 1975 (as shown in Figure 4). The drop in labor productivity was particularly sharp, declining about 15 percent from early 1974 to a low in February 1975, a marked change from the high rates of increase during postwar years.

The sharp declines in sales and production with relatively modest layoffs of regular employees were associated with a sharp drop in corporate profits. Japanese companies tend to have relatively higher levels of fixed costs than in North America, partly associated with the permanent employment practice, but also with the relatively larger reliance on debt and relatively less reliance on internal funds as a source of financing. By early 1975 about half the companies listed on the Tokyo Stock Exchange were experiencing losses, and bankruptcies have been more frequent. Some adjustment in costs could be made by reduced bonuses to employees (which are normally quite large), reduced sub-contracting, or laying off temporary employees. Some of these adjustments shift the impacts of lower sales levels onto smaller firms, and their employees and creditors.

Once economic activity begins to recover, however, the increases can be spectacular, even for an economy with the high rate of long-term growth of Japan. For example, from early 1975 to the end of 1976, the index of industrial production had increased about 22 percent, a spectacular increase for a 20-month period, even though activity was still below the previous peak. Labor productivity in manufacturing had increased more than 20 percent by late 1976 and surpassed the previous peak reached about two years earlier by the spring of 1976 as shown in Figure 5. The recovery in hours worked has been much less complete,

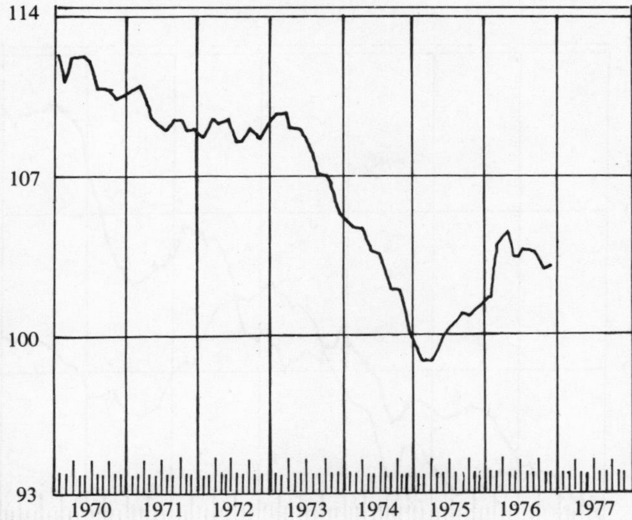

Fig. 4. Index of hours worked per month (1975 = 100)
Manufacturing, regular hours, Japan

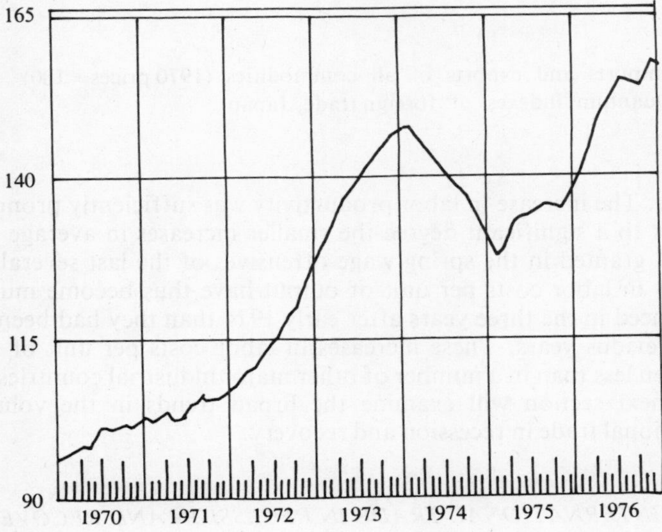

Fig. 5. Labor productivity indexes (1970 = 100)
Manufacturing, Japan

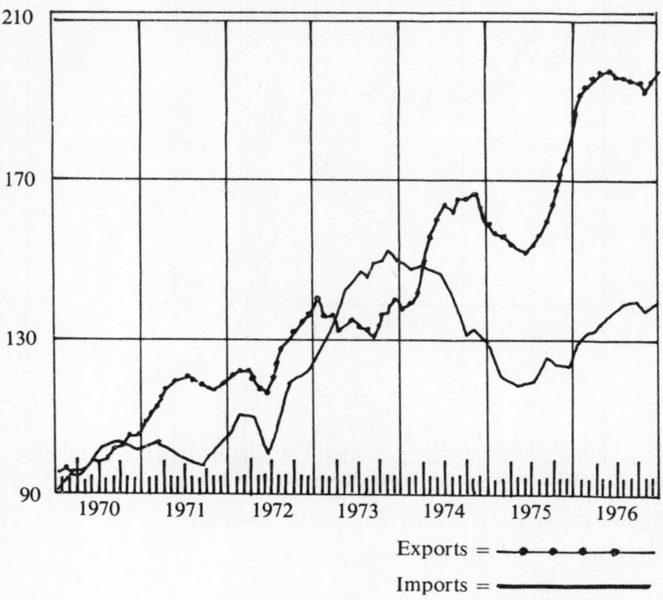

Fig. 6. Imports and exports of all commodities (1970 prices = 100)
Quantum indexes of foreign trade, Japan

however. The increase in labor productivity was sufficiently pronounced to offset to a significant degree the smaller increases in average hourly earnings granted in the spring wage offensives of the last several years. Changes in labor costs per unit of output have thus become much less pronounced in the three years after early 1974 than they had been in the three previous years. These increases in labor costs per unit of output have been less than in a number of other major industrial countries.

The next section will examine the broad trends in the volume of international trade in recession and recovery.

IV. INTERNATIONAL TRADE IN RECESSION AND RECOVERY

Changes in the tempo of Japanese economic activity of the magnitudes discussed in the last section were bound to have an impact on the levels of imports. The volume of imports dropped about 24 percent in 19 months from the peak in November 1973 to a low in the spring of 1975, roughly comparable to the decline in the index of industrial production. This is a significant contrast from the increases of 14 percent per year compounded from 1963 to 1973. Imports were temporarily swollen in the later stages of the expansion by increasing inventories (such as wool) and were depressed in 1975 by the beginnings of some inventory liquidation. As discussed previously, the levels of import volume would have been significantly higher from 1974 to 1976 if the Japanese economy had not experienced a sharp recession domestically. This would have been associated with an enlarged balance of payments deficit, with implications for the exchange rate, and perhaps other aspects of domestic policy.

Before considering developments in exports, it may be useful to consider the competitive position of Japanese manufacturing in relation to North America. In 1967, the level of output per person in total manufacturing in Japan was about half the U.S. level, although the increases since the latter part of the 1950s had brought a small but growing number of individual industries above the U.S. level.[18] The annual growth rates in output per person from 1967 to 1973 were significantly greater in Japan (13.0) than in the United States (2.8) or in Canada (4.5).[19] By 1973, output per man year was higher than in Canada and not too much below the U.S. level. Part of this difference reflected the much longer hours of work per year in Japan than in North America. Although no data are available for manufacturing on the relative levels of capital stock per person employed in Japan and North America, the levels of nonresidential structures and equipment for all industries in Japan were about 60 percent of the U.S. level in 1970.[20] Even with the sharp increases in average hourly earnings in Japan in the early 1970s, the levels of average hourly earnings in Japanese manufacturing are less than two-thirds the levels in Canada and the United States.[21] This is an important factor in costs when labor income is such a major share of net national income both for the economy as a whole and manufacturing.

A number of points about Japanese manufacturing in the context of international trade can be pointed out. For one thing, Japanese exports

[18] Kenzo Yukizawa, *Japanese and American Manufacturing Productivity: An International Comparison of Physical Output Per Head* (1975), Table 3, p. 17, based on Japanese price weights. *See* also Kenzo Yukizawa, "Relative Productivity of Labor in American and Japanese Industry and Market Size, 1958–1972" in this volume. However, Kazuo Sato notes in his paper, "Did Technical Progress Accelerate in Japan?" in this volume that the ratio of value added to gross value of production is lower in Japan than in the U.S. He suggests an adjustment to Yukizawa's estimate of the Japanese level for total manufacturing from .78 to .60 of the U.S. level for 1972 (using U.S. weights).

[19] Philip A. Klein, *Business Cycles in the Postwar World* (1976), Table 10, p. 36.

[20] MITI, *Japan's New Energy Policy* (1976), p. 78.

[21] D.J. Daly, "Productivity and Costs of Canadian Manufacturing" (1976), p. 8.

of manufactured products have become a much larger share of the world market for manufactured products. Furthermore, exports of manufactured products in volume have tended to grow in relation to the volume of GDP, but that primarily reflects the growing role of real manufacturing output in the Japanese economy, rather than exports becoming a larger share of manufacturing.

On the competitive position of Japanese manufacturing, the rapid increases in Japanese wage rates during the inflationary period in the early 1970s and the appreciation in the foreign exchange value of the Japanese yen have eroded the strong competitive position present in 1970.[22] Cost advantages have probably become more selective by 1976 in spite of the more rapid increases in output per person and per man hour in Japanese manufacturing than in the other major industrialized countries.

The volume of exports from Japan was inevitably affected by economic developments in other countries. During the 1960s there was little correspondence between cycles in the United States and eight other major industrialized countries.[23] However, "the recession of 1973–75 affected all the major industrialized market-oriented economies, and in virtually all it was the most severe setback since World War II."[24]

The volume of exports from Japan declined for seven or eight months from late 1974 to a low in June 1975, but the declines were shorter and milder than what had occurred in imports, as can be seen in Figure 6. The greater mildness of the decline in exports reflects the fact that the recessions in domestic activity in the major industrialized countries to which Japan was selling were milder than the recession in Japan. Furthermore, Japan exports to a wide range of countries, and the timing of weakness in foreign markets, although widespread, was not uniform in the various major markets.[25]

Most observers would regard the renewed expansion in the major in-

[22] JETRO (1976), pp. 25–27. This contains comparisons with the U.S. and the Federal Republic of Germany, and the comparisons of 1970 and 1973 with 1975 allow for the changes in prices of primary products and wage costs in the three countries. However, the use of unchanged input-output coefficients would not appear to have incorporated the differential changes in productivity among the countries over the period, and thus overstate the increase in costs in Japanese manufacturing.

A paragraph on p. 27 recognizes that since a reduction in production costs in Japan relative to other countries could be offset by a further appreciation of the yen, it states "To stabilize Japan's export expansion in the future, non-price competitiveness will have to be strengthened along with efforts to further consolidate its overseas information network and to develop new markets." However, insofar as this emphasis on non-price competition leads to a larger trade and basic balance of payments surplus in the future, it seems just as likely to lead to an appreciation of the yen as increased price competitiveness.

[23] Klein (1976), p. 42.

[24] Moore and Klein (1976), p. 55.

[25] Moore and Klein (1976), pp. 56 and 58.

dustrialized countries as hesitant and sluggish compared to the early stages of previous postwar expansions, and rates of capacity utilization were still low relative to the 1950s and 1960s.

The increases in the volume of exports from Japan were significant, however. By late 1976, export volume was about 30 percent above the low reached in June 1975 and more than 15 percent above the previous peak reached in late 1974. These changes provide a significant contrast to the movement in imports shown also in Figure 6 and discussed previously, with exports being above their previous peak and showing significant longer-term growth since 1970. The volume of imports was still below earlier highs, however, and has increased less in volume (but at a much higher price level as discussed earlier). After the first quarter of 1976, the terms of trade recovered somewhat, giving further assistance to the improving balance of trade, as pointed out by Ippei Yamazawa.

It is of interest that Japanese exports to the United States increased more in the five previous downswings in Japanese growth cycles than during the intervening upswings, and the differences in magnitude were striking.[26] It appears that the weakness in the domestic market encouraged Japanese manufacturers to search out export markets more aggressively during this, as in previous periods of slack activity.

This pattern of rapid growth in export volume during periods of slack domestically and abroad that has been observed during the mid-1970s is fully in accord with earlier experience as noted previously in references to studies by Shinohara, Hollerman, Patrick, and the Ohkawa-Rosovsky volume. The sharp decline and then recovery in manufacturing productivity was similar in nature to what had been previously observed in studies by the staff at the National Bureau of Economic Research and elsewhere, but the magnitudes of the changes were greater than anything normally experienced in North America. By late 1976, the levels of productivity in manufacturing were sufficiently above the recession lows to more than offset the increases in wage rates and leave labor costs per unit of output below their previous highs of late 1974 and early 1975, a significant contrast with the experience of the other major industrialized countries. This was leading to renewed increases in exports from Japan and increased concern from manufacturers in other countries. These developments in Japan were broadly similar to comparable previous periods. However, Japan was a growing share of the world market for manufactured products during the postwar period, making perhaps for a greater impact, and the extent and duration of underutilization in manufacturing has probably been greater during this hesitant recovery. On the whole, however, the developments within Japan, and in export markets, have been markedly similar to previous experience, apart from the magnitudes of the decline and subsequent recovery to the end of 1976.

[26] Klein (1976), Table 10, p. 36.

The depressed level of Japanese imports has had a deflationary impact on countries normally exporting to Japan, such as exporters of primary products and some of the developing countries in Asia. Furthermore, the usual response of Japanese manufacturers to increase exports during a period of slack has been a factor in an increase in protectionist sentiment in many industrialized countries and increased trade frictions between Japan and her trading partners. This is an unfortunate development coming as it does at a key time in the current negotiations in Geneva, especially since the Japanese government has been an active supporter of these negotiations, which are called the Tokyo Round (the initial meetings having been held in Tokyo). More expansionary policies earlier would have raised imports and eased the pressure to increase exports by manufacturers.

V. LONGER-TERM MEASURES

This paper has emphasized economic developments at the aggregate measures level and the associated macro policies. Martin Bronfenbrenner correctly mentioned additional measures, beyond those emphasized here, in his oral remarks on this paper concerning the composition of investment and relations with other countries, which are important policy responses to the higher raw material and energy prices beyond the macro policies emphasized in this paper.[27]

The impact of increased prices for raw materials and energy has also affected the thinking on the longer term. The *White Paper on Japan's Future Industrial Structure* provides for a significant increase in the extent of direct investment in foreign sources of supply of raw materials. This would normally be in the form of a joint venture between a Japanese firm (or group of firms) and an affiliated firm in the host country, who would undertake a major responsibility in management and provide a significant part of the initial financing. In addition, a considerable degree of effort is going into major longer-term plans on the energy side, an area not emphasized in this paper. A summary of Japan's new energy policy concluded "Japan has been obliged to change her energy policy from an oil-oriented one aiming at an abundant, cheap and stable supply of energy from oil, to a future policy giving the highest priority to a steady supply of energy by diversifying energy sources.... We further suggested that...major objectives of energy policy for the coming decade should be (1) reduction of dependence on oil and diversification of energy supply, (2) the ensuring of a stable supply of oil, (3) the promo-

[27] This paragraph has been added to clarify the emphasis in the paper, and the author would agree on the importance of the additional policies mentioned by Professor Bronfenbrenner. Sueo Sekiguchi raised further interesting questions in this area, but these will have to be deferred in this paper.

tion of energy conservation, and (4) the promotion of research and development of new sources of energy. We emphasized that these problems should be solved systematically from a long-range and international viewpoint, using all possible means."[28] Part of this longer-term research is the Sunshine Project, which is exploring new longer-term energy sources.

Although this paper has emphasized the shorter-term developments to the end of 1976, it is clear that the longer-term issues of potential vulnerability in an interdependent world economy to higher prices for raw materials and vulnerability to supply interruptions (such as petroleum from the Middle East) are being given serious consideration in the thinking within Japan on domestic policy in an international environment.

REFERENCES

Balassa, Bela A. 1965. "Tariff Protection in Industrial Countries: An Evaluation," *Journal of Political Economy*, vol. 73, December, pp. 573–94.

Bank of Japan, Research Division. 1975. "The Importance of Money Supply in the Japanese Economy," *Chosa Geppo [Monthly Report on Research].* Tokyo: Bank of Japan.

Bronfenbrenner, Martin and Minabe, Shigeo. 1977. "Book Review of Asia's New Giant: How the Japanse Economy Works," *Journal of Japanese Studies*, vol. 3, Winter.

Daly, D.J. 1976. "Productivity and Costs of Canadian Manufacturing: Some International Comparisons," York University, June. Mimeo.

Denison, Edward F. and Chung, William K. 1976. *How Japan's Economy Grew So Fast: The Sources of Postwar Expansion.* Washington, D.C.: Brookings Institution.

——— and Poullier, Jean-Pierre. 1967. *Why Growth Rates Differ: Postwar Experience in Nine Western Countries.* Washington, D.C.: Brookings Institution.

Hollerman, Leon. 1967. *Japan's Dependence on the World Economy: The Approach Toward Economic Liberalization.* Princeton, N.J.: Princeton University Press.

Hout, Thomas. 1973. *Japan's Trade Policy and the U.S. Trade Performance.* New York: Japan Information Service.

Hultgren, Thor. 1960. *Changes in Labor Cost during Cycles in Production and Business.* Occasional Paper 74. New York: National Bureau of Economic Research.

——. 1965. *Cost, Prices and Profits: Their Cyclical Relations.* Studies in Business Cycles, No. 14. New York: National Bureau of Economic Research.

Japan External Trade Organization (JETRO). 1976. *White Paper on International Trade.* Tokyo: JETRO.

Japan, Ministry of International Trade and Industry (MITI). 1976. *Japan's New Energy Policy.* Bulletin Bl-18. Tokyo: MITI.

——. 1977. *Statistics on Japanese Industries, 1976.* Tokyo: MITI.

Kanamori, Hisao, ed. 1976. *Recent Developments of Japanese Economy and its Differences from Western Advanced Economies.* Tokyo: Japan Economic Research Center.

——— and Sekiguchi, Sueo. 1977. *The 1974–75 Recession and Economic Stabilization Policy—Japanese Experience.* Tokyo: Japan Economic Research Center.

[28] MITI (1976), p. 78.

Klein, Philip A. 1976. *Business Cycles in the Postwar World: Some Reflections on Recent Research.* Washington, D.C.: American Enterprise Institute for Public Policy Research.

Kravis, Irving B., *et al.* 1975. *A System of International Comparisons of Gross Product and Purchasing Power.* Baltimore: Johns Hopkins University Press for the World Bank.

Mitchell, Wesley Clare. 1951. *What Happens During Business Cycles: A Progress Report.* Cambridge, Mass.: Riverside Press for the National Bureau of Economic Research.

Moore, Geoffrey H. 1961. "Business Cycles and the Labor Market," in *Business Cycle Indicators. Vol. 1: Contributions to the Analysis of Current Business Conditions.* Princeton, N.J.: Princeton University Press for the National Bureau of Economic Research.

——— and Klein, Philip A. 1976. "Recovery and Then?" *Across the Board: The Conference Board Magazine,* no. 10, October.

——— and Shiskin, Julius. 1967. *Indicators of Business Expansions and Contractions.* New York: Columbia University Press for the National Bureau of Economic Research.

Ohkawa, Kazushi and Rosovsky, Henry. 1973. *Japanese Economic Growth: Trend Acceleration in the Twentieth Century.* Stanford, Calif.: Stanford University Press.

Patrick, Hugh J. 1965. "Cyclical Instability and Fiscal Monetary Policy in Post-war Japan," in *The State and Economic Enterprise in Japan: Essays in the Political Economy of Growth.* Edited by William W. Lockwood. Princeton, N.J.: Princeton University Press.

——— and Rosovsky, Henry, eds. 1976. *Asia's New Giant: How the Japanese Economy Works.* Washington, D.C.: Brookings Institution.

Saito, Kunio. 1977. "The Japanese Economy in Transition," *Finance and Development,* vol. 14, June, pp. 36–59.

Sekiguchi, Sueo. 1976. "Recent Development in Japan's Foreign Trade and Investment," in *Recent Developments of Japanese Economy and its Differences from Western Advanced Economies.* Edited by Hisao Kanamori. Center Paper No. 29, Tokyo: Japan Economic Research Center, pp. 18–25.

Shinohara, Miyohei. 1962. *Growth and Cycles in the Japanese Economy.* Tokyo: Kinokuniya Bookstore.

Shiskin, Julius. 1961. *Signals of Recession and Recovery.* New York: National Bureau of Economic Research.

Walters, Dorothy. 1968. *Canadian Income Levels and Growth: An International Perspective.* Ottawa: The Queen's Printer.

Yukizawa, Kenzo. 1975. *Japanese and American Manufacturing Productivity: An International Comparison of Physical Output per Head.* Kyoto: Kyoto Institute of Economic Research.

Comments

Ippei Yamazawa (Japan)

The big price hike of petroleum and other primary products in 1973–74 gave a big shock to world economy, and the response to it by individual countries is a challenging subject of research in which basic characteristics of individual economies will be revealed. I would like to congratulate Professor Daly on having taken up this subject, and I hope the following comments of mine will be complementary to Professor Daly's analysis.

Professor Daly put focus of his analysis on the deterioration in the terms of trade from 1973 through 1975, and he discussed possible responses to it in *Section* II. However, the *deterioration* in the terms of trade would be regarded as reflecting the *response of Japanese economy to the price increase of imported resources rather than the cause of response itself.* Import prices increased most in Japan among major industrialized countries because of her high dependence on resource imports but Japan's export price increased least, thereby resulting in the biggest deterioration in the terms of trade. The smaller increase in export prices reflected the deflationary response by Japanese economy where the *increase in domestic wholesale price was repressed.* The yen was also depreciated against the U.S. dollar by 20% (from 254 yen per dollar in July 1973 to 307 yen in December 1975), which tended to repress the increase in dollar price of export and the recovery in the terms of trade.

Let me add an analysis of change in the terms of trade after the third quarter of 1976. The export price increased by 10 percentage points by April 1977, which, with a small decline of import price, tended to improve the terms of trade. But it is still at the low level of 78 (with the year 1970 as the base) and it is a long way to recover the pre-1973 level of 108, which partly explains the increasing balance of trade surplus today. The yen rate has turned upward since December 1975 and has been appreciating rapidly since March 1977, but it will take three or six months at least for the appreciated yen rate to be reflected in higher dollar price of Japanese exports and lower sale price of imports at home.

The terms of trade is the first measure of Japanese response to resource price increases but it misses some of the major impacts of the latter. One of them is their impact on the structure of comparative advantage. Some resource-processing industries cannot shift increased material costs onto their output prices under severe competition with foreign exporters. The yen appreciation has further aggravated their competitive position. They have maintained their sales at constant dollar prices so far, but they cannot continue it for long and eventually their price will be raised and their sales will decrease. We will see changes in commodity composition of exports and imports in five years or so.

The deflationary response of the Japanese economy to resource price increases and the delayed recovery today have also been affected by such long-run factors as labor shortage, environmental pollution, and too much dependence on resource imports. They were aggravated during the rapid growth in the 1960s and tended to constrain it, and the change to slower growth was suggested by some economists in early 1970s before the oil price increase. This certainly has discouraged investment in plant and equipment and delayed recovery from depression. If oil price had increased in the mid-1960s when Japan was still in the middle of rapid growth, the business sector would have responded more vigorously to it, the Japanese economy would have been less deflationary, the deterioration in the terms of trade would have been much smaller, and the recovery from the oil shock would have been much quicker.

Finally I would like to comment on *Japan's high dependence on resource imports.* Professor Daly listed factors causing increased import dependence on primary products by Japan by the early 1970s. They are "limited domestic availability," "high domestic demand growth," and "the reduction of tariff and non-tariff barriers." The first two were major factors, but not the third. Except foodstuffs, both mineral and agricultural resource product had been imported free either of duty or of quota restriction since 1895. Crude oil was an exception and on its import was 6–7% *ad valorem* equivalent duty, which, however, has not been changed much since 1960.

Impacts of the first two factors should be understood carefully. Increased imports of resources in the 1960s basically resulted from the growth of the Japanese economy with high requirements for imported resources, but it was also affected by its export structure concentrated in products with high resource content, such as steel, synthetic fiber material, plywood, and so on. A large proportion of increased imports of resources during the 1960s was required for the expansion of exports. That is, they were required to be input *both directly and indirectly to the production for export expansion.* The figures were 34.3% for nonferrous metal ore, 26.7% for iron ore, 25.5% for coal, 13.5% for petroleum, and 13.2% for lumber. Of course we cannot do without those resource imports for export totally because those exports are needed to pay for imports of resources required for the production for domestic demand, and Japan is responsible for supplying her overseas customers with those products. In the future, however, *with export structure more concentrated in products with less resource content* and high value-added, together with slower growth of industrial production and demand for resource input, Japan will be able to reduce her dependence on imported resources.

Comment

Martin Bronfenbrenner (U.S.A.)

The first point which I want to raise arises from Professor Yamazawa's remarks rather than the main paper by Professor Daly. He refers to Japanese "deflation" as a response to the oil crisis. I found this hard to understand in view of the performance of the Japanese price indexes.

Passing to the main paper, I should like to add two points, one historical and the other political. (a) Professor Shinohara's study of *Growth and Cycles in the Japanese Economy* finds a similar phenomenon (falls in the terms of trade) as a major cause of recovery from recessions and depressions as far back as the middle of the Meiji Era. (2) The obverse or "dual" of the fall in Japan's terms of trade is the widespread change of "dumping" level against Japanese exports.

Professor Daly's second point relates to labor productivity. He is completely right in explaining the recession decline on the basis of "hoarding" of excessive labor permanently employed with nothing to do. But perhaps he should go on to explain some part of the subsequent increase in productivity on the reverse basis of "dishoarding" this same labor. As a result, Japan is currently concerned with increasing unemployment rates.

Japanese reactions to the oil shock have also induced many policies not reflected in Professor Daly's statistical series. Let me merely mention four such policies here.

(1) Expansion of nuclear power.
(2) Improving relations with Arab states (un-cooperation with the U.S. anti-OPEC activities).
(3) Developing alternative sources of energy on Siberia and West China.
(4) The so-called "Sunshine" project for developing solar energy, geothermal power, etc.

Urbanization and Land Prices: The Case of Tokyo

Yuzuru Hanayama *Tokyo Institute of Technology*

In this paper the author describes the price mechanism of residential land in the Tokyo area. He stresses the psychology of landowners: the landowners who were small-sized farmers originally and who are not familiar with the entrepreneurial way of managing their lands are apt to select the course of hoarding lands and simply waiting for the future rise of land prices without doing anything. The author thinks that this is the cause of the confusion of the land use and.the astonishingly high speed of land price inflation. He examines some policies and concludes that the heavier property tax levying, especially the abolishment of preferential assessment for farmland in urbanized areas, is the most effective and operational measure to regulate the rise of land prices.

I. INTRODUCTION

A striking characteristic, which distinguishes Tokyo from most Western cities, is the poor quality of housing: low, dense, and spreading without bounds. Comparing the available data internationally, the floor space per capita of Tokyo is 8.5 square meters, while that of Detroit is 34.5, East Berlin is 25.5, and Prague is 11.5 sq.m. Detroit does not seem especially rich among American cities, and East Berlin and Prague were both destroyed during the war as Tokyo was. According to a survey conducted by the Ministry of Construction, 40 percent of all the households in Tokyo, some one million families, live in wooden apartment houses; 65 percent of them are one room apartments, 56.4 percent do not have the exclusive use of a toilet, and 91.3 percent do not have a bathroom. A large number of these dwellings are unhealthy because of lack of adequate sunshine and ventilation, and they are very vulnerable to fire. Almost all of these wooden apartment houses mushroomed in the early 1950s on the ruins left by air raids and are entirely superannuated today.

In Tokyo (a collection of 23 wards) there has not been any large change in population during this decade; every year about six hundred thousand people arrive in Tokyo to work and live, and almost the same number depart to live in other places. This influx of new residents consists mainly of younger and lower income people who have migrated from their home towns or villages, while the efflux is primarily senior and higher income people who have saved some money while working in the city. The former find their habitations in the old wooden apartment houses in the city, while the latter find their new dwellings in the suburbs of the city. Is the environment of the suburbs good? Unfortunately not. On account of the rise in land prices, the builders of houses today can afford rather poorer homes than they could have bought 10 or 20 years earlier, in spite of the growth in income. According to a survey in Higashimurayama, one of the cities in the western suburbs of Tokyo that was newly developed 20 years ago, the average size of a housing lot was 250 sq.m. and the percentage of those smaller than 100 sq.m. was 1 percent in 1962. But the average declined to 200 sq.m., and the number of smaller lots increased to 10 percent in 1968. In 1972 the average lot and the smaller lots became 150 sq.m. and 33 percent, respectively.

The sprawl of housing areas begins from the railroad stations in the suburbs first, and the construction of roads cannot adequately keep up with the housing spread. The existing roads are filled with cars and trucks, which go to the construction sites of the subdivisions. All the commuters who rush to the station, the farmers who go to their farms still left in the suburbs, the housewives who go shopping, and the students who go to schools are continually threatened by traffic accidents. They are showered with mud on rainy days and dust on fine days. The drinking water in these areas is gotten from the ground water at first, but it soon becomes polluted with viruses leaked from septic tanks. It quickly becomes necessary to build a water supply, a very costly undertaking because it requires long and inefficient pipelines connecting housing lots that are spread over a sparse but extensive area. At the same time, sewerage from the houses pollutes the irrigation ponds and channels of neighboring farms, discouraging the farmers from maintaining them. The water becomes contaminated and emits a bad odor. The sewer system requires much more money than the water supply. Every time a new subdivision is built, roads are laid out with water supply and sewerage lines, which add to the community expenses and disturbs the traffic. In addition, mass transit companies do not want to run buses in these sprawling suburban areas because the population is too scattered there to make a profit. As a result the number of private cars increases rapidly.

The degradation of the living environment both in the city and in the suburbs must be attributed to the lack of an appropriate policy on land development. In the Tokyo area, land prices have been soaring continuously for more than two decades since the restoration began after the ruins of the war. The only exception has been these three past years of recession after the oil shock. The rate of land price inflation has not been under 20 percent a year. Under such circumstances farmers who owned land in the suburbs were inclined to keep their land instead of selling it, and if any situation forced them to sell, they sold only a small portion. But on the demand side, the workers continued to come into the Tokyo area, and they have wanted to acquire their own houses and lots before land prices got too far ahead of their incomes. The confused and boundless sprawling of housing land and general waste of land resources seem consequences of the failure of adequate land policy rather than the result of rapid centralization of population in the urban area.

II. MECHANISM OF THE RISING PRICES OF LAND

Hearing of such a desperate housing situation, some may think that land is much too limited to meet the demand of the increasing population in the metropolitan area. But in fact there is a large potential supply of residential land within this area, for if we swing a circle with a radius of 40 kilometers centered at the Tokyo railroad station, we will have an area of about 4000 square kilometers (excluding water) within the circle, and we will find that we use only 1500 square kilometers of it as building area. As the circle can be regarded as the outer limits of residential land for the people who commute to the central part of the metropolis, it can be said that we have 2500 square kilometers left for potential housing provision. Why then cannot this potential be realized?

On account of the rapid concentration of capital and population in the

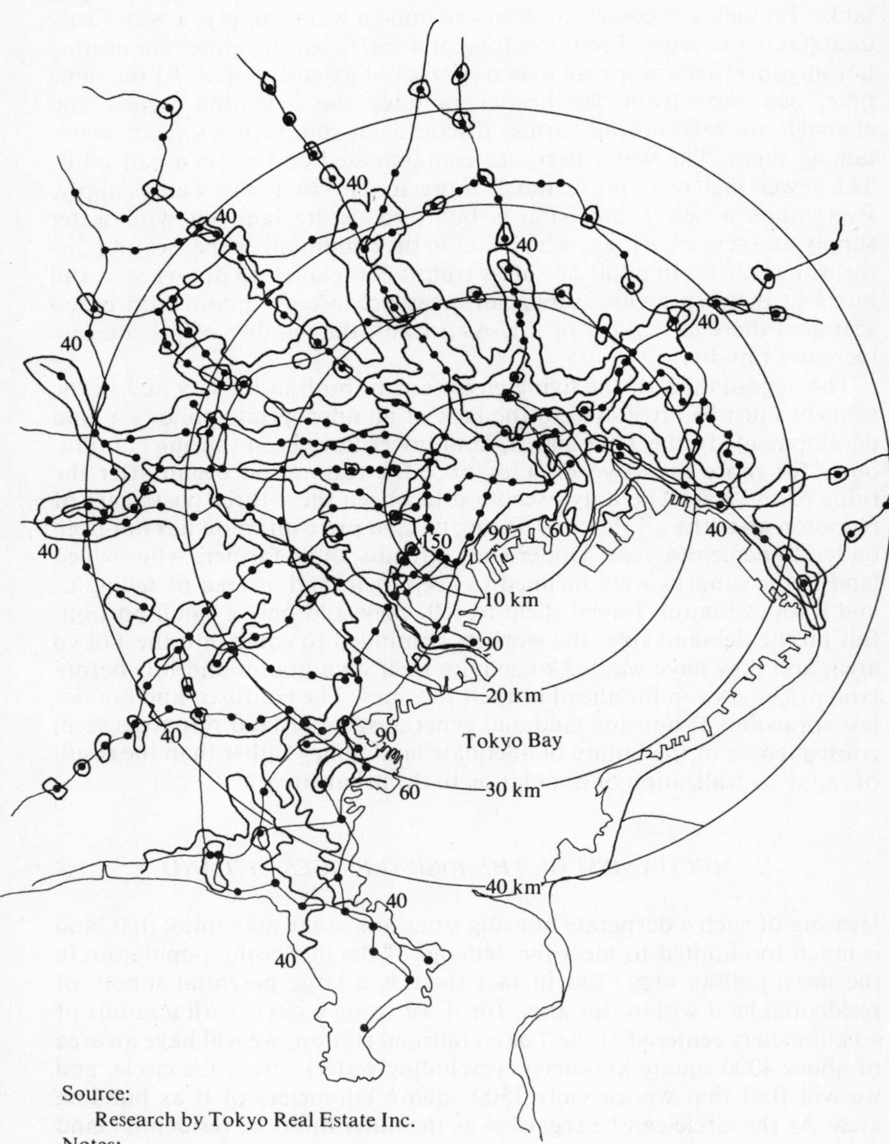

Source:
Research by Tokyo Real Estate Inc.
Notes:
(1) Contour line figures indicate 10,000 yen/m².
(2) Networks of dotted lines are railways.

Fig. 1: The Land Price Topography within Tokyo Metropolis, 1970

TABLE 1

THE LAND PRICE OF A LOCATION IN THE WESTERN PART OF HIGASHIMURAYAMA CITY, AT SELECTED TIME INTERVALS

Period (Month, Year)	*Actual Price (Yen/m²)*	*Rate of Price Rise*
March 1961	2,500	(500/2,500/9) × 12 ≒ .27
December 1961	3,000	(500/3,000/3) × 12 ≒ .67
March 1962	3,500	(700/3,500/3) × 12 ≒ .40
September 1962	4,200	
April 1964	8,000	(5,800/4,200/20) × 12 ≒ .83
May 1964	10,000	
July 1965	10,000	(2,000/10,000/41) × 12 ≒ .6
October 1967	12,000	
February 1968	14,000	
May 1968	13,000	(4,000/12,000/14) × 12 ≒ .29
December 1968	16,000	

existing urban centers resulting from the high growth rate of the Japanese national economy during the second half of the fifties period, the rate at which the price of land rose surpassed the prevailing rate of interest and, since then, the rising rate in the price of land has not dropped below the interest rate. Under such circumstances, it is inevitable that the owners of farmland are inclined to hoard land, and accordingly, the land is placed under the seller's option.

Let's look at some actual facts.

Figure 1 shows the distribution of price of residential land. Here we can see some rules of rising land prices: First, the price of land makes a mountain-like topography, as the center is the highest and the farther the distance from the center, the lower the price is, spreading in ridges along the railways; second, this mountain-like topography is always upheaving, but the speed of rising is not the same at every place; but the faster the speed, the farther the place is from the center, while the absolute value can never be reversed and the area of the part above the water level (the level of price that is reasonable to agricultural management) is increasing.

We can adduce Table 1 as a proof of the second rule: in this table we can see how the land price at a spot in Higashimurayama City, located 30 km from the center of the Tokyo metropolis in the western direction, has risen, and we can see that the speed of the rising price accelerated immediately after the spot was incorporated into the border of the residential area (in other words the spot appears above the water level

Note:
Prepared by author from aerial photographs.

Fig. 2: Sprawl of Building Expansion of Built-up-Areas in Higashimurayama City

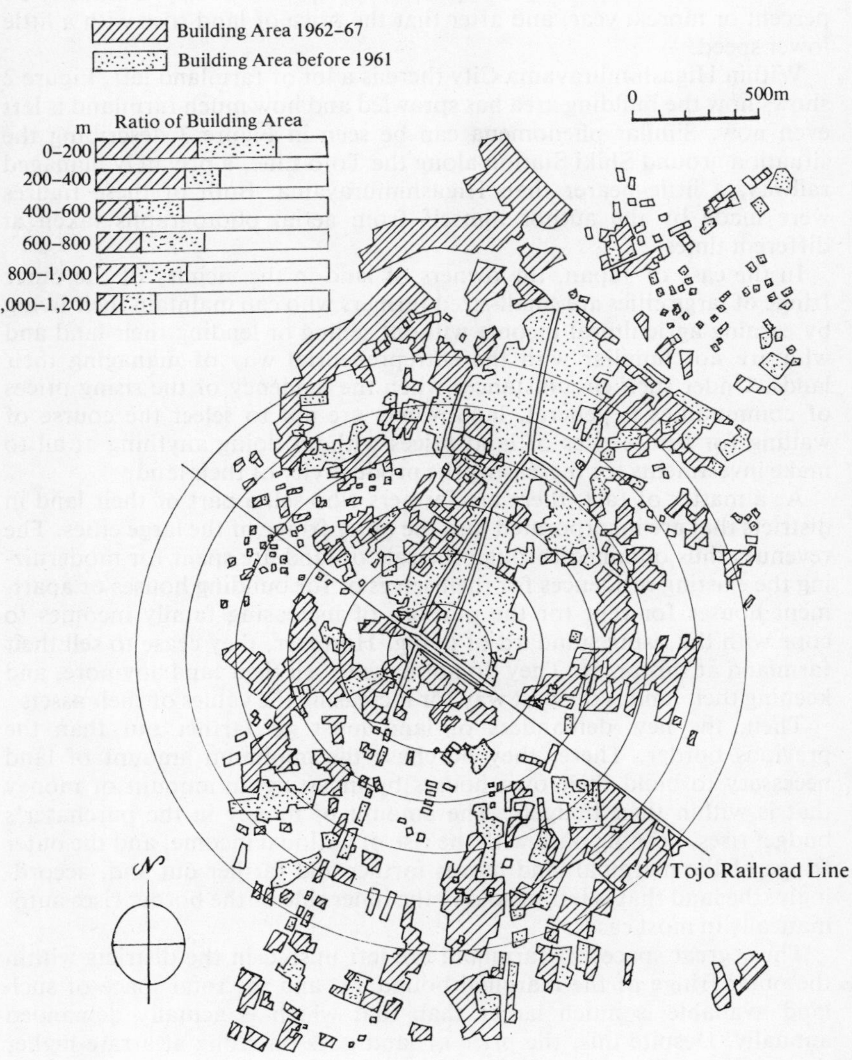

Note:
Prepared by author from aerial photographs.

Fig. 3: Sprawl of Building Area near Shiki Station

mentioned above). During this period, the speed was approximately 30 percent or more a year, and after that the price of land rose with a little lower speed.

Within Higashimurayama City there is a lot of farmland left. Figure 2 shows how the building area has sprawled and how much farmland is left even now. Similar phenomena can be seen in Figure 3 describing the situation around Shiki Station along the Tojo Line, a privately managed railway, a little nearer than Higashimurayama. Both of these figures were made by the author himself from aerial photographs taken at different times.

In the case of Japan, the owners of land in the vicinity of the outer fringe of large cities are small-sized farmers who can maintain their living by earning agricultural income without selling or lending their land and who are not familiar with the entrepreneurial way of managing their lands. Under such circumstances when the tendency of the rising prices of commodities is general, landowners are apt to select the course of waiting for future rises of land prices without doing anything at all to make investments for increasing the productivity of their land.

As a matter of fact, there are farmers who sell a part of their land in districts that are incorporated into the outer fringe of the large cities. The revenues thus obtained through the sale of land are spent for modernizing the existing residences for themselves or for building houses or apartment houses for rent for the purpose of increasing family incomes to cope with the rising standard of living. However, they cease to sell their farmland at this point. They do not have to sell their land anymore, and keeping their land is the best way for increasing the values of their assets.

Then, the new demanders of land must go farther out than the previous border. There, they purchase the minimum amount of land necessary to build their own houses by investing an amount of money that is within their budgets. The amount of money in the purchaser's budget rises, keeping pace with the rise of national income, and the outer fringe of the marginal land moves farther and farther out and, accordingly, the land that is left unsold in the inner side of the border rises automatically in most cases.

Thus, great spaces of farmland are left unsold in the districts within the outer fringe of the marginal boundary, and the total space of such land available is much larger than that which is actually demanded annually. Despite this, the price of land goes on rising at a rate higher than the prevailing rate of interest, and the border is moving farther and farther out every year.

The situation can be described in a specified demand-supply table. An individual farmer is supposed to have a specified purpose, which he wants to realize by selling a part of his land. As he either inherited the land free or bought it very cheap through the land reform immediately after the Second World War, the cost by which he has obtained the land is negligible. Then his supply function can be presented in the equation:

$$pq_i = \pi_i.$$

Here p is the price of land, q is the quantity of land to sell, π is the amount of money he needs in order to realize his specified purpose, and suffix i indicates the order of the farmer. The function is hyperbolic. The integrated table of individual supply functions (within an area) is also hyperbolic-shaped because the total amount of land to sell is obtained by summing up the individual tables.

$$q = \Sigma\, q_i = \frac{1}{p}\sum_i \pi_i.$$

On the other hand, from the demand table for land of new residents, a declining curve just like an ordinary demand curve can be drawn.

While the demand is weak, the price is very low; but if the demand curve has been shifted in a northeasterly direction by increased population resulting in an intersection of the hyperbolic supply curve, the price begins a rapid increase, finally reaching an equilibrium point because the demand surpasses the supply while the price is between P_A and P_B in Figure 4.

In the process, some of suppliers go out of the market because they have realized their purpose by selling land. This means that the hyperbolic supply curve is always shifting towards the southwest corner. The number of demanders might have declined if the population were fixed because some of them also go out of the market after they have acquired housing lots but, as it is, a continuing influx of population keeps the demand level high and an increasing income makes it even higher.

Some parts of demand may move from a place to an outer place, where the number of farmers is larger and the supply level is still high. In Figure 5, *SS* is the supply curve of the outer area, which lies more to the northeast than the supply curve of the inner area shown by *S′S′*. As the buyers can be supposed to pay more for a housing lot of same size if it is located closer to the center of the city, the demand table can be shown in a declining curve.

The price of land in an outer place is lower than that of an inner place. But *SS* also will come closer to the northwest corner if some farmers cease to sell land after they have realized their specified purposes, and the price will go up if the population increases.

III. POLICIES

As the result of soaring of land prices, at least two grave problems have arisen. First, it has made it impossible for newly married couples or young people migrating from their hometowns into large cities to get their own houses. Although it is said that the wages of workers have

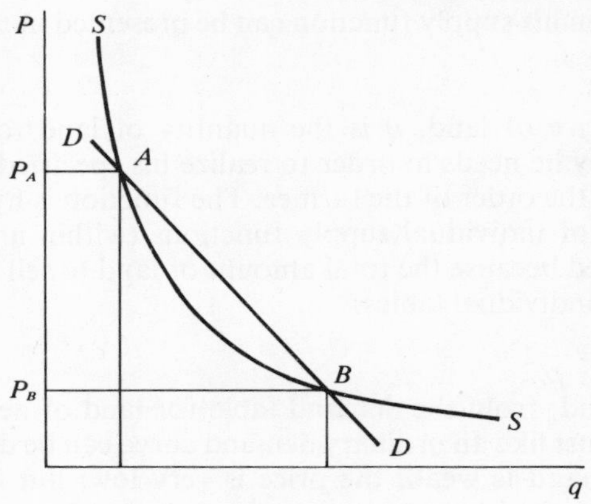

Figure 4

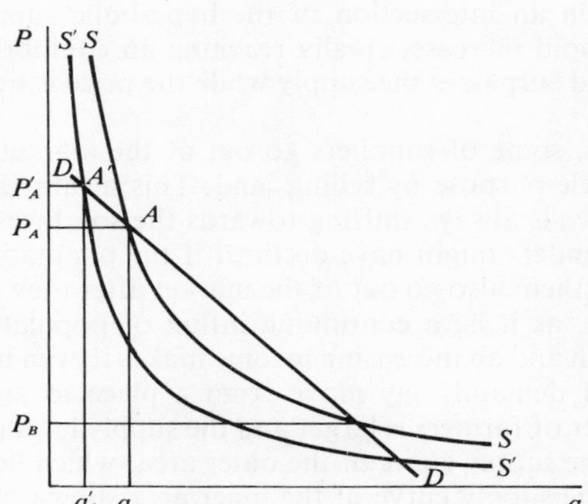

Figure 5

increased by three times in nominal terms and two times in real terms during this decade, for those people who save money in order to purchase a residential lot, the wages actually decline because the rate of land price rise is greater than that of wages. On the other hand, the rising prices of land have brought a large windfall to landowners, almost all of whom were originally farmers on the outskirts of large cities. In short, the rising prices of land have been enlarging the earning differentials and income inequality. Second, it prevents the local governments from providing adequate facilities such as roads, parks, water and sewerage works, schools and nurseries, or public buildings. The share of expenditures for land in public budgets has been increasing on account of the rising prices of land. To add to this, as landowners hold on to the land expecting increased value for it, it is very difficult for the local governments to acquire sites for necessary facilities. In the case of the Tokyo Metropolitan Government, the expenditure for land is above half of that for parks or the road budget, and every year some tenths of the budget are of necessity carried over unused into the following year because of the difficulties in acquiring land. Rising land prices also accelerate the sprawling of building areas into the outskirts of cities and consequent wide scattering of the population, so that city or town governments are required to make a higher amount of public investment than in the case of more intensively developed areas. In this case, the final sufferers are also the citizens who pay heavy taxes while enduring insufficient public services. The soaring of land prices can be said to be one of the gravest problems brought on by the high-paced economic growth in Japan.

As we have seen earlier, as a result of the interactions between the increase of population and the rising prices of land, a lot of farmland and forests is left undeveloped within the border of the metropolitan area, and its amount surpasses that of the building land that is actually developed every year. Therefore, it is not too much to say that the surrounding districts around the built-up area of the Tokyo region have enough potential for providing residences.

As we have also seen, the reason for the failure of this potential supply to manifest itself is that the landowners have ceased to sell land, resulting in further boosting of land prices.

Such being the case, it is necessary for us to hammer out some measures for providing dwellings that would shatter the mechanism raising land prices. From this point of view, levying a heavy property tax is most effective.

The rate of the property tax is legislated by every city, town, or village, and the current rate is 1.4 percent in most cases, but in major cities, a City Planning Tax with a rate of 0.2 percent is added to the ordinary property tax. In both of these taxes every property is legislated to be assessed *ad valorem* or at the market price, but farmland, nevertheless, receives a *preferential assessment.* Even within the 23 wards, for example, there are some hundreds of hectares of farmland, and the average of the assessment value of such farmlands is only 100 yen or less per square meter, while the assessment value of residential lots adjoining those farmlands is 30,000 yen or more per square meter. It means that a salaried man who has just gotten a small lot, say, of 100 square meters, would have to pay a larger amount of tax, besides a payment for loan in most cases, than what a farmer, who owns one hectare or 10,000 square meters of potential land for residence, would pay.

It is said that there are some reasons why the *preferential assessment* should be adopted for farmland, one of the most important being that the actual income gained from the farmland by agricultural use is too small for the farmer to pay property tax. But the latent capital gain is very large; the market value of it may amount to some hundred million yen.

Recently a bill has passed the Diet that requires every autonomy to abolish the *preferential assessment* for farmland within the urbanized area that was settled by the Town Planning Act in 1970. The farmers, however, so strongly opposed this act that the act has been revised with a number of exemptive clauses.

Levying heavier property tax shifts the supply curve in Figure 4 in a northwesterly direction and makes the price lower. So some people may suppose that the income tax that will be levied when the farmers sell the land and get the capital gain actually can function much as a property tax. Progressive income taxation, certainly, may be useful in redistribution of income. But the income tax may cause a lock-in-effect: the landowner may hoard the land and expect an adequate rise in the price of land to offset the additional tax.

We may be able to say, therefore, that the property tax levying, especially the abolishment of the preferential assessment for the farmland in an urbanized area, is the most effective and operational measure to regulate the rise of land prices.

Even eminent domain is weakened in its power because of the soaring of land prices because the Eminent Domain Act orders every public organization to acquire land at the market price.

Under the housing situation that I have described in this paper, the individual housing policies, which may have merits of their own, would function to boost the price of land, which may make the solution of housing and land problems all the more difficult. For example, extension of housing loans by the public to individuals who can build houses on their own lands and the loan extension by large companies to their employees for housing purposes have the effect of strengthening and increasing the demand for building land and accelerating the soaring of land prices because the price of land on the fringe of large cities depends upon the budget of the demanders, and such loan extensions enlarge the budget.

IV. CONCLUSION

As we have seen above, it is clear that land prices were boosted by accumulation of population in urban areas caused by the economic growth of Japan. And the economic growth has profited reversely by the rising of land prices because the land, whose price has risen with a high enough rate to surpass the prevailing rate of interest, is the best security for a loan by a bank for investment in plant and equipment, and it is this investment that has been assuring the economic growth of Japan during the last two decades. But nowadays it has become obvious that the economic growth accompanied by accumulation of population makes land prices go up if land ownership is left beyond policies and that the soaring of land prices is a large obstacle to promoting people's welfare, in that it enlarges the earning unfairness and that it prevents people from enjoying common goods or services that should be provided by the public sector. So the policy should be changed from pursuing pure economic growth to removing the obstacle to promoting people's welfare, even though it may result in restraining the speed of economic growth somewhat, or in depriving a few persons of some kinds of vested interests or anticipation of windfalls. Under conditions of limited resources, it seems that restricting some kinds of private rights, especially of the richer people, is inevitable in order to improve the lives of the poorer people.

There are measures for this. The problem does not seem to exist in methodology, but in decision to adopt them.

Summary of Discussions

On the sessions dealing with the topics of "Environmental and Resources Constraints"

I. GENERAL SUMMARY (Chairman: Ronald Dore, U.K.)

Prud'homme (France), in his paper on "Appraisal of Environmental Policies in Japan," stressed that although Japan has made very considerable success in controlling pollution, there has been much less success in securing high level of amenity, such as quietness, privacy, and natural beauty. Relative failure of land planning was also noticeable. Further, Japanese policy has not followed prescriptions of economists, equating marginal costs and benefits; instead, it has often been framed in terms of absolutes and arbitrary prescriptions. But these in fact proved "rational" in hindsight. Another main point suggested by *Prud'homme* was that simulation models indicated macroeconomic consequences of pollution control expenditures to have been not very serious even for GNP as conventionally calculated, *a fortiori* not so for any concept of net national welfare. *Okishio* (Japan), the official discussant, while agreeing with the effectiveness of political and social pressures in the sphere of environmental problems, took a more critical attitude on the action taken, or lack of action, by the Japanese government in the past, and in particular, he warned against the movement of some polluting industries from Japan to other Asian countries.

Miyamoto (Japan), in his paper on "The Environmental Protection Policy in Japan: Brief History and Appraisal," pointed out that there did occur a change in government policy between 1967 and 1971 toward more effective anti-pollution measures enforced by popular pressure; however, that many of the government measures purporting to control industrialists were in effect designed to protect industry from popular attack. The author was of the opinion that the so-called reform of industrial structure would not fundamentally reduce the level of pollution in Japan. *Godchot* (France), the main discussant, while stressing the importance of public opinion and the need for an anti-pollution educational system, expressed the view that pollution control was relatively quite efficient in Japan and strategy for environmental protection was also successful. He questioned, however, what the author's view was as regards the international aspect of the problem of pollution. In the subsequent course of discussion, participated by *Bronfenbrenner* (U.S.A.), *Wood* (U.S.A.) and *Prud'homme* (France), a number of questions remained unresolved, to wit: (1) If all the market principles are rejected in environmental matters (a view imputed to *Miyamoto*), what standards are we to use? (2) Does or should current stagflation depress expenditures on pollution control ("Can't afford it!") or strengthen the case for it (as demand booster)? (3) Is export of polluting industries to South East Asia an inadmissibly bad international behavior or of some benefit to LDC's?

Yasuba (Japan), in his paper on "Resources in Japan's Development," observed by way of historical review that resources constraint had been deeply felt in the Meiji period and had led to all kinds of resource-saving creative technological adaptation and that recent Japanese industrial development of intensively resource-using type (e.g. exporting of steel to U.S. made from coal from the U.S. east coast and ore from Brazil) was made possible chiefly because of immense reduction of ocean transport cost. The author indicated, however, some considerable reasons for pessimism on account of rising trend for protectionism in other industrial countries and also of the possibility of output curtailment by producers' cartels. This pessimism, he felt, should be no cause for fear that Japan might react in the military fashion as in 1930s. The official discussant on *Yasuba*'s paper was *Narongchai* (Thailand), whose comments are reproduced in the text. *Kirby* (UK) stressed from the floor that because technology was constantly changing, we should speak in terms of quality of resources also. He cited an example of uranium. In concluding remarks *Yasuba* warned once again on the gathering clouds of protectionism all over the world.

Kirby (UK), in his paper on "Resource Potential of Continental East Asia for Japan's Material Needs," called attention to the often neglected point that Russia and China were destined to play a dominant part in Japan's economic environment; in particular, the potential complementarity of Siberia and Japan was the case in point. In general, the author was of the opinion that there should be a more careful calculation of potential contributions to Japan's needs by centrally planned economies. As for Japan's position in South East Asia, he felt that there were likely to be political difficulties, both in securing resources and in finding outlets for investment in polluting industries. As for Japan herself, *Kirby* predicted that even with a "restructured" pattern of production proposed by some economists, she would still require considerable resource imports in metals as well as energy. *Nakauchi* (Japan), the official discussant, was somewhat skeptical of the extent of close ties resource-wise between Japan and Siberia; after all, he said, although "Japan is now walking alone," she has feet in Asia and must seek a new pattern of industrial and social structure based on the international division of labor approach in the Asian scene as a whole. In the course of discussion from the floor it was pointed out that Japanese efforts to develop resources by Japanese capital with long-term purchasing agreements would be less likely to be welcomed in the future; and that if politically-motivated interruption of supplies in the face of Japan's concern for resources were to lead to more accommodating attitude by Japan towards LDCs, it might be a good thing.

II. GENERAL SUMMARY (Chairman: Lynn Turgeon, U.S.A.)

Japanese growth (or lack of growth since 1973) could be explained on a comparative basis. But rather than comparing Japanese growth with that of the United States, France, or Italy (as previous speakers had done earlier), it seems that comparison with Germany would be more instructive. In this connection, it is necessary to look at similarities between Japanese and German developments, both before and after World War II.

It is interesting to note that both Japan and Germany managed to extricate themselves from the Great Depression using basically Keynesian policies. And it is perhaps significant that two of the earliest translations of the *General Theory* were made by the Japanese and the Germans (both in the fall of 1936, in contrast to the French who waited until 1942). Keynes, himself, recognized the potential market for his ideas by writing separate introductions to each translation in which he emphasized particularly the supposed "thirst" of the Germans for some theory, hopefully his own. At any rate, both economies achieved full employment, rapid growth in both output and in labor productivity *(See Yasuba's* paper), and surprisingly little inflation. To achieve this result (which is remarkable in view of the 1937–38 downturn afflicting the remainder of the capitalist world), both economies were cleansed of significant money and banking influences, and real interest rates remained at a very low level, far below any rates ensuing from previously determining market forces.

In the postwar heritage, there are also many similarities between Japan and Germany (that is, the Federal Republic of Germany), as well as a few differences. Both countries experienced devastating destruction of their capital stock, as well as a huge repatriation of labor. In the case of Japan, six million persons came home from areas formerly under Japanese influence (*See Yasuba*'s paper), while a comparable number of Sudeten and other Germans came to the Federal Republic of Germany from Czechoslovakia, Poland, and the Soviet zone of Germany, as well as from the German Democratic Republic after 1949. These new factor proportions established the base for potentially very high rates of return to capital. In both countries, there was a strong ideological commitment on the part of the occupying powers to market forces and a return to conventional pre-Keynesian monetary policy, with balanced budgets and generally impotent fiscal policy.

Initially, German recovery was earlier and more rapid than that of the Japanese due to the Marshall Plan for Europe. Both countries were demilitarized and therefore maintained extremely low defense budgets. In this connection, it is interesting to note that no mention of this factor—military spending—was made in comparing Japanese with either United States or French relative growth the previous day. Both countries paid some reparations: the FRG to Israel and the Jewish people generally, and Japan to various underdeveloped Asian countries; but the German reparations were earlier while the Japanese were later as well as spread out over many years.

The resulting West German economic miracle of the 50s and early sixties under Adenauer and Erhard is well known, and this success was reflected in the first revaluation of the mark in 1961. Trouble signs began to appear, however, as early as the 1966–67 "mini-recession" when, for the first time, West German national product actually declined. This shock produced a change in the postwar FRG Constitution which, in its reaction or backlash to the fascist period, prevented substantial deficit financing. (The previous day's paper by *Zaneletti* pointed out that Italian deficit financing was postponed still later until 1963.) In addition, Willi Brandt began to forge his *Ostpolitik,* which, in its effect, opened up Eastern Europe to West German machinery and other capital exports.

In comparison with the West Germans, Japanese economic policy seems to be even more basically pre-Keynesian. Deficit financing is apparently confined to "special budgets" reminiscent of FDR's early pre-Keynesian period. While some opening moves have been made for greater trade with the Chinese People's Republic, this trade must remain more or less balanced as long as the Maoist tenet of "balanced equivalents" continues to have validity. Thus, in contrast to the huge surpluses to the non-Chinese socialist world engineered by both the United States and the FRG, Japan is forced more or less to balance her trade with China and to depend on Asian underdeveloped countries or advanced capitalist economies for her employment-creating export surpluses.

This weakness of Keynesian policy in Japan showed up in the morning session, when it was reported that Japanese government policymakers—in contrast to the Swedes, who use environmental expenditures in a counter-cyclical manner—seem to have relegated environmental problems to a secondary priority in comparison with the problem of the Great Recession. Although it can be shown that environmental control measures and compensation for victims can be stimulative to both growth and profits (*See Prud'homme*'s comments), these expenditures seem to have weakened in Japan during the most recent period according to *Miyamoto* (*See* his paper). As we see when we come to an examination of the OPEC energy shock in the afternoon's papers, the weakness of Japanese Keynesian policy in fiscal matters has created problems for the Japanese similar to those of the Ford Administration which, in contrast to the Nixon Administration, began to pursue pre-Keynesian policies.

The first afternoon paper, that of *Sakisaka* (Japan), outlined the transition of the Japanese economy to one which was highly dependent on formerly cheap petroleum product imports. The Japanese "Affluent Throwaway Society" was constructed on the basis of cheap oil which evaporated after the OPEC oil shock. The elasticity of oil consumption which had exceeded unity in the 60s dropped to 0.75 in the early 70s, partly as a reflection of the weakness in the industrial sector and partly as a result of the growing citizen's environmental movement. *Sakisaka* emphasized a trade-off between the energy problem and the unemployment or underutilization problem. When there is underutilization, as was the case after 1973, there is a lesser problem of securing sufficient energy or of polluting the environment.

Sakisaka also emphasized the fragility of the Japanese energy base, especially in view of the opposition of the citizen's movement to the expansion or reliance on nuclear power. His recommendations included such things as greater government research and development efforts in the areas of solar energy, nuclear fusion, geothermal energy, etc., since world oil output will peak in 1990. *Uzawa* (Japan), his official discussant, emphasized the need for greater in-depth analyses of consumer behavior, as well as the need for developing new energy-economizing types of social infra-structure. *Derek Healey* (Australia) intervened to suggest that he felt that the paper had given insufficient attention to the role of energy prices. He doubted that there were important obstacles to nuclear power emanating from the citizen's movement and wondered why the importation of Australian uranium (which is now allowed) and coal might not remedy Japan's problems. *Sakisaka* responded by saying that there had been a threefold increase in electric power prices and that the Japanese were unlikely to come up with energy sources other than those of OPEC. Australian coal was considered too polluting and gasification was not considered too useful as far as Japan was concerned.

The second paper, that of *Nishikawa* (Japan), emphasized the increased dependence of Japan on imported raw materials, as well as the giving up of coal mines and the decline in coal reserves. By endorsing GATT principles as early as 1955, the Japanese economy had become more integrated into the world's economy. Thus, by subscribing to free trade principles and comparative advantage, she became more dependent on foreign sources of supply (even for food) than other advanced capitalist countries. As a result of the fourfold increase in oil prices, and the possibility of an embargo, a great deal of anxiety developed. Stockpiling, the moving of plants to foreign shores, and conservation measures generally have only begun in Japan, and some of these steps might aggravate the conflict between north and south. Rather than following the Ministry of International Trade and Industry (which recommends more private overseas investment), *Nishikawa* would favor a more nationalistic rather than an internationalist solution to Japan's energy problems: he would abandon the philosophy of international division of labor and attempt to re-establish the primary sectors of agriculture and fishing.

Hernadi (Hungary), *Nishikawa*'s discussant, doubted that domestic agriculture had been discouraged as much as *Nishikawa* had suggested, since Japan's protectionism was strongest in agriculture. He also wondered how Japanese coal stocks or reserves had been reduced since presumably, at some price, closed coal mines could be reopened. *Hernadi* felt that *Nishikawa* was too autarkic in his approach and that the underdeveloped countries would tend to lose out if the Japanese followed his suggestions. Since Japanese importers seem to have considerable monopsonistic power with regard to their food purchases (due to procurement from a number of sources), he did not seem to find any real basis for *Nishikawa*'s anxiety. *Nishikawa* responded that he did not use the word "autarky" and that only a few products (mandarin oranges, sugar and pineapple) were being supported by government policy. Although 50 percent of the farmers are prosperous (due to non-farm supplementary activity and windfall gains from sales of former agricultural land), Japanese agriculture has been wrecked. *Nishikawa* felt that the FRG had handled this problem in a better way since 40 percent of their energy still comes from coal.

Donald Daly (Canada) tended to deal with the short-term effects of the increase in OPEC prices. Japanese terms of trade were especially hard hit, declining by 50 percent in comparison with a 10 percent improvement in Canada's, since Canada is not a significant net importer of oil. Industrial production fell by 23 percent—the greatest of all advanced capitalist countries. Labor productivity dropped by 15 percent, but on an hourly basis it has increased sharply in 1975 and especially 1976. However, there has been little recovery in hours worked so that considerable underutilization of labor is occurring despite an official unemployment rate amounting to 2.1 percent of the labor force. Because of the more recent increases in labor productivity, unit labor costs have remained low, thus supporting the recent Japanese export surpluses. By early 1977, exports were 2 times higher than the 1970 level, while imports were only 30 percent higher than in 1970.

The official discussant, *Yamazawa* (Japan), did not challenge any of *Daly*'s findings, but rather made a few complementary additions to his analysis. He felt that if the oil price increase had taken place in the mid-60s, the deflationary measures of the government could have been greater and subsequent recovery faster. *Bronfenbrenner* (U.S.A.) questioned the fact that the Japanese government had even pursued any deflationary policy in view of the increases in the money supply reported by *Daly*. The moderator intervened to say that the absence of any stimulation on the part of the Japanese fiscal policy in 1974 might, in some Keynesian sense, be considered deflationary. As Charles Schultze, *et al.* have shown in their Brookings volume on the oil crisis, the OPEC price increase must be considered as if there had been a huge excise tax imposed on oil. To offset this deflationary effect of the tax increase, the appropriate Keynesian policy would be some tax reduction to offset the destimulative effects of the higher oil prices. Thus, a pure case of cost-push inflation (such as the OPEC oil price increase) must be handled by means of a tax reduction (or increased government expenditures) elsewhere. Since Japanese fiscal policy is virtually non-existent, they were extremely vulnerable to the OPEC shock. However, since President Ford, as late as September, 1974, was calling a summit meeting of economists designed to sell an *increase* in taxes to cool off an overheated economy (almost a year after the beginning of the Great Recession), we can hardly hold up the United States policymakers as any sort of role models.

The final paper, that of *Hanayama* (Japan), dealt with the irrational use of urban land in Tokyo, and the burden of the present situation on the working class, since land prices have risen faster than wages. The existing tax system favors the former farmers as they sell their land in a patchwork fashion. There has been a decline in the average size of housing lots as a result of the inflation in land prices, and farmers are hoarding land which should be made available to the urban population. His solution would be something reminiscent of the single tax advocated by Henry George, but obviously this would not be welcomed by the current landowners, so he offered no solution to the "transition" problem. *Stefani* (Italy) suggested that the land now held by farmers be made available for parks and infra-structure, which seems weak in Japan. Although denying any connection with urban planners, he suggested more vertical housing surrounded by green spaces as opposed to the existing generally horizontal pattern. Two Swedish colleagues were critical of proposals using taxation and the market to solve the Japanese (or any other) housing problem.

PART THREE

CONTRIBUTED PAPERS

A Social Indicators Approach to Economic Development*

Kimio Uno *University of Tsukuba*

The paper deals with the evaluation of economic and social growth performance employing a system of social indicators. The social indicator approach is a recent attempt to overcome the recognized analytical limitation of purely economic analysis in appraising a development process. The paper describes the multidimensional changes of the socioeconomic system using empirical observations from the Japanese experience over the past quarter century. The indicators are arranged in an accounting framework, which includes economic accounts, demographic accounts, and environmental accounts. In the accounting framework, the more conventional concepts are expanded to include the quality of life aspects. The social indicators in the paper should be viewed as summary expressions of these accounts, rather than a random selection of the various aspects of socioeconomic activities. The statistical accounts can then be utilized in econometric model-building. The present paper, however, is intended to reveal the overtime changes of social indicators in the course of development, rather than presenting the statistical framework or the model itself.

* The author extends his appreciation for helpful comments by Professor Kenneth E. Boulding and the session chairman Professor Lynn Turgeon. The present paper is a revised summary of the paper of the same title that appeared in a research report *Fukushi Shakai ni Okeru Shotoku Hosho—Riron to Jissho*, Yoshimasa Kurabayashi, ed., Tokei Kenkyukai [The Institute of Statistical Research], 1977. The full text includes a proposed accounting framework from which individual indicators are derived.

I. PURPOSE OF THE PAPER

The purpose of the paper is two-fold. First, it will introduce a socioeconomic statistical framework system from which social indicators may be derived. Second, it will describe the performance of the Japanese development process using a series of social indicators based on the statistical account.

When an economist leaves the artificial world of real economic growth as measured by the increase in national income and asks himself the more fundamental question of what economic growth means to a society as a whole or to the individuals involved in it, he often finds himself at a loss.

The framework of national income accounting has served as a common framework in which the socioeconomic performance of a society can be summarized. Its universal applicability, however, is not undisputed.[1] The national income concept, undoubtedly useful for some purposes, is of a very limited validity for describing a many faceted process of economic development.

The social indicators approach is a recent attempt to overcome the recognized shortcomings of national income framework in its ability to describe the aspects of a society that have more direct bearings on the quality of life of the population.[2] The main point to focus our attention is the disputed compatibility of the quality of life and the continued expansion of productive activities. By integrating the two aspects of the working of a society, *i.e.*, the aspect of production and the aspect of the quality of life, into an integral statistical account, it would be possible to reveal the influence of increased production on the quality of life and *vice versa*.

II. SOME VIEWPOINTS

When we look into the quality of life consequences of economic development, we should remember, first of all, that macroeconomic aggregation

[1] *See*, for example, Simon Kuznets, "Quantitative Aspects of the Economic Growth of Nations: Some Conceptual Problems of Measurement" (1956). As an example of a Japanese reflection, *see* Shigeto Tsuru, "In Place of Gross National Product" (1971).

[2] For a summary review of the theoretical development in this field, *see* Kimio Uno, "Shakai Shihyo—Hatten no Keifu to Makuro Shakai Shihyo Moderu" ["Social Indicators and Macro-Economic Framework"] (1974).

Most active at present are two main conceptualizations of social indicators, one by the Organization for Economic Cooperation and Development and the other by the United Nations. The former has a scheme to define a hierarchy of social concerns. *See* OECD, *Progress Report on Phase II—Plan for Future Activities* (1976). The latter defines social indicators as components of the proposed SSDS. *See* United Nations Department of Economic and Social Affairs, *Towards A System of Social and Demographic Statistics* (1975) and *Draft Guidelines on Social Indicators* (1976).

is a fiction, an economists' invention. Happy figures on the aggregate do not necessarily mean everything is all right inside. To assume so is another kind of laissez faire, expansion on the aggregate automatically fulfilling harmonious results. What exist in reality are the economic actors including households, producers, and the government, and the ultimate beneficiaries and evaluators of the society are the household sector.

Conventional measures of economic development lack the viewpoint of the beneficiaries of the society: existence of the household is of course recognized, but in a piecemeal fashion, being represented by wage income, consumption, housing investment, etc., but not as an integral unit as an actor. The viewpoint has to be turned upside down, and the socioeconomic variables pertinent to the household sector must be so arranged as to allow for sectoral analysis. This implies the need to prepare a statistical base that is compatible with the national income account, but much wider in scope.

Second, it may be necessary to focus on the conflicts of interests among various social groups in the development process. Social groups could be debtors and creditors, farmers and employed workers, or various age groups of the population including future generations. Needless to say, under inflation, the debtors gain and the creditors lose; on the aggregate, however, the gain and the loss cancel out, and it is not stated explicitly. The same holds true in the case of transfers of income by agricultural price policy or by social security measures. What should be stressed here is that those items do not show up in the national income framework, which is on the aggregate. If analysis by social groups is indispensable in growth evaluation, the analytical framework should be expanded to meet the need.

Third, there is another drawback in analyzing on the aggregate. That is the fact that balance among items included is not questioned. Aggregate analysis hides such phenomena as congestion, which is an excessive use of public goods in general. As the income level goes up, the demand for food and clothing is more or less satisfied, and a stage is reached where the quality of life cannot be improved unless public goods, which include institutional arrangements as well as social capital, are supplied together with the private goods. There are numerous such cases: housing and sewerage, and automobiles and highways are some examples. Perhaps we should look into the individual components of social life.

Fourth, the increasing importance of public goods and social security measures indicates that we have to think of the social decision-making process, which takes place outside the market mechanism. It is outside the market, but still within the domains of economic studies. Inclusion of decision-making in our scope means that we have to deal with subjective evaluation of the participants in the social processes.

Also, it is questionable whether it is possible to judge *a priori* on the goodness or badness of products or activities apart from the value judg-

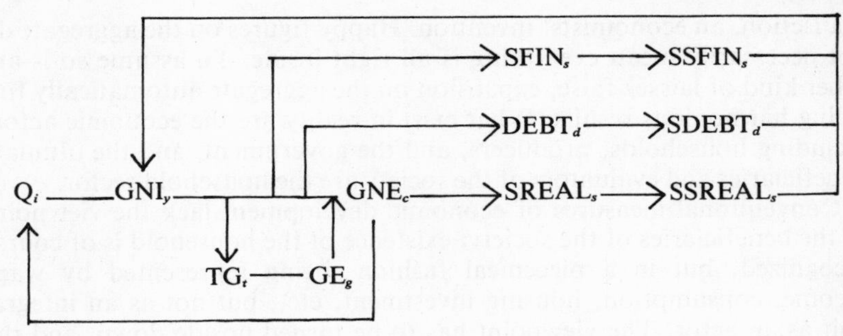

production *distribution* *expenditure* *saving* *assets*

DEBT:	debt	TG:	government revenue
GE:	government expenditure	*c*:	expenditure category
GNE:	gross national expenditure	*d*:	debt category
GNI:	gross national income	*g*:	government expenditure category
Q:	production	*i*:	industry category
SDEBT:	debt outstanding	*s*:	saving category
SFIN:	financial saving	*t*:	tax category
SREAL:	real assets purchases	*y*:	income category
SSFIN:	financial assets		
SSREAL:	real assets		

Fig.1: Economic Aspects

ment of the people concerned. Is the amount of household debt an indicator of deficient income or does it on the contrary signify that more resources are directed to that sector? Does the increased incidence of divorce indicate the collapse of a family system or does it indicate improved status of the female population? Can shorter work hours necessarily be interpreted as increased leisure or are we to infer whether it is voluntary or involuntary, induced by decreasing demand for labor?

Thus, it is only when we include the relation between subjective evaluation and objective situation that we can evaluate the quality of life consequences of economic development. Subjective evaluation is an objective statistic when observed on the society-wide basis, and it is not a random expression of one's ideas. Based on the correspondence between objective situation and subjective evaluation, we should be able to identify the needs of a society. The list of social indicators should include those aspects of the society which are subject to such evaluation.

Finally, the points made above may have given an impression that the production sector is being negated, but the real point is that it is clearly possible to integrate the dual viewpoints in a single framework, *i.e.*, the

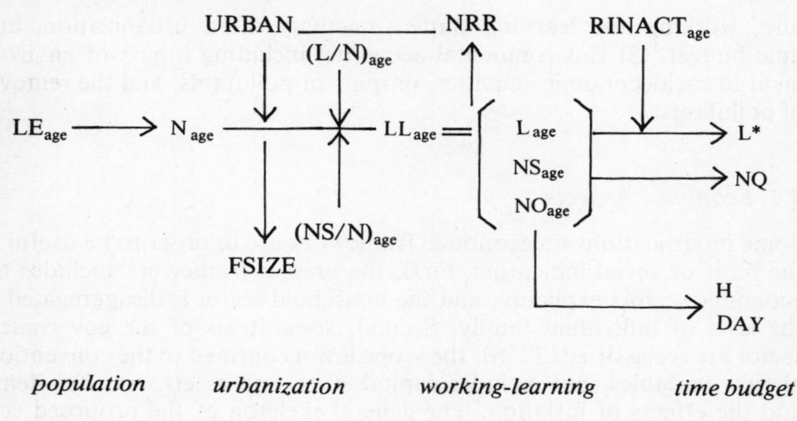

DAY:	working-nonworking days of the year	NRR:	regional relocation of the population
FSIZE:	family size	NS:	learning force
H:	time budget of the day	NS/N:	learning force participation rate
L:	labor force		
L*:	effective employment	NQ:	educational quality of the population
LE:	life expectancy		
LL:	working-learning status of the population	RINACT:	inactivity rates
		URBAN:	urbanization of the population
L/N:	labor force participation rate		
N:	population size	age:	age bracket
NO:	housekeeping and other		

Fig.2: Demographic Aspects

viewpoint of production as represented by national income framework and that of the beneficiaries as represented by quality of life indicators or social indicators.

III. CONCEPTS AND DESIGN

It is perhaps promising to define first the statistical base in a matrix form and then seek a summary expression of a particular matrix. The summary expression may be called a social indicator. The indicators may be arranged in an accounting framework, which includes the following: (1) Economic accounts, including various flow and stock variables, prices, and economic sectors including the household and the government. (2) Demographic accounts, including population, birth and death, age structure, working and learning status, inactivity rates, urbanization, and time budget. (3) Environmental accounts, including inputs of environment to socioeconomic activities, outputs of pollutants, and the removal of pollutants.

3.1 Economic Aspects

Some modifications on economic framework are in order to be useful as the basis of social indicators. First, the present framework includes the economic sectors explicitly, and the household sector is disaggregated to the level of individual family. Second, some items of the government sector are reclassified. Third, the scope is not confined to the conventional flow variables, but includes capital stock and assets, transfer items, and the effects of inflation. The general skeleton of the proposed economic accounts is as shown in Figure 1.

Of the income and expenditure items, those which are pertinent to the household are disaggregated by the age bracket of the household head. From the flow of expenditures, accounts on fixed assets, financial assets and liabilities are formed.

3.2 Demographic Aspects

From the standpoint of the beneficiaries of a society whose concern is with the quality of life, the main concern would be working-learning status and inactivity rates rather than labor force alone; education level rather than skill level; leisure and total time budget rather than working hours; and the adequate provision of real and financial assets in addition to being a consumer of current products.

In Figure 2 on demographic aspects, life expectancy is given first. Life expectancy determines the population size. Second, age structure of the population is provided. Third, working-learning status is entered. In order to derive labor force participation and school attendance by age, we can envisage diagonal matrices having the respective rates on the diagonal. Fourth, the available labor force is not necessarily utilized effectively for productive activities. This aspect is seen in the inactivity matrix, which includes as the inactive factors: labor disputes, industrial accidents, traffic accidents, communicable and other diseases, and crimes, which in the present framework are reflected in the detention rate. Unemployment rate and the percentage of labor force in military service may be included here, depending on the purpose.

Fifth, the cumulative effect of the education process is to improve the educational quality of the population. In the present framework, the educational level is defined in terms of the length of the educational process.

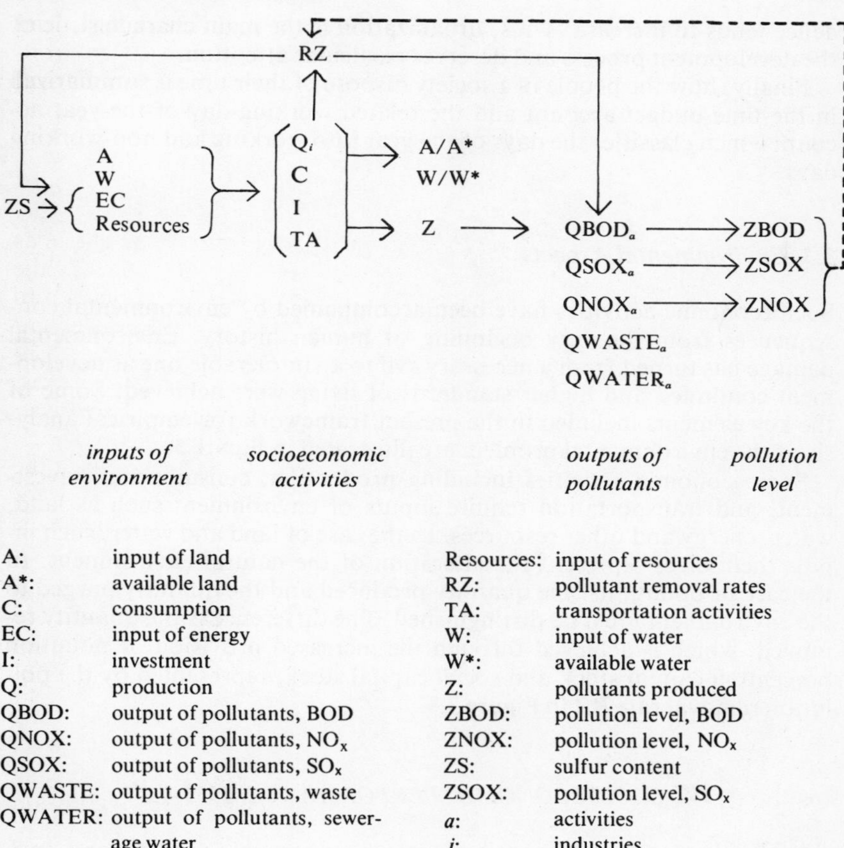

Fig.3 Environmental Aspects

Sixth, urban-rural distinction is made. With the process of urbanization, the market mechanism extends to every socioeconomic activity that has been previously carried out at individual homes. The concept of unemployment becomes relevant, and the mutual help in local communities disappears. The family size tends to become smaller. We have observed one or two members of the rural family maintaining employed jobs outside the agricultural sector year-round, while residing with the rest of the family who work in the agricultural sector. These working arrangements perhaps constitute an important part in the successful urbanization and development. Also, the number of people who change their place of residence tends to increase. Thus, urbanization is the main characteristic of the development process and deserves particular attention.

Finally, how the people in a society dispose of their time is summarized in the time budget account and the related working-day-of-the-year account which classifies the days of the year into working and non-working days.

3.3 *Environmental Aspects*

Socioeconomic activities have been accompanied by environmental consequences from the very beginning of human history. Environmental damage has turned from a necessary evil to an intolerable one as development continued and higher standards of living were achieved. Some of the key elements included in the present framework for empirical analysis of the environmental problem are illustrated in Figure 3.

Socioeconomic activities including production, consumption, investment, and transportation require inputs of environment such as land, water, energy and other resources. In the case of land and water, such inputs themselves represent the alteration of the natural environment. In the case of pollutants, the quantity produced and the quantity purged to the environment must be distinguished. The difference is the quantity removed, which is achieved through the increased provision of pollution prevention capital stock and social capital stock, represented by the pollution removal rate *RZ* in Figure 3.

IV. SOCIOECONOMIC PERFORMANCE, 1955–1975

Social indicators on economic aspects, demographic aspects, and environmental aspects are summarized in the Tables 1 to 3, respectively. Various standpoints can be accommodated in the framework, including national income accounts, the Measures of Economic Welfare (MEW, commonly known as NNW in Japan), and social indicators.

During the two decades under consideration, household total income in real terms (YF/) has increased from 55 thousand yen per month to 142 thousand yen. As for the distributive features, the share of wages (YW/Y) increased from 0.496 to 0.634, and the Gini coefficient (GINI) has shown some improvement for the period where data is available and stands at 0.188 at the end of the period. Increased provision of social security is seen both on a household basis and a macro basis (YSI and GRT, respectively). Also noteworthy is the rapid increase in household debt, especially for housing (DEBTH, DEBTO, and DEBTC on a household basis and LOANH and LOANO on a macro basis).

On the expenditure side, food consumption seems to be approaching a saturation point as far as protein intake is concerned, and the Engel's coefficient (ENGEL) has dropped from 0.445 in 1955 to 0.300 in 1975. On the other hand, expenditure on consumer durables (CD) and other (CO), which include among other things medical care (COMED), private transportation (COCAR), and education expenditure (COEDU), have shown a marked increase. Household income expenditure balance (BF) has improved from 1.089 initially to 1.265 at the end of the observation period, and the surplus is reflected in the increasing saving rate (RSFIN and RS).

As a result, the amount of household financial assets (SSFIN) has increased together with the percentage of households with savings (RSSFIN), which currently stands at 0.993 as compared to 0.935 in 1960. Real assets have also increased; the stock of consumer durables on a macro basis started from an estimated 2727 billion yen in 1955 reaching 22592 billion yen in 1975. Housing capital (KH) has also increased by 45 percent in value during 1955 and 1973, and has more than doubled in the number of housing units (KHUNIT). In the process, floor space per unit (KHFLOOR) has changed from 95.2 square meters to 77.1 square meters in 1973, after hitting the bottom of 73.1 square meters in 1965; the share of owned houses has decreased from 0.679 in 1955 to 0.580 in 1975. On the whole, the housing situation has been improving judging from the ratio of housing units to number of households (KHUNIT/F), which climbed from 0.854 in 1955 to 1.137 in 1975.

The benefit of economic development is not limited to economic aspects. Social indicators on demographic aspects in Table 2 indicate that life expectancy (LE) has improved from 66.2 years in 1955 to 74.4 years in 1975. The birth rate (RB) has dropped only slightly, but the infant death rate (RDB), which is regarded as a good indicator of general sanitary and health conditions of a society, has dropped from 0.040 to 0.010 during the twenty-year period, whereas the death rate of other than infants (RDA) dropped from 0.008 to 0.006. As a result, the ratio of the old aged ((N65+)/N) has increased from 0.053 in 1955 to 0.079 in 1975; the dependency rate (NDEPEND) was initially 0.379, but soared to 0.557 in 1960 reflecting the postwar baby boom, subsequently declining to 0.475 toward the end of the period.

The labor force participation rate (L/N) has steadily declined during the observation period from 0.708 to 0.630, mainly due to the shift of labor force out of the agricultural sector where the female participation rate is relatively high. The shift can be seen in the relative size of primary sector labor force and the non-primary sector in total labor force (L1 and L2). Another reason for declining labor force participation is the increased access to higher education (NSH, NSC, and NSU). The labor market situation is reflected in unemployment rate (U) and job opening–applicant ratio (RLDS); both have steadily improved, and the former has reached 0.012 and the latter 1.41 in 1970, but the trends were reversed

TABLE 1

SOCIAL INDICATORS—ECONOMIC ASPECTS

Indicators	*Symbols*	*Unit*	*Source*	*1955*	*1960*	*1965*	*1970*	*1975*
INCOME								
Household								
Wages and salaries, household head	YWAGE	¥/month	(3)	24065	34051	54111	94632	198316
	YWAGE/	″	cal.	45664	59014	70092	94632	119396
Wages and salaries, wife and other	YWIFE	″	(3)	3015	4134	6628	10836	24140
	YWIFE/	″	cal.	5721	7159	8585	10836	14533
Social security benefits	YSI	″	(3)	332	286	396	659	1586
Returns from assets	YSS	″	″	359	586	697	732	1059
Other income	YO	″	″	788	926	1831	2891	6462
Total income	YF	″	″	29169	40895	65141	112949	236152
	YF/	″	cal.	55349	70875	84380	112949	142175
Debt for houses and land	DEBTH	″	(3)	} 509	} 462	} 843	860	2154
Other debt	DEBTO	″	″				459	685
Installment and credit	DEBTC	″	″	1294	1989	1176	2838	4233
Macro								
Compensation of employee	YW	bil. yen	(4)	3527	6435	14348	30984	79270
Distributive share of wages	YW/Y	ratio	cal.	0.496	0.502	0.560	0.545	0.634
Gini coefficient of distribution	GINI	ratio	cal.	n.a.	n.a.	0.198	0.179	'74 0.188
Income from property	YSSNI	bil. yen	(4)	483	1253	2979	6658	17504
Interest on consumers' debt	R*SDEBT	″	″	16	33	85	241	596
Government transfer to household	GTR	″	″	356	585	1414	3042	10275
Housing loan, new	LOANH	″	(2)	n.a.	n.a.	n.a.	349	2021
Other loan, new	LOANO	″	″	n.a.	n.a.	n.a.	351	398

EXPENDITURE								
Household								
Consumption total	C	¥/month	(3)	23513	32093	49305	82582	166032
	C/	″	cal.	44617	55620	63905	82582	99959
Consumption, food	CF	″	(3)	10465	12440	17858	26606	49828
	CF/	″	cal.	17767	20228	22129	26606	27051
Consumption, non-durables	CC	″	(3)	4929	6910	10247	15718	30284
	CC/	″	cal.	10762	13288	13847	15718	17178
Consumption, durables	CD	″	(3)	n.a.	1715	2469	5396	9635
	CD/	″	cal.	n.a.	1612	2462	5396	7599
Consumption, other	CO	″	(3)	7568	11028	18761	34862	76285
	CO/	″	cal.	17807	27751	25115	34862	44146
Medical care	COMED	″	(3)	506	687	1221	2141	3957
Private transportation	COCAR	″	″	n.a.	n.a.	477	2077	5975
Education	COEDU	″	″	713	933	1753	1894	3686
Income and other taxes	TY	″	″	2365	2006	3274	5099	10787
Social security contribution	TSIF	″	″	762	1115	2169	4067	9514
Property purchases, net	SREAL	″	″	n.a.	n.a.	670	2288	7054
Savings, net	SFIN	″	″	754	2120	4576	9471	24137
Total expenditure	YE[a]	″	″	26786	35280	54919	91897	186676
Protein intake	—	g./day	(16)	69.7	69.7	71.3	77.8	80.0
Engel's coefficient	ENGEL[b]	ratio	cal.	0.445	0.388	0.362	0.322	0.300
Tax burden	TY/YF	ratio	cal.	0.081	0.049	0.050	0.045	0.046
Saving rate, household basis	RSFIN	ratio	(3)	n.a.	10.5	10.1	15.0	17.3
Household income expenditure balance	BF[c]	ratio	cal.	1.089	1.159	1.186	1.229	1.265
Macro								
Personal consumption	CNI	bil. yen	(4)	5529	8823	18098	36287	82307
	CNI/	″	cal.	10492	15291	23443	36287	49553

TABLE 1 *(Continued)*

Indicators	*Symbols*	*Unit*	*Source*	*1955*	*1960*	*1965*	*1970*	*1975*
Housing capital formation, private	IHP	bil. yen	(4)	256	617	1811	4759	10521
government	IHG	″	″	29	48	131	370	839
total (real)	(IHP + IHG) /	″	cal.	626	1191	2642	5129	6852
Housing capital formation, floor space per unit	IHFLOOR	square m.	(13)	58.2	59.0	58.9	68.1	82.9
Housing capital formation, number of units	IHUNIT	thousand	(13)	257.4	424.2	842.6	1484.6	1356.3
Savings	SPNI	bil. yen	(4)	853	1864	3884	9054	27329
Saving rate, macro basis	RS[a]	ratio	cal.	0.120	0.145	0.152	0.159	0.219
ASSETS								
Household								
Amount of savings	SSFIN	1000 yen	(3)	n.a.	359	764	1603	3168
Ratio of household with savings	RSSFIN	ratio	″	n.a.	0.935	0.975	0.989	0.993
Macro								
Stock of consumer durables	KCD	bil. yen	est.	2727	4517	7318	13935	22592
Diffusion of consumer durables:								
Motor vehicles	—	ratio	(4)	n.a.	'62 0.051	0.091	0.221	0.412
Washing machines	—	″	″	n.a.	'62 0.581	0.685	0.914	0.976
Refrigerators	—	″	″	n.a.	'62 0.280	0.514	0.891	0.912
Air conditioners	—	″	″	n.a.	'62 0.007	0.020	0.059	0.172
TV sets: black & white	—	″	″	n.a.	'62 0.794	0.903	0.902	0.487
color	—	″	″	n.a.	'62 0	0	0.263	0.903

Stereo sets	—	ratio	(4)	n.a.	'62 0.072	0.135	0.312	0.521
Tape recorders	—	″	″	n.a.	'62 0.054	0.179	0.308	0.516
Pianos	—	″	″	n.a.	'62 0.033	0.034	0.068	0.118
Housing capital stock	KH	bil. yen	est.	35936	35364	37353	45746	'73 52251
Housing capital stock, floor space per unit	KHFLOOR	square m.	est.	95.2	82.6	73.1	75.2	'73 77.1
Housing capital stock, number of units	KHUNIT	thousand	est.	16180	18354	22630	27540	33315
Share of owned houses	—	ratio	(3)	0.679	0.645	0.596	0.582	0.580
Housing unit–household ratio	KHUNIT/F	ratio	cal.	0.854	0.860	0.951	1.035	1.137
LIABILITIES								
Household								
Debt outstanding, housing	SDEBTH	1000 yen	(3)	n.a.	69.2	130.8	171.0	579.0
Debt outstanding, other	SDEBTO	″	″	n.a.			112.8	271.0
Ratio of household with debt	RSDEBT	ratio	″	n.a.	0.310	0.356	0.406	0.443
Macro								
Housing loan outstanding	SLOANH	bil. yen	(2)	n.a.	n.a.	29.3	622.9	5806.1
Other loan outstanding	SLOANO	″	″	n.a.	n.a.	30.5	317.6	408.2
PRICES								
Personal consumption deflator	PC	1970 = 100	(4)	52.7	57.7	77.2	100.1	166.1
Consumer prices, food	PCF	″	(3)	58.9	61.5	80.7	100.0	184.2
Consumer prices, non-durables	PCC	″	cal.	45.8	52.0	74.0	100.0	176.3
Consumer prices, durables	PCD	″	(3)	110.6	106.4	100.3	100.0	126.8
Service price	PS	″	(3)	42.5	50.7	74.7	100.0	172.8
Housing investment deflator	PIHP	″	(4)	45.5	55.9	73.5	100.0	165.8
Land price	PA	″	(10)	6.82	19.1	50.1	100.0	213.4

TABLE 1 *(Continued)*

Indicators	*Symbols*	*Unit*	*Source*	*1955*	*1960*	*1965*	*1970*	*1975*
GENERAL GOVERNMENT								
Social capital formation	IG	bil. yen	(4)	334	676	1574	3330	8183
	IG/	"	"	547	970	1936	3340	5307
Share of social capital formation	IG/V	ratio	cal.	0.038	0.044	0.050	0.047	0.056
Social capital stock	KG	bil. yen	(4)	14013	17632	26626	43219	69711
Government consumption expenditure	GC	"	(4)	894	1382	2949	5827	16204
Military expenditure	GM	"	(15)	136	164	303	594	1386
Research and development expenditure	R&D	"	(3)	n.a.	n.a.	438	1065	2975
Government transfer abroad	GAID	"	(4)	9	29	34	64	83
GENERAL ECONOMY								
GNP	V	bil. yen	(4)	8622	15487	31953	70709	145446
Real GNP	V/	"	"	16907	25388	40861	70638	91996
Real growth rate	—	ratio	cal.	0.101	0.133	0.051	0.109	0.026
Per capita real GNP	(V/)/N	1000 yen	cal.	189.4	271.8	415.8	681.0	821.9
Investment in plant and equipment	I2/V	ratio	cal.	0.103	0.188	0.159	0.201	0.142
Import ratio	MNI/V	ratio	cal.	0.105	0.111	0.100	0.106	0.140
Trade balance	BT[e]	ratio	cal.	1.082	1.035	1.114	1.105	0.995
Exchange rate	X	¥360 = $1 as unity	cal.	1.000	1.000	1.169	1.358	1.200

Symbols are in general common to the econometric model of social indicators. The sources of data in this and the following tables are indicated by the numbers that correspond to the numbers in the separate *Sources of the Data* list. Estimated figures are indicated by est. and calculated figures by cal., respectively. The figures marked by / are real values, deflated by appropriate deflators provided elsewhere in the framework. All figures are annual, unless otherwise stated.

Definitions in the table are as follows: [a]YE = C + TY + TSIF. [b]ENGEL = CF/(CF + CC + CD + CO). [c]BF = YF/YE. [d]RS = SPNI/Y. [e]BT = ENI/MNI.

TABLE 2
SOCIAL INDICATORS—DEMOGRAPHIC ASPECTS

Indicators	*Symbols*	*Unit*	*Sources*	*1955*	*1960*	*1965*	*1970*	*1975*
LIFE EXPECTANCY	LE[a]	year	(16)	66.2	67.9	70.7	72.0	74.4
POPULATION SIZE DETERMINATION								
Population size	N	thousand	(16)	89276	93419	98275	103720	111934
Birth rate	RB	ratio	"	0.019	0.017	0.019	0.019	0.017
Death rate, infants	RDB	"	"	0.040	0.031	0.018	0.013	0.010
Death rate, other than infants	RDA	"	"	0.008	0.008	0.007	0.007	0.006
AGE STRUCTURE OF THE POPULATION								
65 years of age and over	N65+	thousand	(3)	4747	5350	6181	7331	8857
15 years of age and over	N15+	"	"	69878	65352	73109	78897	84747
Less than 15 years of age	N−(N15+)	"	"	19798	28067	25166	24823	27187
Ratio of the old	(N65+)/N	ratio	cal.	0.053	0.057	0.063	0.071	0.079
Dependency rate	NDEPEND	ratio	cal.	0.379	0.557	0.468	0.449	0.475
WORKING-LEARNING STATUS								
Labor force	L	thousand	(3)	41940	45110	47870	51530	53230
Primary sector	L1	"	"	15280	13310	11040	8860	6580
Non-primary sector	L2	"	"	25620	31060	36260	42070	45650
Unemployed	LU	"	"	1050	750	570	590	1000
Self-employed	—	ratio	(3)	0.262	0.221	0.196	0.194	0.174
Family workers	—	"	"	0.344	0.240	0.195	0.163	0.127
Employed	—	"	"	0.392	0.540	0.607	0.642	0.698

TABLE 2 (Continued)

Indicators	*Symbols*	*Unit*	*Sources*	*1955*	*1960*	*1965*	*1970*	*1975*
Not in labor force								
Attending school, 15 years and older	NS15+	thousand	(3)	n.a.	4540	7330	7350	7590
Housework and other	NO	"	"	n.a.	10210	12020	13790	16110
Labor force participation rate	L/N	"	"	0.708	0.692	0.657	0.654	0.630
Unemployment rate	U	ratio	cal.	0.018	0.011	0.008	0.012	0.019
Job opening–applicant ratio	RLDS	ratio	(19)	0.22	0.59	0.64	1.41	0.61
Military service ratio	RM[c]	ratio	cal.	0.007	0.008	0.008	0.007	0.007
School attendance rate	RNS[d]	ratio	cal.	n.a.	0.070	0.101	0.093	0.090
INACTIVITY RATES								
Industrial accidents	RAI	ratio	est.	0.00829	0.00601	0.00442	0.00252	'74 0.00140
Traffic accidents	RAT	ratio	est.	0.00009	0.00017	0.00019	0.00035	'74 0.00022
Crime rate (detention rate)	RCRIME	ratio	est.	0.00140	0.00105	0.00086	0.00060	'74 0.00054
Infectious diseases	RDIS	ratio	est.	0.00138	0.00116	0.00108	0.00049	'74 0.00019
Labor disputes	RSTRIKE	ratio	est.	0.00056	0.00054	0.00054	0.00033	'74 0.00062
EDUCATIONAL ATTAINMENT								
Junior high school graduates	NSJ	thousand	(14)	1663	1771	2360	1667	1581
Senior high school graduates	NSH	"	"	716	934	1160	1403	1327
Junior college graduates	NSC	"	"	28	30	56	115	141
College and university graduates	NSU	"	"	95	120	162	241	313
Quality of the population	NQ	index	est.	1.136	1.151	1.167	1.191	1.212

URBANIZATION								
Urbanization of the population:								
Major urban areas	URBAN	ratio	cal.	0.391	0.408	0.445	0.476	0.490
Share of the agricultural household with nonagricultural employment	—	ratio	(12)	0.275	0.321	0.418	0.507	0.621
Regional relocation of the population	NRR	%	(3)	2.5	2.6	3.7	4.1	3.3
FAMILY STRUCTURE								
Family size	FSIZE[e]	persons	(3)	4.71	4.38	4.13	3.90	3.82
Working members of a household	FL[e]	persons	(3)	1.45	1.52	1.53	1.55	1.50
Number of household	F	thousand	cal.	18955	21329	23795	26595	29302
TIME BUDGET								
Hours of the day (male, weekdays only)								
Working hours	H	min./day	(19)	491	503	490	489	479
Commuting hours	HC	"	(7)	n.a.	37	50	51	59
Time consumption	HL	"	"	n.a.	908	900	904	895
Sleeping hours	HO	"	"	n.a.	495	490	485	486
Days of the year								
Working days	DAY	days/month	(19)	23.8	24.2	23.6	22.9	'74 22.0

[a] average of male and female. [b] NDEPEND = [N − (N15 +) + (N65 +)]/[(N15 +) − (N65 +)].
[c] RM = LM/L. [d] RNS = (NS15 +)/(N15 +). [e] household other than primary.

TABLE 3
SOCIAL INDICATORS—ENVIRONMENTAL ASPECTS

Indicators	*Symbols*	*Unit*	*Sources*	*1955*	*1960*	*1965*	*1970*	*1975*
LAND UTILIZATION								
Sources of increase in the land use								
Forest and waste land: agriculture	—	square km	(12)	n.a.	187.1	307.4	438.3	410.0
manufacturing	—	"	(18)	n.a.	7.5	9.5	22.9	['74]23.5
Reclaimed: agriculture	—	"	(12)	n.a.	6.6	8.1	35.9	38.6
manufacturing	—	"	(18)	n.a.	4.8	13.7	8.6	['74]12.1
From agricultural land: manufacturing	—	"	(18)	n.a.	20.4	13.0	25.3	['74]14.7
Current use								
Primary sector (agriculture only)	A1	square km	(12)	59810	60710	60040	58070	55720
Non-primary sector: manufacturing	A2	"	(18)	429	572	802	1086	['74]1241
railway	ARAIL	"	(20)	754	756	770	780	781
road	AROAD	"	(13)	3527	3527	3768	4116	4664
Residential land	AH	"	(17)	5249	5572	6774	8014	10302
Rate of cultivation of arable land	RA	ratio	cal.	1.364	1.332	1.238	1.089	0.964
Rate of land utilization	A/A*	ratio	cal.	0.808	0.824	0.836	0.835	0.842
WATER UTILIZATION								
Sources of increase in the water supply								
Plain water	—	mil. cubic m.	(18)	n.a.	n.a.	341	2744	934
Sea water	—	"	"	n.a.	n.a.	600	1568	0
Recycling	—	"	"	n.a.	n.a.	772	2167	1459

Current use								
Non-primary sector: manufacturing	W2	mil. cubic m.	(18)	n.a.	'[62]14998	19972	33850	'[74]43944
Household (incl. commercial, etc.)	WH	″	(16)	n.a.	n.a.	5047	7987	'[74]9962
Rate of water utilization	W/W*	ratio	est.	n.a.	n.a.	0.037	0.062	0.080
Ratio of recycled water	—	ratio to plain	(18)	n.a.	'[62]0.278	0.363	0.517	'[74]0.648
Ratio of sea water	—	ratio to total	est.	n.a.	'[62]0.286	0.303	0.310	'[74]0.279
Diffusion of waterworks[a]	—	ratio	(16)	n.a.	n.a.	0.694	0.808	'[74]0.867
Social capital, waterworks	(KG)	bil. yen	(4)	258	423	795	1373	2544
Diffusion of sewerage system[b]	RZWATER	ratio	(11)	n.a.	'[61]0.193	0.238	0.266	'[74]0.347
Social capital, sewerage system	(KG)	bil. yen	(4)	284	366	542	968	2020
Water pollution level: BOD	ZBOD	ppm	(5)	n.a.	n.a.	'[66]9.4	6.6	3.6
WASTE								
Waste treatment capacity	(RZWASTE)	ton/day	(5)	n.a.	8717	20736	53998	'[74]111228
Social capital, waste treatment	(KG)	bil. yen	(4)	3	14	67	143	345
ENERGY								
Sources of increase in the energy supply								
Charcoal and wood	—	10^{10}Kcal	(1)	27	(−)127	(−)167	(−)330	(−)57
Coal	—	″	″	334	5564	2560	2391	(−)3518
Hydro	—	″	″	682	(−)790	1828	789	276
Crude oil	—	″	″	1361	9386	15307	34950	(−)16521
Nuclear	—	″	″	0	0	9	857	1330
Current consumption								
Total	EC	10^{10}Kcal	(1)	50583	81452	135512	253556	310124
Primary sector	—	″	″	1351	2110	3413	6642	7719
Non-primary sector	—	″	″	34574	51929	83967	159486	181842
Transportation	—	″	″	7975	12491	20820	36592	47711
Household	—	″	″	11192	14922	27308	50836	72852

TABLE 3 *(Continued)*

Indicators	*Symbols*	*Unit*	*Sources*	*1955*	*1960*	*1965*	*1970*	*1975*
Air pollution level								
SO_x: Sulfur content	ZS	%	(9)	n.a.	n.a.	'68 2.32	1.93	1.43
Sulfur removal rate	RZS	1000 b/day	(9)	n.a.	n.a.	'67 40	370	1270
SO_x output	QSOX	1970 = 100	(5)	n.a.	n.a.	n.a.	100	53
SO_x pollution level	ZSOX	ppm	(5)	n.a.	n.a.	0.057	0.043	0.021
NO_x: NO_x output	QNOX	1970 = 100	(5)	n.a.	n.a.	n.a.	100	108
NO_x pollution level	ZNOX	ppm	(4)	n.a.	n.a.	0.030	0.041	n.a.
			(5)	n.a.	n.a.	n.a.	0.028	0.030
Lead: lead content in gasoline	—	cc/gal.	(5)	0.49	0.49	0.49	0.49	0.00
Lead pollution level	—	$\mu g/m^3$	(5)	n.a.	n.a.	n.a.	0.59	0.21
POLLUTION PREVENTION MEASURES								
Private investment for pollution prevention								
Air pollution prevention	—	bil. yen	(6)	n.a.	n.a.	'66 9	84	312
Water pollution prevention	—	″	″	n.a.	n.a.	'66 19	89	296
Waste treatment	—	″	″	n.a.	n.a.	'66 5	21	73
Share of pollution prevention investment	IZ/IP	″	″	n.a.	n.a.	0.031	0.053	0.172
Government expenditures								
Pollution prevention expenditures	—	bil. yen	(5)	n.a.	n.a.	n.a.	75.8	333.0
Environmental protection expenditures	—	bil. yen	(5)	n.a.	n.a.	n.a.	7.1	42.1

[a] diffusion in terms of population. [b] diffusion in terms of the area served.

following the "Oil Shock" and subsequent policy failures.

As for inactivity rates, the industrial accident rate (RAI) dropped from 0.00829 in 1955 to 0.00140 in 1974, and the infectious disease rate (RDIS) from 0.00138 to 0.00019 during the same period; but the traffic accident rate (RAT) has increased from an estimated 0.00009 to 0.00035 in 1970 when the trend was reversed, and the figure has declined to 0.00022 in 1974. The labor dispute rate (RSTRIKE) had been decreasing through the entire period until the labor relations were disrupted by the worsening labor market situation in recent years.

Increased educational opportunities have resulted in the improvement in the quality of the population index (NQ) as measured by the years of the education received. The index has been constructed taking junior high school level (9 years) as unity; thus senior high school graduates (12 years) are counted as 1.33, college graduates (16 years) as 1.77, etc. The index for the whole population started from 1.136 in 1955 and reached 1.212 in 1975.

Economic development accompanied rapid urbanization (URBAN) and regional relocation of the population (NRR). The share of the agricultural households with nonagricultural employment has more than doubled from 0.275 in 1955 to 0.621 in 1975. Family size (FSIZE) has become smaller, from 4.71 persons in 1955 to 3.82 persons in 1977.

The benefit of increased productivity may be received in the form of increased leisure, which is summarized in the time budget account. In the early period under study, namely from 1955 to 1960, both daily work hours and monthly work days (DAY) were lengthened, reflecting the sharing of the limited job opportunities, which eased as the economic development continued. Since then, both working hours and working days have had declining trends. We observe that a part of the shorter work hours is eaten up by longer commuting hours (HC), which is clearly a cost accompanying urbanization.

Increased use of land and water for socioeconomic activities had environmental consequences, although increase in residential land (AH) was almost indispensable to accommodate increased housing capital, plant sites (A2) and road construction (AROAD). As a result, the rate of land utilization has increased from 0.808 in 1955 to 0.842 in 1975. However, agricultural land (A1) and the rate of cultivation of arable land (RA) had declining trends during the period. Water resources were consumed in larger quantity, both for industrial purposes and household use (W2 and WH).

Increased production and improved levels of living required larger energy consumption (EC), of which a larger portion was met by crude oil, causing air pollution.

It took some time before any effective measures were taken to prevent environmental pollution. It was only in the latter half of 1960s when any pollution statistics were officially collected and made available, as can be seen in Table 3. Water pollution level as measured by biochemical oxygen demand, BOD (ZBOD), air pollution level as measured by sulfur oxides, SO_x (ZSOX), nitrate oxides, NO_x (ZNOX), and lead content in the air indicate that, in general, considerable improvement has been achieved in the pollution level in recent years, although some reservations are necessary in the case of NO_x. It should be remembered, however, that pollution phenomena are not exhausted by the indicators listed here.

Increased effort to prevent pollution is reflected in such indicators as the diffusion of sewerage systems (RZWATER), social capital for sewerage systems and waste treatment (component of KG), waste treatment capacity (RZWASTE), decreased sulfur content (ZS), increased sulfur removal rate (RZS), and decreased lead content in gasoline. The magnitude of pollution prevention measures is most clearly reflected in the share of pollution prevention investment in total private investment in plant and equipment (IZ/IP), which rose from 0.031 in 1965 to 0.172 in 1975. This was accompanied by increased government effort to prevent pollution and to protect environmental quality.

V. CONCLUDING REMARKS

The present paper introduces an attempt at systematizing social and economic statistics into a system of socioeconomic accounts, then deriving a series of social indicators from that data base.[3] Needless to say, the actual items to be picked up would depend on the issue at hand, which varies from society to society and from time to time. The important point is that, being expressed in an accounting framework, every single variable is explicitly related to other aspects of socioeconomic activities. Finally the social indicators can be built into an econometric model.[4]

With the aid of this kind of data base and modeling, monitoring of economic and social policies with respect to various aspects of the society will become possible.[5] Current economic statistics and macroeconomic modeling are perhaps too limited in scope when we realize the very com-

[3] This approach is in line with the one taken by the United Nations Economic and Social Council, although the scope and definitions are quite different.

[4] For an econometric model of social indicators, *see* Kimio Uno, "An Econometric Model of Social Indicators and Its Application to Social Policies: A Japanese Experience" (August, 1977).

[5] Undoubtedly, there are other aspects to be taken into consideration in order to evaluate a development process, including the changing regional characteristics in the course of economic development and urbanization. On this point, *see* Kimio Uno, "Shakai Shihyo to Toshika—Inshibunseki ni yoru Yobiteki Kohsatsu" ["Social Indicators and Urbanization—A Preparatory Consideration by Factor Analysis"] (1976).

plex nature of the interdependence among various aspects of the development process. In this sense, the social indicators approach could be a helpful analytical tool in recognizing what economic development means to a society.

There have been attempts to synthesize the social indicators into a scalar magnitude, which supposedly reflects the general welfare level of the population in a society. There can be two ways of convincingly achieving this. First is the use of the subjective weights that people attach to the various aspects of social life. Apart from the subjective evaluation, however, objective relations may have to be obtained to be useful as an analytical tool. Second is the imputation in monetary terms of various aspects of the society. This is basically the method used in the Measures of Economic Welfare (MEW) approach.[6] This procedure, however, necessarily involves the loss of dimensions when limiting its scope to the aspects that are subject to monetary evaluation. The variety of aspects in the social indicators approach cannot be preserved.

No attempt is made in the present paper to aggregate the social indicators into a single aggregate measure.

As for individual aspects, we observe that through the twenty-year period from 1955 to 1975 the availability of goods and services has increased, assets of the people and the society as a whole have increased, access to educational facilities has improved, and the people are healthier and live longer. The Japanese economy achieved full employment in early 1960s and now is achieving some improvement in the prevention of environmental pollution in the first half of 1970s. The judgment of success or failure of the development process would depend on what to focus on for which period. If the performance of the society survives a series of tests on various aspects, including the increased availability of goods and services, equality in income distribution and of economic and social opportunities, and improvement in the negative side effects accompanying the development on both an objective and a subjective basis, it would be possible to judge the performance as a whole. Improvement in environmental pollution reduction in recent years strengthens the case for success.

REFERENCES

Kuznets, Simon. 1956. "Quantitative Aspects of the Economic Growth of Nations: Some Conceptual Problems of Measurement," *Economic Development and Cultural Change*, vol. 5, October, pp. 5–94.

Nordhaus, William D. and Tobin, James. 1972. "Is Growth Obsolete?" in *Economic Growth*. National Bureau of Economic Research, General Series No. 96. New York: Columbia University Press for NBER, pp. 1–80.

[illegible]e William D. Nordhaus and James Tobin, "Is Growth Obsolete?" (1972).

Organization for Economic Cooperation and Development. 1976. *Progress Report on Phase II—Plan for Future Activities*. Paris.

Tsuru, Shigeto. 1971. "In Place of Gross National Product," in *Kutabare GNP [Down with GNP]*. Tokyo: Asahi Shimbun.

United Nations Department of Economic and Social Affairs. 1975. *Towards a System of Social and Demographic Statistics*. ST/ESA/STAT/SER. F/18. New York.

———. 1976. *Draft Guidelines on Social Indicators*. E/CN. 3/488. New York.

Uno, Kimio. 1974. "Shakai Shihyo—Hatten no Keifu to Makuro Shakai Shihyo Moderu," ["Social Indicators and Macro-economic Framework"], *Journal of Japan Economic Research*, no. 3, March.

———. 1976. "Shakai Shihyo to Toshika–Inshibunseki ni yoruYobiteki Kohsatsu"["Social Indicators and Urbanization—A Preparatory Consideration by Factor Analysis"], in *Toshika to Fukushi [Urbanization and Welfare]*. Edited by Miyohei Shinohara. Tokyo: The Institute of Statistical Research. Reissued as Japan Economic Research Center Discussion Paper No. 18, January, 1977.

———. 1977. "An Econometric Model of Social Indicators and Its Application to Social Policies: A Japanese Experience," *Journal of Japan Economic Research,* no. 6, August.

SOURCES OF DATA

(1) Agency for Natural Resources and Energy.
(2) Bank of Japan.
(3) Bureau of Statistics, Office of the Prime Minister.
(4) Economic Planning Agency.
(5) Environmental Agency.
(6) Japan Association of Machinery Producers.
(7) Japan Broadcasting Association (NHK).
(8) Japanese National Railway.
(9) Japan Petroleum Association.
(10) Japan Real Estate Institute.
(11) Japan Sewerage Association.
(12) Ministry of Agriculture and Forestry.
(13) Ministry of Construction.
(14) Ministry of Education.
(15) Ministry of Finance.
(16) Ministry of Health and Welfare.
(17) Ministry of Home Affairs.
(18) Ministry of International Trade and Industry.
(19) Ministry of Labor.
(20) Ministry of Transportation.

Ocean Resources: An Analysis of Conflicting Interests

Tomotaka Ishimine *California State University*

From time immemorial, the ocean has provided food, adventure, and inspiration to humanity. In recent years, the nations began to recognize the ocean as an important source of resources. The immense potential of the ocean in providing food and nutrition, particularly protein, began to be reexamined. However, the ocean also contains a seed of conflict among nations, since claims over ocean resources are overlapping. Attempts to establish the law of the sea have failed to reach an accord with regard to the definition of territorial waters and economic zones. At stake are the freedom of navigation, the right of fishing, and claims over mining deep-seabed resources. It is imperative to examine the conflicting claims over ocean resources and to foresee the possible outcome of the law of the sea to avoid scrambles over the ocean resources among nations.

I. THE LAW OF THE SEA AND OCEAN RESOURCES: BACKGROUND

The First United Nations Decade of Economic Development ended with a mixed blessing. The average annual growth rate of less developed countries (LDCs) during the 1960s surpassed that of the 1950s and even that of developed countries (DCs) in the same decade. But the population growth was such that the growth rate for per capita income for LDCs lagged behind that of DCs. As a result, the gap in the standard of living between LDCs and DCs was actually widened in the decade of the 1960s.[1]

Disillusioned by the policy of import substitution in the 1950s and the early 1960s, many LDCs turned their attention to export efforts. For example, tariff preferences are to be granted by DCs on manufactured or semimanufactured products produced by LDCs. Although there has been some progress in this regard, the extent of preferences accorded by DCs has been much too small, and the pace of the preferences provided was much too slow to accommodate the LDCs' need as they saw it. In addition, their dissatisfaction with the amount of foreign aid received, and the tendency towards a growing economic consciousness and nationalism have led LDCs to look upon their natural resources as an effective economic and political weapon to be aimed at DCs.

Due to pressures by LDCs, the United Nations adopted a resolution in 1966 that allowed nations sovereign rights over natural resources within their domain.[2] The pace of nationalization of natural resources quickened, and many nations, including Chile, Algeria, and Jamaica, nationalized resource industries by the early 1970s. The potential of natural resources as an economic and political weapon was effectively dramatized in the fall of 1973 when the Organization of Petroleum Exporting Countries (OPEC) quadrupled petroleum prices.[3]

The waves of sovereignty over natural resources have finally reached the area of ocean resources. The oceans, which have largely been outside the domain of national sovereignty, contain vast amounts of marine and mineral resources, as well as water, energy, and space resources.[4] Meanwhile, the growth in ocean technology, population, and economic and political awareness, coupled with such recent discoveries as the large manganese nodule deposits in the South Pacific, have led the LDCs to influence the United Nations' declaration in 1970 that the high seas and

[1] Gerald E. Meier, *Problems of Cooperation for Development* (1974), pp. 4–6.

[2] United Nations General Assembly, Resolution 2158 (1966).

[3] Edith Penrose, "The Development of Oil Crisis" (1976), pp. 39–57.

[4] Statement by John L. Mero in "National Oceanographic Program Legislation," Hearings before the Subcommittee on Oceanography of the House Committee on Merchant Marine and Fisheries (1965b), pp. 599–600. Also, for estimates of various ocean minerals, *see* John L. Mero. *The Mineral Resources of the Sea* (1965a).

their resources belong to the "common heritage of mankind."[5] The declaration also stated that no nation can unilaterally develop the seabed and that the benefits reaped from the seabed must be "equitably" distributed among the nations.[6]

LDCs' consciousness of the importance of the oceans as a source of natural resources is also reflected in their demand for enlarging the size of territorial waters and the setting up of an economic zone beyond territorial waters. To the present, the main interest of LDCs has been in marine resources, while that of DCs has encompassed consideration of the military, freedom of navigation, and mineral resources. Herein lies the source of potential conflict among interested nations, since ocean resources and activities related to ocean resources are overlapping in many areas in the three dimensional space.[7] For example, mining oil from the ocean deposits in the continental shelf might interfere with the migration of fish in the sea and navigation of ships on the surface of the sea.

There have been various attempts to coordinate the diverse interests of nations and to formulate international laws. The United Nations International Law Commission produced four basic laws in 1958 concerning oceans and ocean resources.[8] These are: (1) The law on territorial waters, which recognized the coastal nations' jurisdiction over incidents occurring within their territorial waters and confirmed the principle of "innocent passage" of seagoing nations within the territorial waters of other nations. However, the law failed to define the limit of territorial waters due to lack of consensus among the nations. (2) The law on high seas, which confirmed freedom of navigation on high seas and reaffirmed the traditional jurisdiction of the "nation of flag" for incidents involving ships of seagoing nations. (3) The law on marine resources, which set rules for preservation of marine resources in the international waters. (4) The law for the continental shelf, which recognized the right of coastal nations over mineral resources on the continental shelves extending to 200 meters. The first three of the laws represent, in a large part, a consolidation of customs and conventions that had existed for years regarding the use of the seas. Many LDCs consider these existing laws as accommodating the need and interest of DCs and as not necessarily reflecting the need and interest of LDCs. It is noteworthy that, within a little more than a decade, the need for reexamination and revision of the existing laws of the sea has become apparent.[9] On the other hand, a greater

[5] For an interesting historical account on the "common property of mankind" in Western civilization, *see* George R. Geiger, *The Theory of the Land Question* (1936), pp. 119–47. Also, *see* Fritz Heichelheim, "Ancient Land Tenure" (1933), pp. 77–82. The latter article is not contained in the 1968 edition.

[6] United Nations General Assembly, Resolution 2750 (1970).

[7] Francis T. Christy, Jr., "New Dimension for Transitional Marine Resources" (1970).

[8] United Nations Conference on the Law of the Sea (1958a, 1958b, 1958c, 1958d).

number of nations are drifting from the traditional three-mile limit for territorial waters and demanding, in addition, an exclusive economic zone beyond territorial waters.

In 1960, another conference was convened by the International Law Commission to discuss the limit of territorial waters, but nations again failed to reach consensus.[10]

A turning point came when Malta's ambassador to the United Nations, Arvid Pardo, spoke before the General Assembly in 1967.[11] He pointed out the recent development of ocean technology, which has uncovered vast deposits of resources on the seabed, including petroleum, natural gas, and other mineral resources; he also warned of possible conflicts among nations over deposits, of disruption of the freedom of the sea, and of damages to the ocean environment. Pardo proposed that the ocean resources in international waters be declared as common property of all nations and that an international administrative agency be established to exploit the resources and to apportion them to all nations, especially to LDCs. His proposal was enthusiastically accepted by LDCs; and the United Nations declared in 1970 that international waters and their resources are the "common heritage of mankind."[12] The United Nations General Assembly also adopted a moratorium for the development of seabed resources until an international organization was established to administer the mineral resources.[13]

The interest in the ocean proliferated, however, beyond the mineral resources and culminated in the Third Conference on the Law of the Sea, held in Caracas, Venezuela, in May 1974, and in Geneva in March 1975, under the auspices of the United Nations Commission for Peaceful Utilization of the Seabed.[14] The Third Conference turned out to be the largest international conference, encompassing 156 nations, including nonmember nations of the United Nations. The 1973 United Nations General Assembly, which proposed the Third Conference, also requested that the new law of the sea should be a package law, unlike existing ones. Aside from its advantages, the attempt to produce a package law has resulted in intense bargaining, maneuvering, and even threatening among nations and blocs of nations in an effort to incorporate their own interest before the law was adopted by the majority vote.[15]

[9] This statement applies for the United States as well. *See* Louis Henkin, *Law for the Sea's Mineral Resources* (1968), pp. 32–36.

[10] For a concise account of the 1958 and 1960 Conferences on the Law of the Sea, *see* Sayre A. Swartztrauber, *The Three-Mile Limit of Territorial Seas* (1972), pp. 209–18.

[11] United Nations General Assembly, Document A/6695 (1967).

[12] United Nations General Assembly, Resolution 2750 (1970).

[13] U.N., Resolution 2750 (1970).

[14] Ann L. Hollick and Robert E. Osgood, *New Era of Ocean Politics* (1974), pp. 41–44.

[15] Before the 1974 Caracas and 1975 Geneva sessions, a preparatory session was held in New York in 1973.

The 1974 and 1975 sessions of the Third Conference remained largely as bargaining sessions and no concrete result emerged.[16] However, various subcommittees produced unofficial texts, which participants brought back to their countries for further study, and on the basis of which the 1976 conferences, *i.e.,* the fourth and fifth sessions of the Third Conference, were convened in March and August, 1976.[17] The conferences again failed to come to an accord, mainly because of disagreement as to how the international agency, to be created to regulate ocean mining, was to function. There are a sense of urgency, both on the part of LDCs and DCs, and the feeling that this is the final year before nations begin scrambling for ocean resources unless a new law is adopted. It is, therefore, useful to survey the issues involved and to explore economic and noneconomic implications in order to foresee the direction in which the matter is moving.

II. TERRITORIAL WATERS AND SPACE RESOURCES

As stated earlier, there has been a tendency to drift away from the traditional three-mile limit for territorial waters. The origin of the three-mile limit itself is obscure.[18] At any rate, the narrow definition of territorial waters has benefited traditional seagoing nations by according them the freedom of navigation. However, as of 1974, it is estimated that only 25 nations uphold the three-mile limit. Another 14 nations are for a four-to-ten-mile limit; 55 nations are for a 12-mile limit; and 21 nations are for up to a 200-mile limit.[19] Through the Caracas Conference, a consensus

[16] For extensive evaluations and discussions of various topics of the Law of the Sea and of the 1974 Caracas session, *see* Francis T. Christy, ed. *Caracas and Beyond: Proceedings* (1975).

[17] *Wall Street Journal,* September 20, 1976.

[18] Sayre A. Swartztrauber, *The Three-Mile Limit of Territorial Seas* (1972), pp. 53–56.

[19] The tabulation is as follows. For a three-mile limit: Australia, Bahrain, Barbados, Belgium, Cuba, Denmark, Fiji, East Germany, West Germany, Guyana, Ireland, Japan, Jordan, Maldives, Monaco, Netherlands, New Zealand, Qatar, Singapore, Taiwan, United Arab Emirates, United Kingdom, United States, Vietnam (South), Western Samoa. For a four-to-ten-mile limit: Dominica, Finland, Greece, Iceland, Israel, Italy, Ivory Coast, Malta, Norway, Poland, South Africa, Spain, Sweden, Yugoslavia. For a 12-mile limit: Albania, Algeria, Bangladesh, Bulgaria, Burma, Canada, China, Colombia, Congo, Costa Rica, Cyprus, Dahomey, Egypt, Equatorial Guinea, Ethiopia, France, Guatemala, Haiti, Honduras, India, Indonesia, Iran, Iraq, Jamaica, Kenya, Khmer, Korea (North), Kuwait, Liberia, Libya, Malaysia, Mauritius, Mexico, Nauru, Oman, Pakistan, Portugal, Romania, Saudi Arabia, Senegal, Soviet Union, Sri Lanka, Sudan, Syria, Thailand, Togo, Tonga, Trinidad and Tobago, Tunisia, Turkey, Venezuela, Vietnam (North), Yemen (Aden), Yemen (San'a), Zaire. For an 18-to-130-mile limit: Cameroun (18 mi.), Gabon (100 mi.), Gambia (50 mi.), Ghana (30 mi.), Guinea (130 mi.), Madagascar (50 mi.),

for a 12-mile limit has developed. Nations that are inclined towards a more than a 12-mile limit have made the acceptance of 12 miles conditional upon satisfactory solutions on other matters such as economic zones, passage of straits, and the question of archipelagos.

At stake is the desire of many nations, especially smaller ones, to expand their jurisdictions to extended waters. At stake are exclusive rights to exploit marine and offshore mineral resources on the part of coastal nations. And at stake is the desire for uninterrupted passage of merchant marines of seagoing nations and warships of naval powers. The traditional concept of innocent passage in territorial waters has conferred upon seagoing nations the freedom of navigation in the territorial waters as long as the passage does not disturb the peace and security of coastal nations. But the concept has been subject to varying interpretations among countries. Thus, seagoing nations are requested to provide advance notification for passage of warships, and submarines are allowed to travel only on the surface of the water. It is possible that overflights of aircrafts and passage of giant tankers are forbidden under the pretext of violation of innocent passage. The accessibility of nations to place disguised intelligence-gathering vessels near the coast of other nations may be seriously hampered. It is not a coincidence that both the United States and the Soviet Union initially opposed expansion of territorial waters beyond the three-mile limit.[20] The seagoing nations such as Japan and England also opposed expansion of territorial waters, for it will put serious limitations on fishing activities and passage of straits that may fall within the territorial waters.[21] Indeed, it is mainly the passage of straits and its strategic and navigational implications that the naval powers and seagoing nations worried about most. When territorial waters are expanded to 12 miles, straits with a width of 24 miles would fall within the territorial waters of either one of the two nations. Such straits number 116, including such important straits as Malacca, Gibraltar, Dover, and Bering, which are international waters at present.[22]

Mauritania (30 mi.), Morocco (70 mi.), Nigeria (30 mi.), Tanzania (50 mi.). For a 200-mile limit: Argentina, Brazil, Chile, Ecuador, El Salvador, South Korea (20–200 mi.), Panama, Peru, Sierra Leone, Somalia, Uruguay. Unknown: Lebanon, Nicaragua, Philippines. Adopted from John R. V. Prescott, *The Political Geography of the Oceans* (1975), pp. 226–28.

[20] For an account of changing U.S. positions, *see* Hollick and Osgood (1971), pp. 17–50. For overall policy recommendations, *see* Lewis M. Alexander, ed., *The Law of the Sea: National Policy Recommendations* (1970) and Marine Technology Society, *Law of the Sea Reports: A Year of Crisis, February 19, 1971; Geneva Report, October 18, 1971* (1972).

[21] Prescott (1975), pp. 106–10.

[22] Robert D. Hodgeson and Terry V. McIntire, "Maritime Commerce in Selected Areas of High Concentration" (1973), pp. 1–18.

Sensing strong demand by many nations, the naval powers and the seagoing nations began to soften their stand. Most of these nations are now ready to admit a 12-mile limit on the condition that an international sea lane must be set for free passage in the straits and that a guarantee of freedom of navigation is accorded to vessels traveling international waters. However, many nations are reluctant to accept an international lane that would allow unconditional passage of ships, which may include submarines, atomic-powered warships, nuclear-carrying vessels, and pollution-prone tankers.[23] On the other hand, the majority of nations would not object to guaranteeing them the right of innocent passage that goes with the territorial waters.[24] The question here is not one of freedom of passage or of no passage, but one of unconditional passage or conditional passage. However, the demand of seagoing nations is so strong that the other nations will perhaps have to accept the international lane in exchange for a recognition of the 12-mile limit, provided that safeguards against pollution and damages are made. The provisions governing the passage of straits are likely to be different from the innocent passage within territorial waters, for sea lanes in the straits are international lanes and as such subject to international negotiations.

The limit for territorial waters is still more complicated for nations consisting of archipelagos such as the Philippines, Indonesia, and the Bahamas. They have put forward the idea of drawing lines for territorial waters connecting 12 miles off the outer island of the archipelagos.[25] An unofficial text distributed by the United Nations Seabed Committee, however, restricts the ratio of water to land to no more than nine to one, and states that the total length of the strait lines shall not exceed 80 miles, with some exceptions calling for 125 miles. All ships passing through the international waters of archipelago nations must go through designated passing lanes; if no lanes are designated, the customary lanes that have been used in the past will be used.

There are also provisions against pollution in the aforementioned unofficial text. The rules governing pollution must be internationally agreed upon and, in order to eliminate interference by a third country, only nations of flag or of destination are allowed to prosecute violating ships. According to the provisions, all nations have an obligation to enforce international rules on their own ships and must prosecute violating ships. When requested in writing by nations incurring damages from violating ships, the nations of flag must investigate their own ships and bring them to the court when violation is evident. When foreign ships that are suspected of violations enter ports, the host countries must investigate immediately and notify nations of the flag. When foreign

[23] Hollick and Osgood (1974), pp. 75–131.

[24] Swartztrauber (1972), pp. 211–72 for the existing innocent passage provisions.

[25] Prescott (1975), pp. 128–41.

ships that are anchored in their own ports emit polluting materials, or when a third nation claims that the foreign ships emitted polluting materials, they can be prosecuted by the anchoring nations provided that no appropriate measure was taken by the nation of flag. These provisions represent a compromise between the demand of seagoing nations that only nations of flag have jurisdiction over their own ships and the demand of coastal nations that wish to exercise sole jurisdiction over polluting ships near their water. These regulations against ocean pollution are likely to be tied in with the existing international agreement against ocean disposal of harmful materials such as mercury and cadmium, and the agreement to regulate the disposal of lead, copper, and arsenic.[26]

Recently oil pollution has become a serious problem as the size of tankers has increased so much that 200,000 to 300,000-ton tankers have become common. Aside from leakage of oil from tankers due to collision, stranding, and seepage, tankers are required to fill empty tanks with water in order to maintain balance during return trips. As the tankers take on cargo in the intermediate ports, the water is drained into the ocean to maintain the balance. Occasionally the emptied tanks are cleansed during the return trip, and the water is drained into the sea. It is understandable that affected nations want to regulate ships that are pollution prone, while the existing international law only permits prosecution of such ships by the nation of flag. Since the aforementioned provisions attempt to exclude involvement of a third country with a passing (but polluting) ship, the coastal nations are likely to resist the provisions that limit jurisdiction only to the nation of flag and the visiting nation.

III. ECONOMIC ZONE AND MARINE RESOURCES

The next question of importance is the economic zone. According to established customs, the ocean beyond the territorial waters, except for continental shelves, is theoretically free to all nations for passing, fishing, and mining. But as LDCs see it, freedom is open only to those advanced seagoing nations that have the technology and means of transportation.[27] These nations are capable of sending their fishing fleets into the offshore waters of other nations, where they are able to deplete LDCs' fish resources with their efficient fishing techniques. LDCs may have wanted to make a unilateral declaration announcing an expansion

[26] Anatoly Andreev, "Activities of the Intergovernmental Maritime Consultative Organization in the Field of Prevention and Control of Operational and Accidental Pollution Emanating from Ships" (1973), pp. 29–47.

[27] Prescott (1975), pp. 117–25.

of territorial waters to a vast area. However, the cost of enforcing such a unilateral declaration and of watching violators in a vast area would have far exceeded the benefit of an exclusive fish catch, the volume of which is limited by the state of their fishing technology.[28]

With a growing awareness of the importance of oceans as a source of food and protein on the one hand, and with the gap in the fishing technology of DCs and LDCs on the other,[29] an increasing number of the latter began to demand the "economic zone."[30] Subsequently the 200-mile economic zone has been gaining momentum not only among LDCs but among some DCs as well.

Originally, the 200-mile economic zone was proposed by Kenya in 1972 as a compromise to the claim of some countries such as Chile, Ecuador, and Peru for the 200-mile territorial waters.[31] Fishing nations such as Japan and England also have opposed the proposal, as it would seriously restrict their fishing activities.[32]

Most nations in favor of the 200-mile economic zone (188 miles beyond the territorial waters of 12 miles) want to have an exclusive right not only for fish but also for mineral resources as well. At the present time the countries that initially opposed the economic zone are largely resigned to the idea. For unless the economic zone is set, there is a prospect that impatient countries may resort to unilateral actions, and lawlessness in the sea may prevail. And it has gradually become clear that establishment of the economic zone may not necessarily exclude fishing nations that have traditionally operated in the area. Nevertheless, coastal nations will have the right to set limits in fish catch, method of catching, catching seasons, antipollution provisions, and so forth; and as a result, the activities of fishing nations will be constrained. Some countries may make it conditional for fishing nations to provide them with fishing technology and to set up processing plants or research facilities on their shores. At any rate, after some advanced countries, such as Canada, Australia, and New Zealand, as well as Ireland, joined this group, the trend for the 200-mile economic zone gained momentum.

It is ironic that the United States, which strenuously opposed creation of the economic zone, gave impetus to the idea by unilaterally declaring, in the 1945 Truman proclamation, exclusive rights to mineral resources on the continental shelf.[33] In 1953, Congress enacted the Outer Continental Shelf Land Act and implemented the proclamation. Other

[28] Kenneth W. Clarkson, "International Law, U.S. Seabed Policy and Ocean Resource Development" (1974).

[29] Prescott (1975), pp. 117–25.

[30] United Nations, Document A/8028 (1970).

[31] Hollick and Osgood (1974), pp. 75–131.

[32] Hollick and Osgood (1974), pp. 75–131.

[33] Presidential Proclamation No. 2667 (1945).

countries followed with similar acts. In 1958, an international treaty for mineral resources on the continental shelf was signed under the auspices of the United Nations International Law Commission. Some countries, such as Iceland, find it arbitrary that mineral resources on the continental shelf are under sovereignty while marine resources are not.[34] In the light that a few countries demand extension of territorial waters to 200 miles and that countries endowed with continental shelves already have the right to mineral resources, setting 200 miles as an economic zone, rather than as territorial waters, seems to be a practical compromise. At any rate, the limit of economic zones is likely to be either 200 miles or the extent of the continental shelf, whichever is larger.

The United States has conflicting interests in recognizing the 200-mile economic zone. Both the Department of Defense and the State Department have opposed it because of adverse military and political implications. The fishermen of the southwestern United States who net tuna and shrimp off the west coast of South America oppose the 200-mile economic zone.[35] The fishermen of the northeastern states, on the other hand, have pressed for just such a law to exclude or limit fishing by other nations off the coast of New England.[36] Foreseeing the eventuality and pressing for an early international agreement, Congress passed a bill that establishes the 200-mile economic zone for the United States. The President signed the bill on 13 April 1976; it has become a law effective 1 March 1977.[37] The new law provides the United States with 2.5 million square miles of exclusive fishing zone, which is approximately 70 percent of its land area; other nations wishing to engage in fishing in this zone are subject to licensing regulations of the United States.

Beyond control of fish and mineral resources in the economic zone, there are still a number of conflicts among the nations as to what more "rights" coastal nations may acquire.[38] They include the right to control ocean pollution, scientific exploration, underwater cables and pipelines, as well as the traditional freedom of navigation and air passage. DCs are against any restriction beyond mineral and marine rights. One question yet to be solved concerns what privileges should be extended to inland or land-locked nations, and nations that have traditionally engaged in

[34] Morris Davis, *Iceland Extends its Fisheries Limit* (1963), pp. 77–94.

[35] Senator Mark Hatfield on U.S. Fishing Industry (1971) and Tom Alexander, "Dead Ahead of a Bounded Main" (1974).

[36] Hatfield (1971) and Alexander (1974).

[37] Public Law 94-265 (HR 200) April 13, 1976, "Fishery Conservation Act of 1976" (1976), pp. 331–61. For legislative history, *see* House Reports: No. 94-445, No. 94-448; Senate Reports: No. 94-416, No. 94-459, No. 94-515 all comprising § 961 and No. 94-711. Also, *Congressional Record,* Vol. 121 (1975): October 9, December 19; Vol. 122 (1976): January 19–28, March 29 and 30. *See* also *Weekly Compilation of Presidential Documents,* Vol. 12, No. 16 (1976), April 13, Presidential Statement.

[38] Prescott (1975), pp. 13–31.

fishing in the distant waters that are now to become economic zones.[39]

The so-called Evensen Proposal of the Second Subcommittee, dealing with territorial waters and economic zones, is indicative of the likely outcome.[40] According to the proposal, the coastal nations are obligated to preserve marine resources and to strive for "optimum utilization" of marine resources. The amount of permissible fish catch within the economic zone is to be regulated by the coastal nations. However, when the coastal nations do not have the capacity to utilize permissible catch, they must permit fishing to other nations. In doing so, consideration must be made for minimizing economic disruption of the countries that have customarily engaged in the fishing in the economic zone. The optimum utilization of migratory fish that travel in and out of an economic zone (*e.g.,* tuna, mackerel, anchovy) must jointly be determined by the coastal nations and fishing nations. The preservation of ocean fish that spawn in fresh water (*e.g.,* salmon and trout) is to be determined by the nations where spawning waters originate; but consideration must also be made to minimize economic disruption of nations that have traditionally caught these fish in the high seas. The inland nations will have the right to participate in fishing within the economic zone of the adjacent coastal nations. The coastal nations will have the right to check, search, arrest, and bring to the court those parties that violate the laws in the economic zone. However, the arrested ships and their crews must be released on bail and cannot be detained in the prosecuting country.

IV. INTERNATIONAL WATER AND MINERAL RESOURCES

International waters that are beyond the 200-mile economic zone are common property of all nations and as such free for all nations to navigate and catch fish as they have for centuries, although the size of international waters will be vastly decreased. A problem exists, however, in the matter of mineral resources in the deep seabed. At a time of ever increasing need for resources on the one hand and their depletion on the other, the potential of the oceans as a source of mineral product has stirred intense interest. In the 1950s, a United States Navy exploration team discovered an immense deposit of manganese nodules on the seabed of the South Pacific near Tahiti.[41] Manganese nodules, containing minerals such as manganese, nickel, copper, and cobalt, are potato-

[39] Martin I. Glassner, *Access to the Sea for Developing Land-Locked States* (1970), pp. 29–35, 205–18, and Hollick and Osgood (1974), pp. 1–73.

[40] United Nations Conference on the Law of the Sea (3rd, 3rd Sess.) 2nd Committee (1975). Also, *see* Prescott (1975), pp. 222–25.

[41] David C. Brooks, "Deep Sea Manganese Nodules: From Scientific Phenomenon to World Resources" (1968), pp. 406–07.

shaped nuggets that lie on the surface of the deep seabed.[42] Since no ground mining is required, the main technological problem involves dredging the nodules, bringing them to the surface, and processing them efficiently. Three methods of dredging have been devised and are known as continuous-path dredging (a giant vacuum cleaner), fixed-area dredging (movable arms with carriages), and continuous-line-bucket dredging.[43] Experiments have already been attempted by some firms, including Deepsea Ventures (a subsidiary of Tenneco), Summa Corporation (the late Mr. Howard Hughes), Kennecott Ocean Resources (California-based), and International Nickel (Canadian-based).[44] However, technology of efficient processing requires substantial capital investment and is still at the development stage. Although knowledge of manganese nodules has existed for over 100 years, it has only been in the 1950s and 1960s that large deposits of nodules have been located. According to one estimate, the Pacific Ocean alone contains as much as 1.6 trillion tons of nodules with concentration of up to 5,000 tons per square mile. Two factors are crucial for successful commercialization of the nodules: increase in world market price of the metals contained in nodules and development of more efficient technology of dredging, surfacing, and processing. When sufficiently processed and marketed, the metals from the seabed in turn are likely to affect the existing price structure of these metals and substitutes.[45] In fact some DCs see a hidden utilitarian motivation behind the pressure exerted by LDCs to declare the seabed resources as a common heritage of mankind. It is apparent that LDCs would like to establish an international agency to control development of the seabed resources, since some of these countries are themselves producers of these minerals (Chile, Peru, Zaire, Zambia).

Although it was agreed upon by DCs and LDCs that the seabed resources are a common property of mankind, the question unanswered is who is to develop these resources.[46] There is a consensus among nations of the need for an international authority to regulate development of the seabed resources. However, opinions vary as to how development is actually to take place. LDCs want the international authority to retain

[42] Mero (1965), pp. 127–241.

[43] Mero (1965), pp. 242–72. *See* also Arnold J. Rothstein, "Deep Ocean Mining: Today and Tomorrow" (1971), pp. 43–50.

[44] United Nations Secretary General, *Progress Report* (1973), pp. 10–14; "Tapping the Lode of the Ocean Floor," *Business Week* (1974); Tom Alexander, "Dead Ahead Toward a Bounded Main" (1974).

[45] Mero (1965), pp. 273-79 and Philip E. Sorensen and Walter J. Mead, "A Cost-Benefit Analysis of Ocean Mineral Resource Development: The Case of Manganese Nodules" (1968), pp. 1611–20.

[46] Richard J. Sweeney, Robert D. Tollison, and Thomas D. Willett, "Market Failure, the Common-Pool Problem, and Ocean Resource Exploitation" (1974), pp. 179-92.

control of all phases of development, *i.e.*, exploration, production, refining, transportation, marketing, and price determination. Private firms play only a passive role through "participation," that is, by acting as agents of the controlling authority. DCs, on the other hand, want a system in which private firms play active roles in the aforementioned activities through contracts with the international authority. However, LDCs are fully aware that there can be no development without capital and technology supplied by DCs, and for this reason, the outcome will probably be greatly reflected by the demands of DCs. In the United States, the American Mining Congress went so far as to introduce a bill in 1971 through Representative Thomas Downing of Virginia and Senator Lee Metcalf of Montana.[47] The bill would grant the United States government the authority to stake out mining blocks in international waters, lease them to private mining firms, and compensate their losses if the succeeding international authority takes over at an unfavorable time. The bill has alarmed many countries, and the United States government itself opposed the bill. The bill has since been shelved.[48]

Another problem associated with the establishment of an international authority is the method of contract with private firms. After mining blocks are staked out, each block can be auctioned off to the highest bidder with clear transfer of property rights. It is argued that this method of allocating licenses ensures efficiency from an economic point of view. Other methods such as a first-come, first-served basis where ownership rights are not transferred may lead to inefficient allocation.[49] The problems of ownership and its effect on economic efficiency are far from settled; and they involve noneconomic judgments as well.

However, there is a consensus among the nations that no matter what authority the international agency assumes and regardless of what method of licensing is adopted, royalties must be paid to the international authority which in turn distributes them to all nations, including land-locked nations. The heart of the matter ultimately lies in the

[47] T. S. Ary, "Statement of American Mining Congress to the Department of the Interior with Respect to Working Paper of the Draft United Nations Convention on the International Seabed Area" (1971).

[48] United States Senate, Committee on Interior and Insular Affairs, *Outer Continental Shelf.* Hearings, Part 2 (1971a), p. 463 and *Outer Continental Shelf,* Committee Report (1971b), p. 25. For interdepartmental differences of opinion, *see* Deborah Shapley, "Law of the Sea: Energy, Economy Spurs Secret Review of U.S. Stance" (1974), pp. 290–92. *See* also *Congressional Record,* daily ed., January 23, 1974, S255–66.

[49] Ross D. Eckert, "Exploitation of Deep Ocean Minerals: Regulatory Mechanisms and United States Policy" (1974), pp. 143–77. The U.S. position in this regard at the UN Conference of the Law of the Sea is presented in UN Doc. A/AC 138/25; A/AC 138/SC, II/L.35; and A/AC 138/22.

"equitable" distribution of wealth.[50]

In summary, the concept of freedom of navigation in the traditional sense is at a turning point. It will be a constrained freedom with certain obligations attached to the navigating ships. Territorial waters will be extended to 12 miles and the economic zone will be set up for 200 miles off coastal nations. Nations customarily engaged in fishing in the 200-mile zone will have to pay license fees and must be subject to various restrictions imposed by the coastal nations. These restrictions include amount, kind, method, and seasons of catch, as well as pollution controls and the tranquility of the coastal nations. Mineral resources within the 200-mile zone will be administered by the coastal nations and in remaining international waters by an international agency. In short, all ocean resources will be divided into two categories, one controlled by coastal nations, the other by an international body.

The above study also reveals the importance of the matter of sovereignty in future economic analyses, especially in the field of natural resources, multinational corporations, and economic development of LDCs. Consideration and integration of national sovereignty in economic analyses have been grossly neglected in the past. Nations endowed with raw materials, through exercise of sovereignty and control of resources, will press for commodity agreements to stabilize supply and prices and will use resources as political and economic leverage for economic development. DCs, despite their reluctance, will be increasingly compelled to accept commodity agreements as a price to be paid for access to the supply of vital resources.

COMMENTS ON THE IMPLICATION OF THE LAW OF THE SEA FOR THE JAPANESE ECONOMY

The outcome of the law of the sea will undoubtedly have significant effects on the Japanese economy, for her interests are closely related to ocean resources. However, unlike the current opinion prevailing in Japan, my assessment is, overall, on the positive side. The reasons are as follows.

The first is concerned with territorial waters and space resources. Japan, as an archipelago nation, will obviously benefit from the expansion of the territorial waters. Although, as a maritime nation, her freedom of navigation will be somewhat curtailed by the expansion of the territorial waters of other nations, the setting of the international sea lanes in the straits seems to be a practical compromise. The fear of coastal nations of oil spill and other forms of damage is real. One needs only to recall the Showa Maru incident of 1974 in which an estimated

[50] Roger L. Miller, *Economics Today* (1976), pp. 626–30.

thousand tons of crude oil was spilled in the Strait of Malacca. The expansion of territorial waters may even lead to positive results if the maritime nations become more aware of the concerns of the coastal nations in preserving their coastal environment and are made accountable for their own actions.

Secondly, with regard to the economic zone and marine resources, there is no doubt that the fishing and fishery industries in Japan will be seriously affected if the economic zone is established. This is especially true because Japan yields 45 percent of her annual fish catch within 200 miles of other coastal nations.

However, an obvious fact is often overlooked: the creation of her own economic zone will benefit Japan as well. Though there will be adverse effects from the limitation of fishing in the other coastal nations, especially in the short run, Japan, with its highly advanced fishing and fishery technology, can utilize her economic zone more efficiently to create a greater quantity and variety of marine resources through ocean farming. Meanwhile, the government and the industry can engage in a more vigorous campaign to change consumers' tastes from those fish caught in distant waters to those caught in near shore or offshore waters. The campaign, if effectively combined with changes in the price structure, can tide over a short-run difficulty, which may arise in that industry. In a global sense, placing restrictions on the wasteful utilization or depletion of marine resources may eventually benefit all of humanity.

Furthermore, one should not overemphasize the role of the fishing and fishery industry in the Japanese economy. This industry employs 676,000 members of Japan's population or only 1.02 percent of the total labor force of 52,235,000. Though the estimates of the contribution of this sector to the entire economy vary, my own estimate indicates that it contributes less than four-tenths of 1 percent of the 500 billion dollar economy.

In addition, the contribution of the fishing and fishery industry to the national economy has been steadily declining in the postwar period. As for employment, the number of people engaged in this sector has been decreasing absolutely. This trend is likely to continue regardless of the outcome of the economic zone. It can be considered that the establishment of the economic zone is a welcome signal to hasten the structural change that has already been taking place for decades.

Thirdly, when one looks into a potentiality created by the problems involved in the international water and marine resources, the long-term gains to Japan in terms of further development of ocean mining and machinery industries will certainly offset whatever temporary loss Japan may suffer in the fishing and fishery industry. When one looks into a large spectrum of ocean-related industries, such as oil drilling gears, ocean reclamation, and antipollution devices, the gains loom particularly large. Added to these prospects are the gains from forthcoming development and opportunity in the field of ocean mining technology, floating atomic power plants, and utilization of ocean energy such as currents, tides and waves. Japan has already developed buoys and lighthouses that generate power from the waves on a self-sustaining basis. My preliminary estimates show that the contributions of these ocean-related industries have already reached the $10 billion mark on an annual basis.

In conclusion, there will be some adverse effects of the outcome of the forthcoming law of the sea sectorally, but in terms of the impact on the entire economy, the effect is at best minimal. In the long run, Japan, with her ever expanding economic base, will even benefit by transforming her industrial structure from fishing to ocean-related industries. It is encouraging that despite an outward sign of despair, Japan is already moving in that direction. Witness the energy and expenditure Japan spent on the 1975 International Ocean Exposition.

One final note is in order. It has been a well-known pattern of behavior for Japan to be extremely agitated by serious or not-so-serious events and then to devote her energy singlemindedly to solve the problem. The current controversy concerning the law of the sea and ocean resources in Japan fits this pattern very well. If agitation eventually produces something constructive, as it usually has in the past, so much the better for Japan. However, even in the midst of the controversy, disinterested economists must be able to tell the fact from the fiction.

REFERENCES

Alexander, Lewis M., ed. 1970. *The Law of the Sea: National Policy Recommendations: Proceedings of the Fourth Annual Conference of the Law of the Sea Institute, June 23–26, 1969.* Kingston: University of Rhode Island Press.

Alexander, Tom. 1974. "Dead Ahead Toward a Bounded Main," *Fortune,* vol. 90, October, pp. 129–31.

Andreev, Anatoly. 1973. "Activities of the Intergovernmental Maritime Consultative Organization in the Field of Prevention and Control of Operational and Accidental Pollution Emanating from Ships," in *Hazards of Maritime Transit: Law of the Sea Institute Workshop, Nassau, The Bahamas, May 1973.* Edited by Thomas A. Clingan, Jr., and Lewis M. Alexander. Cambridge, Mass.: Ballinger, pp. 29–47.

Ary, T.S. 1971. "Statement of American Mining Congress to the Department of the Interior with Respect to Working Paper of the Draft of United Nations Convention on the International Seabed Area," Washington, D.C., January.

Business Week. 1974. "Tapping the Lode of the Ocean Floor." 19 October, pp. 130–34.

Christy, Francis T., Jr. 1970. "New Dimensions for Transnational Marine Resources," *American Economic Review,* vol. 60, May, pp. 109–13.

———, ed. 1975. *Law of the Sea, Caracas and Beyond: Proceedings.* Kingston: University of Rhode Island Press.

Clarkson, Kenneth W. 1974. "International Law, U.S. Seabeds Policy and Ocean Resource Development," *Journal of Law and Economics,* vol. 17, April, pp. 117–42.

Davis, Morris. 1963. *Iceland Extends its Fisheries Limit.* Copenhagen: Universitetsforlaget.

Eckert, Ross D. 1974. "Exploitation of Deep Ocean Minerals: Regulatory Mechanisms and United States Policy," *Journal of Law and Economics,* vol. 17, April, pp. 143–77.

Geiger, George R. 1936. *The Theory of the Land Question.* New York: Macmillan.

Glassner, Martin I. 1970. *Access to the Sea for Developing Land-Locked States.* The Hague: Martin Nijhoff.

Hatfield, Mark. 1971. "Statement on the U.S. Fishing Industry," U.S. Senate, *Congressional Record,* 20 November.

Heichelheim, Fritz. 1933. "Ancient Land Tenure," *Encyclopedia of the Social Sciences,* vol. 9.

Henkin, Louis. 1968. *Law for the Sea's Mineral Resources.* Institute for the Study of Science in Human Affairs, Monograph No. 1. New York: Columbia University Press.

Hodgeson, Robert D. and McIntire, Terry V. 1973. "Maritime Commerce in Selected Areas of High Concentration," in *Hazards of Maritime Transit: Law of the Sea Institute Workshop, Nassau, The Bahamas, May 1973.* Edited by Thomas A. Clingan, Jr., and Lewis M. Alexander. Cambridge, Mass.: Ballinger, pp. 1–18.

Hollick, Ann L. and Osgood, Robert E. 1974. *New Era of Ocean Politics.* Studies in International Affairs Series, No. 22. Baltimore: Johns Hopkins University Press.

Marine Technology Society. 1972. *Law of the Sea Reports: A Year of Crisis: February 19, 1971; Geneva Report, October 18, 1971.* Washington, D.C.: Author.

Meier, Gerald M. 1974. *Problems of Cooperation for Development.* New York: Oxford University Press.

Mero, John L. 1965a. *The Mineral Resources of the Sea.* New York: Elsevier.

———. 1965b. "Statement," in *National Oceanographic Program Legislation.* Hearings before the Subcommittee on Oceanography of the House Committee on Merchant Marine and Fisheries. 89th Congress, 1st Session. Washington, D.C.: U.S.G.P.O., pp. 599–600.

Miller, Roger L. 1973. *Economics Today.* San Francisco: Canfield Press.

Penrose, Edith. 1976. "The Development of Oil Crisis," in *The Oil Crisis.* Edited by Raymond Vernon. New York: Norton, pp. 39–57.

Prescott, John R. V. 1975. *The Political Geography of the Oceans.* Baltimore: Halsted Press.

Rothstein, Arnold J. 1971. "Deep Ocean Mining: Today and Tomorrow," *Columbia Journal of World Business,* vol. 16, January-February, pp. 43–50.

Shapley, Deborah. 1974. "Law of the Sea: Energy, Economy Spurs Secret Review of U.S. Stance," *Science,* vol. 183, 25 January, pp. 290-92.

Sorenson, Philip E. and Mead, Walter J. 1968. "A Cost-Benefit Analysis of Ocean Mineral Resource Development: The Case of Manganese Nodules," *American Journal of Agricultural Economics,* vol. 50, December, pp. 1611–20.

Swartztrauber, Sayre A. 1972. *The Three-Mile Limit of Territorial Seas.* Annapolis: Naval Institute Press.

Sweeney, Richard J.; Tollison, Robert D.; and Willett, Thomas D. 1974. "Market Failure, the Common-Pool Problem, and Ocean Resource Exploitation," *Journal of Law and Economics,* vol. 17, April, pp. 179–92.

United Nations Conference on the Law of the Sea. 1958a. *Convention on the Territorial Sea and the Contiguous Zone.* A/Conf. 13/L. 52. 28 April.

———. 1958b. *Convention on the High Seas.* A/Conf. 13/L. 53. 29 April.

———. 1958c. *Convention on Fishing and Conservation of the Living Resources of the High Seas.* A/Conf. 13/L. 54. 28 April.

———. 1958d. *Convention on the Continental Shelf.* A/Conf. 13/L. 55. 28 April.

United Nations General Assembly. 1966. *Permanent Sovereignty over Natural Resources.* Resolution 2158. A/6316, GAOR, 21st Session, Suppl. No. 16. 20 September–20 December.

———. 1967. *Request for the Inclusion of a Supplementary Item in the Agenda of the 22nd Session.* A/6695. 17 August.

———. 1970. *Reservation Exclusively for Peaceful Purposes of the Seabed and the Ocean Floor.* Resolution 2750. A/8028, GAOR, 25th Session, Suppl. No. 28. 15 September–17 December.

———. 1975. Third Conference, Third Session, 2nd Committee. A/Conf. 62/SR. 54. Geneva, 17 March–9 May.

United Nations Secretary General. 1973. *Progress Report.* A/AC 138/90. 3 July. *United States Code Congressional and Administrative News.* 1976. "Fishery Conservation and Management Act of 1976." 25 May, pp. 331–61.

U.S. President. 1945. *Presidential Proclamation No. 2667, September 28, 1945. Federal Register,* Vol. 10.

United States Senate, Committee on Interior and Insular Affairs. 1971a. *Outer Continental Shelf.* Hearings. Part 2. Washington, D.C.: U.S.G.P.O.

———. 1971b. *Outer Continental Shelf.* Committee Report. Washington, D.C.: U.S.G.P.O.

Wall Street Journal. 1976. 20 September.

Name Index

Entries in **bold** type under the names of participants in the conference indicates their papers of discussions of their papers. Entries in *italic* type indicates contributions of participants to the discussions.

Subject Index